FRONTIERS IN OPTICAL TECHNOLOGY

MATERIALS AND DEVICES

FRONTIERS IN OPTICAL TECHNOLOGY

MATERIALS AND DEVICES

P. K. CHOUDHURY
O. N. SINGH
EDITORS

Nova Science Publishers, Inc.
New York

Library of Congress Cataloging-in-Publication Data
Frontiers in optical technology : materials and devices / P.K. Choudhury and Onkar N. Singh, (editors).
p. cm.
Includes index.
ISBN 13 978-1-60021-084-6
ISBN 10 1-60021-084-8
1. Optoelectronic devices. 2. Optical materials. I. Choudhury, P.K. II. Singh, Onkar N.
TA1750.F74 2006
621.36--dc22 2006006148

Published by Nova Science Publishers, Inc. ✢New York

CONTENTS

Contributors		**vii**
Foreword	*Akhlesh Lakhtakia*	**ix**
Preface		**xiii**
Introduction		**xv**
Chapter 1	Towards shaping of pulsed plane waves in the time domain via chiral sculptured thin films *Joseph B. Geddes III*	**1**
Chapter 2	One-dimensional bigyrotropic magnetic photonic crystals: non-defect and defect structures *N.N. Dadoenkova, I.L. Lyubchanskii, M.I. Lyubchanskii, E.A. Shapovalov, and Y.P. Lee.*	**23**
Chapter 3	Integrated optics on silicon *Ulrich Hilleringmann*	**73**
Chapter 4	Photonics applications of nano-silicon *Lorenzo Pavesi*	**97**
Chapter 5	Combining energy efficiency with aesthetic appeal using advanced optical materials *Geoffrey B. Smith*	**125**
Chapter 6	High-speed optical modulators and photonic sideband management *Tetsuya Kawanishi*	**159**
Chapter 7	Lateral biasing in *p-n* junction light emitters: HELLISH devices *J. Y. Wah and N. C. Balkan*	**191**

Chapter 8	Techniques and analyses of coupling between laser diode and fiber-based microlens *Faidz Abd. Rahman*	**213**
Chapter 9	An overview of optical sensors and their applications *Pankaj Kr. Choudhury and Onkar N. Singh*	**235**
Chapter 10	Optical current sensors *Toshihiko Yoshino and Masayuki Yokota*	**273**
About the Editors		**289**
Index		**291**

CONTRIBUTORS

Prof. N. Balkan
Department of Electronic Systems Engineering, University of Essex, Colchester, UK.

Prof. Pankaj Kr. Choudhury
Faculty of Engineering, Multimedia University, Cyberjaya, Malaysia.

Dr. N. N. Dadoenkova
Donetsk Physical and Technical Institute, National Academy of Sciences of Ukraine, Donetsk, Ukraine.

Dr. Joseph B. Geddes III
Computational and Theoretical Materials Science Group, Department of Engineering Science and Mechanics, The Pennsylvania State University, University Park, USA.

Prof. Ulrich Hilleringmann
Institut für Elektrotechnik und Informationstechnik, Universität Paderborn, Paderborn, Germany.

Dr. Tetsuya Kawanishi
National Institute of Information and Communications Technology, Koganei, Tokyo, Japan.

Prof. Y. P. Lee
Quantum Photonic Science Research Center, Physics Department, Hanyang University, Seoul, Republic of Korea.

Prof. I. L. Lyubchanskii
Donetsk Physical and Technical Institute, National Academy of Sciences of Ukraine, Donetsk, Ukraine.

Dr. M. I. Lyubchanskii
Donetsk Physical and Technical Institute, National Academy of Sciences of Ukraine, Donetsk, Ukraine.

Prof. Lorenzo Pavesi
Dipartimento di Fisica, Università di Trento, Trento, Italy.

Dr. Faidz Abd. Rahman
Centre for Photonics Research, Innovation and Application, Faculty of Engineering, Multimedia University, Cyberjaya, Malaysia.

E. A. Shapovalov
Donetsk Physical and Technical Institute, National Academy of Sciences of Ukraine, Donetsk, Ukraine.

Prof. Onkar N. Singh
Department of Applied Physics, Institute of Technology, Banaras Hindu University, Varanasi, India.

Prof. Geoff Smith
Department of Applied Physics and Institute of Nanoscale Technology, University of Technology, Sydney, Australia.

Dr. J. Y. Wah
Faculty of Engineering, Multimedia University, Cyberjaya, Malaysia.

Dr. Masayuki Yokota
Department of Electrical and Electronics Engineering, Faculty of Engineering, Gunma University, Kiryu, Japan.

Prof. Toshihiko Yoshino (Retd.)
Retired Professor of the Department of Electrical and Electronics Engineering, Faculty of Engineering, Gunma University, Kiryu, Japan.
Presently with The Kaisei Academy, Nishi-Nippori, Arakawa-ku, Japan.

FOREWORD

For eight centuries prior to the expulsion of the Moors from Spain, Islamic science and technology was dominated by optics. Understanding of the operation of the human eye was put on the right track, the laws of reflection and refraction were discovered and design principles for magnifying glasses and mirrors were enunciated and tested. Observed though the telescope of history, the progress in optics then made throughout the Islamic crescent appears breath-taking to me.

But even more spectacular are the optical achievements of the mid-to-late 20th century. These have taken place in my lifetime: I was barely a toddler when the laser was invented. Spurred by a myriad of applications in industry, medicine, and military thereafter, the pace of innovation in optics has sped exponentially. Today, it is difficult to decide where the frontiers of optics exactly are, in part due to the immense diversity of optics research and application. My fellow researchers' remarks often echo the Red Queen in Lewis Carroll's *Through the Looking Glass*: "Now, *here*, you see, it takes all the running *you* can do, to stay in the same place."

Drs. Choudhury and Singh and all contributors to this volume, therefore, have done a signal service to researchers by mapping out the frontiers of optical technology in two major subdisciplines: optical materials and optical devices. The optical materials and material architectures covered include nanostructured silicon, chiral sculptured thin films, magnetic photonic crystals, and switchable materials for efficient lighting and decorative optics. The optical devices addressed include silicon waveguides for integrated circuitry, high-speed electro-optic modulators, laser diodes coupled with fiber-tip lenses, and optical sensors. Reading the ten chapters, either altogether or piecemeal, one would get a virtually up-to-date review of the state-of-the-art. Therein lies the value of this volume, as I can attest from personal experience.

Much of my research is on the fundamentals of electromagnetic field behavior in complex materials. While going through the chapter by Geoff Smith on optical coatings for buildings, I was struck by the complexity of the problems considered by paints researchers in relation to hiding power, particulate dimensions, energy efficiency, and lightpiping from the outside to the interior of a building. When I next visit Sherwin Williams or some other paints store, I am sure to look at their products with new appreciation. I will also feel like attending a session on paints at the next meeting of SPIE. Perhaps, I shall even feel compelled to fruitfully apply the principles outlined in Smith's chapter to my theoretical research on homogenization of particulate composite materials by incorporating the source spectrum as well as the spectral response of a detector (which could be the human eye).

As another example of potential cross-fertilization across subdisciplines, let me cite the chapter of Choudhury and Singh on fiber optic sensors. Some of my research has been on optical sensing of fluid concentration as well as genomic material through sculptured thin films. While reading the chapter, and with a recent fatal disaster in a West Virginia coal mine at the back of my mind, I thought that one could coat the tip of an optical fiber by a series of chiral sculptured thin films, which, being porous, could be used to sense a variety of gases in a mine or some other underground chamber.

Optical sensing of electrical current, described by Drs. Yoshino and Yokota in the last chapter of this volume, turned out to be a revelation for me. I have known of magneto-optical effects for many years, and the Ampère circuital law for even longer, but had never thought that the two concepts could be put together to optically detect the magnitude of an electric current. Somehow, I feel, the Yoshino-Yokota chapter has sown in my head a seed that is bound to germinate sooner or later.

Seeing possible benefit to my own future research, I am confident that other researchers, both novices and experts, will put this volume to a variety of uses. In appreciation of these potentialities, I raise my glass of water.

I am delighted that the editors have dedicated this volume to Prof. Prasad Khastgir, who inspired a love of physics in several generations of students at the Institute of Technology, Banaras Hindu University. Teachers like him are rare, and his students – including me – are very fortunate.

Akhlesh Lakhtakia

Akhlesh Lakhtakia was born in Lucknow, India on July 1, 1957. He obtained a Bachelor of Technology degree in Electronics Engineering from the Banaras Hindu University, Varanasi, India in 1979; Master of Science and Doctor of Philosophy degrees in Electrical Engineering from the University of Utah, Salt Lake City in 1981 and 1983, respectively. Thereafter, he joined the faculty of the Pennsylvania State University, where he was elevated to the rank of Distinguished Professor of Engineering Science and Mechanics in January 2004. From 2004 to 2007 he also holds the rank of a Visiting Professor of Physics at Imperial College, London. He has published more than 500 journal articles; has contributed chapters to 14 research books and encyclopedias; has edited, co-edited, authored or co-authored 11 books and 6 conference proceedings; has reviewed for 75 journals; serves on the editorial boards of four electromagnetics journals; and was the Editor-in-Chief of the international journal *Speculations in Science and Technology* from 1993 to 1995. He served as an international lecturer for the International Commission for Optics and the Optical Society of America; was twice a Visiting Professor of Physics at Universidad de Buenos Aires, a Visiting Professor of Physics at the University of Otago, and a Visiting Fellow in Mathematics at the University of Glasgow; headed the IEEE EMC Technical Committee on Nonsinusoidal Fields from 1992 to 1994; and is a Fellow of the Optical Society of America, SPIE-International Society for Optical Engineering, and the Institute of Physics (UK). He also served as the 1995 Scottish Amicable Visiting Lecturer at the University of Glasgow. He was awarded the PSES Outstanding Research Award in 1996 and the PSES Outstanding Advising Award in 2005. For his research, he received the Faculty Scholar Medal in Engineering in 2005. His current research interests lie in the electromagnetics of complex materials, sculptured thin films, chiral nanotubes, and negative refraction of light due to gravitational fields. At Penn State, he co-developed a course on green engineering for undergraduate engineering students, as well as a course on fundamentals of engineering principles and design for pre-service elementary schoolteachers.

PREFACE

This book focuses research details in several areas of optical technology reported by pioneering scientists from ten different countries. Preeminent researchers themselves report their recent research results in the emerging areas of optics and photonics, and review the recent developments. The book provides readers a solid foundation to understand the rapidly growing area of optical technology. There are ten chapters in the book, each focusing on special aspects of optics/photonics. The topics covered in the book are chiral sculptured thin films, magnetic photonic crystals, silicon integrated optics on silicon, nano-silicon photonics, advanced optical materials and their aesthetic appeal, high-speed optical communications, efficient optical sources and their fiber couplings, optical sensing, and their industrial applications. Through this book we aim to provide (i) novice researchers, such as Ph.D. students, an understanding of exciting new research advances that may fuel their own research areas as well as provide thesis topics; (ii) expert researchers, an overview of advances in contiguous research fields; and (iii) cross-disciplinary researchers, a set of topics whereby optics plays the role of enabling advances in other research areas. Finally, it is expected that the book will provide a sound understanding and the technological know-how of the rapidly growing areas of optics and photonics technology to the academicians belonging to various categories ranging from students to expert researchers.

We are thankful to all the authors who agreed to contribute for the book. Their deep interest, hard work and cooperation made it possible to bring their valuable research results under one cover in the form of this book. Our sincere thanks to Nova Science Publishers (USA) for appreciating the central idea of the book, and for accepting this edited volume for publication. The completion of this volume would not have been possible without help and support received from our respective families; we greatly appreciate their patience and forbearance. We owe Dr. Akhlesh Lakhtakia, Distinguished Professor of the Pennsylvania State University, a debt of gratitude for providing the idea of this timely volume, and for kindly accepting to write the foreword for the book. Finally, we are immensely grateful to our revered teacher and esteemed colleague Professor (Retired) Prasad Khastgir of the Banaras Hindu University, intellectual discussions with whom have always been the source of inspiration for us; this volume is dedicated to him.

P. K. Choudhury
O. N. Singh

INTRODUCTION

Optoelectronics is now a billion dollars per year industry, and it continues to grow. This has laid the foundation for technology of the information age enabling all functions of the information industry and the RandD works at universities, and industries are playing significant role in this evolution. This subject deals with the electronics devices used for the emission, modulation and transmission of light signals. However, today the subject has widened its meaning by incorporating technologies related to photonics, electro-optics and magneto-optics too. Photonics deals with the properties and applications of photons as a medium for the transmission of information, whereas electro- and magneto-optics, respectively, deal with the effects of electric and magnetic fields on light passing through a material.

So far as the optoelectronics industry is concerned, truly speaking, there is no single such industry, and it is, in fact, a spectrum of several industries supported by common technology and common infrastructure such as optical communication, integrated optics, silicon photonics, material science, sensing etc. Today internet and data services are showing their rapidly increasing dominance, and for that, high-speed and broadband transmissions are required at the lowest possible cost. Specific attention towards the communication bandwidth fuelled the amazing growth rate of optical communication systems. Wavelength division multiplexing (WDM) system is the one that gave a major boost after its commercialization. However, many other practical and commercial issues are also responsible to keep up the growth.

A drastic reduction in the cost of optoelectronic components is required for high volume access networks. It is noteworthy that today the access systems like internet, WDM and vertical cavity surface emitting lasers (VCSELs) promise to lower the cost and facilitate the integration of electronic and photonic components, thereby rendering optical networks more affordable. It is rather difficult to predict on the growth of market volume of specific optoelectronic components; however, an estimation of this growth can always be made depending on the latest technology developments and inputs from industry sources and analysts. Today an overdriving requirement of these components results in their improvement in performance as well as cost reduction. One should expect a steady improvement in the quality performance of optoelectronic components with simultaneous cost reduction which, along with other activities, will surely fuel the market growth of optics industry.

This book delineates several important advances occurring at the forefront of optics and optical technology. In equal measures, the two focuses of this book are (i) integrated optics and (ii) applications of classical optics in new settings engendered by advances in materials sciences. The book emphasizes time-domain research on chiral sculptured thin films, recent advancements in magnetic photonic crystals and their applications, silicon technology in integrated optics and nano-silicon photonics, advanced optical materials and their applications, photonics in communication, and optical sensing. These topics are of increasing demand in the present information age, and the RandD workers at universities and industries are playing significant role in their evolution. It is now the appropriate time to collect several such advancements under one cover to enable researchers to use the book as springboard for their own ideas.

Rotationally nonhomogeneous thin films comprising parallel nanowires with tailored shapes are a new class of materials that have been addressed as sculptured thin films (STFs). These nanoengineered materials were recognized as one of the seventeen promising classes of nanomaterials by NASA Nanotech Briefs in 2005. In **Chapter 1**, Joseph B. Geddes III of the Pennsylvania State University (USA) reviews the interactions between pulsed plane waves of picosecond and femtosecond durations and chiral STFs. An examination of the spatiotemporal anatomy of the circular Bragg phenomenon is carried out, and the pulse-bleeding phenomenon is characterized. Efforts are put to explain the importance of this phenomenon for the shaping of ultrashort pulses.

One of the most exciting recent developments in the field of photonics has been the realization by E. Yablonovitch (in 1987) that photons in dielectric mediums could be guided just like electrons in semiconductors or more generally in crystals. The basis for the guidance of photons in dielectric mediums is the same as that for electrons in crystals, i.e. Bragg diffraction due to a periodic lattice. If electrons can be diffracted by a periodic nuclear potential, photons could also be equally well diffracted by a periodic modulation of the relative permittivity or refractive index of the medium. This idea has given birth to many new artificial photonic materials or optical microstructures. **Chapter 2** is devoted to a new class of photonic structures, namely, magnetic photonic crystals (MPCs). These are artificial periodic multilayer structures with either magnetic-nonmagnetic or magnetic-magnetic unit cells, with periods comparable with the wavelengths of electromagnetic radiation. In this chapter, I. L. Lyubchanskii, N. N. Dadoenkova, M. I. Lyubchanskii, and E. A. Shapovalov of Donetsk Physical and Technical Institute of the National Academy of Sciences of Ukraine (Ukraine), and Y. P. Lee of the Hanyang University (Korea) paid special attention to MPCs containing bigyrotropic magnetic materials, which are characterized by the off-diagonal components of both permittivity and permeability tensors. The application of transfer matrix method for polar, longitudinal and transverse magneto-optical configurations is described, and the reflection/transmission spectra for normal and oblique incidence of electromagnetic radiation are presented. Recently published results of theoretical and experimental investigations of MPCs as well as possible applications of such structures are reviewed.

The subject of silicon technology is of great importance as it finds multifarious applications in photonics. Silicon waveguides, viz. rib waveguides, channel waveguides, strip loaded film waveguides etc., form the basic building blocks of the present day integrated optics technology. In **Chapter 3**, Ulrich Hilleringmann of the Universität Paderborn (Germany) thoroughly reviews different types of waveguides, layers used for waveguide integration, various deposition techniques, the etching of ribs, and the characteristics of waveguides. Different types of couplings of waveguides and photodetectors, viz. butt coupling, leaky wave coupling, the coupling by integrated mirrors etc., are also discussed.

Monolithic integration of waveguides and metal oxide semiconductor circuits is also described in the context of combined integration technique and modular processing. Micromechanic add-ons are discussed, and the different applications, e.g. pressure sensor with optical read-out etc., are also reviewed.

Very fascinating and new optical properties of materials appear with the change in the dimensionality of silicon. In particular, light emission starts to be a process very efficient in nanostructured silicon. In this context, highly efficient light emitting diodes (LEDs) have been fabricated. Optical amplification has also been observed when silicon nanocrystals are embedded in a dielectric matrix material, which renders this system a potential candidate for observing a laser action. In addition to variations in the electronic properties, nanostructured silicon can also be used as a nanodielectric material, where changes in the refractive index can lead to trapping or slowdown of photons. In **Chapter 4**, Lorenzo Pavesi of the Università di Trento (Italy) reviews such interesting photonic aspects of nanostructured silicon. Some examples are also discussed where photonic Bloch oscillations, photon Zener tunneling and high quality microcavities are demonstrated.

Developments of new materials have opened up exciting new opportunities for reducing the energy demands, while at the same time creating more appealing interiors and greater decorative options. In designing optical materials, consideration must be given to tailoring spectral and directional properties for specular or beam radiation and visual clarity. When light scattering is desired, skylights and LED color mixing energy efficiency must limit backscattering and absorption at specific wavelengths, and directions must be controlled. The latter includes a new approach to white light and arbitrary colors from red-green-blue LED arrays. Glare reduction, light spreading and optical homogenization are central issues in the deployment of bright LEDs for lighting and display. In **Chapter 5**, Geoffrey B. Smith of the University of Technology, Sydney (Australia) discusses the optical properties and structure of a range of materials, pigments and thin films that can meet these goals, and outlined the associated underlying optical science. The chapter presents a review of the important general role for engineered nanostructured surfaces in energy efficiency. Progress in optically switchable materials is also outlined; these include electrically switchable materials (electrochromics, gasochromics and suspended particle systems) and thermally switchable materials (thermotropics and thermochromics).

Optical communications has been one of the most important areas in the optoelectronics technology. In such communication systems, one has to use some interactions with unmodulated lightwaves, in order to produce and read optical signals having some information. There are two types of interactions – (i) interaction between electric and optical signals and (ii) interaction between two or more optical waves. Recently, many types of all-optical systems using interaction among optical waves have been proposed to enhance the processing performance at nodes. However, this development does not decrease the importance of the interaction between electric and optical signals, because mostly the information, which can be accessed by human beings, is described by electric signals, at this moment. One has to use the interaction between electrical and optical signals at terminals of the optical communications system, while all-optical interaction can be used at the nodes. In **Chapter 6**, Tetsuya Kawanishi of the National Institute of Information and Communications Technology (Japan) presents a review of the recent results in high-speed optical modulation and their applications. Some mathematical expressions for optical phase modulators are given to discuss the performance of various types of modulators comprising the phase modulators.

Discussions are made of the electrode structures for high-frequency operations, and the design schemes for resonant electrodes. Descriptions are presented of various types of optical modulators consisting of phase modulators viz. intensity modulator, single-sideband (SSB) modulator, carrier suppressed SSB modulator, frequency-shift-keying (FSK) modulator, and optical quadrature phase modulator. A new concept of optical sideband modulation technique, i.e. photonic side-band management (PSBM), is also reviewed in this chapter with emphasis on various PSBM techniques, which enable us to manipulate optical signals by electric control signals, while most of modulators are just to convert the electric signal to the optical one. Basically, PSBM does not use optical oscillation, so that there are no chaotic phenomenons, such as mode-hopping and mode-competitions, leading thereby to stable and agile manipulation of optical signals in frequency and time domains.

In optical communications and/or other applications, the use of suitable light sources cannot be exaggerated. VCSELs have been the most important class of such inexpensive, tiny and radiation hard light source at wavelengths of 750 nm to 1000 nm. These involve the proper blend of group III-V materials, and their spacing determines the emission wavelength. VCSELs essentially work on the principle of distributed Bragg reflectors (DBRs), i.e. a narrow range of wavelengths is reflected depending on the thickness and refractive indexes of the composing materials. The refractive index can be controlled by properly biasing the *p-n* junction in VCSELs. However, conventional vertical biasing of *p-n* junctions has induced several problems when carriers are injected through the DBRs. In **Chapter** 7, J. Y. Wah of the Multimedia University (Malaysia) and N. Balkan of the University of Essex (UK) describe a series of lateral biasing devices, namely HELLISH (hot electron light emission and lasing in semiconductor heterojunction) – a novel hot electron surface emitter. The injection of hot electron-hole pairs into the quantum well is achieved via tunneling and thermionic emission processes. The structure has been then evolved and developed to ultra-bright LED, VCSEL, VCSOA, wavelength converter, high efficiency LEDs/lasers etc. The chapter reviews these novel structures.

The use of laser diodes as the signal source in communications requires that the emission be coupled into other optical components such as optical fibers, optical amplifiers, waveguides or lenses. In **Chapter 8**, Faidz Abd. Rahman of the Multimedia University (Malaysia) reviews the relevant concepts in respect of optical sources, and examined the coupling efficiency and alignment tolerance between laser diodes and single-mode fibers. The review describes an approach based on the integrated lens. A model based on overlap integral and transmission matrix is discussed considering conical and hemispherical types of microlenses. Also, examples of applying the model are taken up, and the results are compared with experimental investigations.

Fiber optic sensing has been the focus area of research and development for over twenty five years. In this period, varieties of sensors have been devised by the investigators for multifarious range of applications. In the developed nations of the world, optical sensor industry is a multimillion-dollar business. In **Chapter 9**, P. K. Choudhury of the Multimedia University (Malaysia) and O. N. Singh of the Instutute of Technology, Banaras Hindu University (India) review the fundamentals of fiber optic sensing techniques, different types of optical sensors and their applications in different industries. Descriptions are made of intensity based sensors, distributed sensors, and interferometric sensors, their different types and specific applications. Examples of some more forms of fiber optic sensors, and their use in the context of environment are also reviewed. In addition, discussions are made of the role of integrated optic technology in sensing applications. Photonic crystals and sculptured thin films have been the emerging areas of research in the present day; the use of such materials and/or mediums in the area of optical sensing is also taken up.

Optical current sensing has been one of the demanding areas in the sensor market, and such sensors are widely used in power industries. In the last chapter (**Chapter 10**), Toshihiko Yoshino, formerly with the Gunma University (Japan) and presently with The Kaisei Academy (Japan), and Masayuki Yokota of the Gunma University (Japan) comprehensively discuss on the optical current sensors based on Faraday effect. The review describes two different types of optical current transformers (OCTs), viz. (i) the all-optical type consisting of an optical fiber itself or a bulk solid Faraday element and (ii) the hybrid type consisting of a combination of a magnetic core for the field concentrator and a bulk solid Faraday element. Ring-core (hybrid) type, flint glass block type, and fiber (all-optical) type sensors are reviewed in this chapter. It has been shown that the all-optical type sensor can offer a precise measurement of current as it can rigorously fulfill the homogeneous isotropic closed optical circuit (HICOC) condition in the Faraday cell. On the contrary, the hybrid-type OCT can only approximately fulfill the HICOC condition, but has some practical advantages over the all-optical type. The types of current sensors described in the chapter received great appreciation by the sensor market, and find many industrial applications.

It is the hope of the editors that the book should be immensely useful chiefly for doctoral students, relevant engineers and faculty members at universities, as well as for industrial researchers as they would find key insights into the current trends and techniques of optics/photonics likely to shape future directions of this increasingly important topic. Finally, it is needless to say that these are exciting times for the RandD community devoted to optoelectronics research to work in the key areas of interest.

P. K. Choudhury
O. N. Singh

In: Frontiers in Optical Technology: Materials and Devices ISBN 1-60021-084-8
Editors: P.K. Choudhury and O.N. Singh, pp. 1-21

Chapter 1

TOWARDS SHAPING OF PULSED PLANE WAVES IN THE TIME DOMAIN VIA CHIRAL SCULPTURED THIN FILMS

Joseph B. Geddes III
CATMAS—Computational and Theoretical Materials Sciences Group
Department of Engineering Science and Mechanics
The Pennsylvania State University
212 Earth and Engineering Sciences Building
University Park, PA, 16802–6812, USA

Abstract

We review our investigations concerning the propagation of ultrashort optical pulsed plane waves across chiral sculptured thin films (STFs) in the time domain. The phenomenon of pulse bleeding—i.e., the time domain manifestation of the circular Bragg phenomenon (CBP)—is explained, and its importance for the shaping of ultrashort pulses is investigated.

1. Introduction

As the generation of ultrashort optical pulses becomes more routine, efforts to more efficiently and inexpensively shape and manipulate such pulses rise in importance. In this chapter, we describe time–domain calculations that show how chiral sculptured thin films (STFs) could be used to sort and alter such ultrashort pulses in ways that depend on their amplitude, shape, duration, carrier polarization, and carrier phase. Chiral STF–based devices could be especially useful in manipulating pulses whose carrier plane wave is circularly polarized.

STFs are assemblies of parallel nanowires affixed to a substrate. They are fabricated via physical vapor deposition—either evaporation or sputtering—in vacuum, and their constituents can be semiconducting oxides, metals, or even polymers. The nanowire shapes, which can be bent and twisted during deposition into both continuous and discontinuous curvilinear shapes, endow these nanoengineered materials with their most useful optical properties. Patterning of substrates before deposition can also affect the shapes of the growing nanowires; such patterning raises the possibility of creating three dimensional STF architectures for photonic crystals and waveguides [1, 2]. The intensive research of the past ten years on STF morphology and optics has been recently summarized in book form [3].

In chiral STFs, the nanowires are helicoidal in shape. They are fabricated by evaporating material at an oblique angle onto a relatively cool substrate that rotates about an axis normal to its surface. The nanowire helices grow normal to the substrate, and their pitch depends on the deposition rate, rotation rate, and deposition angle. An electron micrograph of a chiral STF is presented in Figure 1.

When the pitch of the helices is constant or nearly so, chiral STFs exhibit the circular Bragg phenomenon (CBP). Suppose light is incident normally on a structurally right–handed chiral STF. The wavelengths of light that lie within a certain range, known as the Bragg regime, are largely reflected from the chiral STF if they are right circularly polarized (RCP). Left circularly polarized (LCP) light, in contrast, is largely refracted into the film, where it is either subsequently transmitted or absorbed, depending on the film properties. The CBP is the phenomenon that makes possible many of the chiral STF–based devices that have been designed and tested, including circular polarization handedness inverters, circular polarization filters, and spectral hole filters [4, 5, 6]. Advances in deposition technology have resulted in improved chiral STF properties [7], including increased optical activity [8].

Almost since the invention of the laser, researchers have sought to discover ways to use that device to create optical pulses of ever shorter duration [9]. Such pulses are used for many applications, e.g. in telecommunications and scientific research [10, 11]. Ultrashort pulses are modeled as a carrier plane wave, which we take to be of unit amplitude, multiplied by a pulse envelope. The envelope determines, in general, the pulse's total energy, shape, and duration, while the carrier plane wave sets the pulse's dominant wavelength and its polarization.

This chapter summarizes the key results—including the discovery of the spatiotemporal manifestation of the CBP and its effect in shaping of pulsed optical plane waves—obtained so far in using chiral STFs to manipulate pulsed plane waves in the time domain. Section 2. contains a description both of the chiral STF constitutive equations, the initial boundary value problem to be solved, and a finite–difference (FD) algorithm to solve it numerically. The following two sections describe the spatiotemporal evolution of optical pulsed plane waves. Section 3. describes this evolution when pulses are incident on chiral STF half–spaces, and Section 4. when pulses are incident on chiral STF slabs of finite thickness. Finally, we tie these results—many of which have been previously collected in a thesis [12]—together and suggest several applications for the work in Section 5.. The work concentrates on the phenomenology of reflection, refraction, and transmission of pulses by chiral STFs; quantification of pulse shaping effects will be the topic of future investigations.

2. Theory

In order to study the spatiotemporal evolution of an ultrashort optical pulse as it propagates through a chiral STF, mathematical models of both the pulse and the film are needed. In this section, we present the constitutive relations for a dielectric chiral STF, formulate

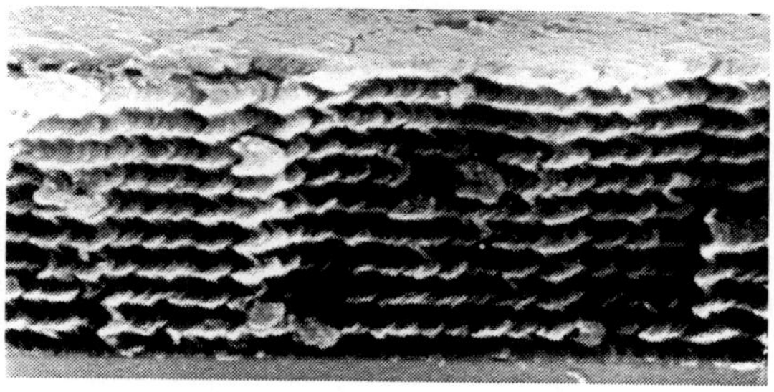

Figure 1. Scanning electron micrograph of a SiO chiral STF showing a cross section of its 10–turn helicoidal microstructure. The film is approximately 4 μm thick. *Courtesy: Russell Messier, The Pennsylvania State University*.

the Maxwell equations in matrix form, and then discretize those equations for numerical solution.

At time $t = 0$, a pulsed plane wave, consisting of a monochromatic carrier plane wave amplitude–modulated by a pulse envelope, is launched from the plane $z = 0$ toward a chiral STF slab. The film occupies the region $z_0 \leq z \leq z_1$ (note that $z_0 > 0$). The pulse traverses the vacuum half–space $z \leq z_0$, excites the chiral STF, whereupon it splits into reflected pulses in the half–space $z \leq z_0$ and transmitted pulses in the half–space $z \geq z_1$, also vacuous.

2.1. Chiral STF Constitutive Relations

As mentioned in Section 1., chiral STFs exhibit a continuously twisted, unidirectionally periodic microstructure, in this case along the z axis. Their nanowires, whose constituents are dispersive in bulk form in accord with causality requirements, possess cross sections that are nominally elliptical, and thus exhibit locally biaxial dielectric properties. The time–domain chiral STF constitutive relations encompass all these properties in the equations

$$\underline{D}(\underline{r},t) = \epsilon_0 \left(\underline{\underline{\epsilon}}_r * \underline{E}\right)(\underline{r},t) \,, \quad \underline{B}(\underline{r},t) = \mu_0 \, \underline{H}(\underline{r},t) \,, \tag{1}$$

where ϵ_0 and μ_0 are the permittivity and the permeability of free space, respectively; $\underline{r}$ is the position vector; the electromagnetic fields $\underline{E}(\underline{r},t)$, $\underline{D}(\underline{r},t)$, $\underline{H}(\underline{r},t)$, and $\underline{B}(\underline{r},t)$ have their usual meanings [13]; while the operation $*$ denotes convolution with respect to time as follows:

$$\left(\underline{\underline{\epsilon}}_r * \underline{E}\right)(\underline{r},t) = \int_0^t \underline{\underline{\epsilon}}_r\left(\underline{r}\cdot\underline{u}_z, t'\right)\cdot \underline{E}\left(\underline{r}, t-t'\right)\, dt' \,. \tag{2}$$

Equations (1) contain the tacit assumption that chiral STFs exhibit only dielectric properties at optical wavelengths [14], an assumption that can be relaxed in future work. The relative permittivity dyadic is defined as

$$\underline{\underline{\epsilon}}_r(z,t) = \begin{cases} \underline{\underline{I}}\,\delta(t)\,, & z \notin [z_0, z_1] \\ \underline{\underline{S}}_z(z-z_0)\cdot\underline{\underline{S}}_y(\chi)\cdot\left[\epsilon_a(t)\,\underline{u}_z\underline{u}_z + \epsilon_b(t)\,\underline{u}_x\underline{u}_x + \epsilon_c(t)\,\underline{u}_y\underline{u}_y\right] & \\ \quad \cdot\underline{\underline{S}}_y^{-1}(\chi)\cdot\underline{\underline{S}}_z^{-1}(z-z_0)\,, & z \in [z_0, z_1] \end{cases} , \tag{3}$$

where the structural handedness of the chiral STF is described by the rotation dyadic

$$\underline{\underline{S}}_z(z) = \underline{u}_z\underline{u}_z + \left(\underline{u}_x\underline{u}_x + \underline{u}_y\underline{u}_y\right)\cos\left(\pi z/\Omega\right) + h\left(\underline{u}_y\underline{u}_x - \underline{u}_x\underline{u}_y\right)\sin\left(\pi z/\Omega\right) \quad (4)$$

with 2Ω as the structural period, and $h = +1$ for a structurally right–handed chiral STF, and $h = -1$ for a structurally left–handed one. The identity dyadic is denoted as $\underline{\underline{I}}$, while $\delta(t)$ is the Dirac delta function and $\underline{u}_x$, $\underline{u}_y$, and $\underline{u}_z$ are the cartesian unit vectors. The tilt dyadic

$$\underline{\underline{S}}_y(\chi) = \underline{u}_y\underline{u}_y + \left(\underline{u}_x\underline{u}_x + \underline{u}_z\underline{u}_z\right)\cos\chi + \left(\underline{u}_z\underline{u}_x - \underline{u}_x\underline{u}_z\right)\sin\chi \quad (5)$$

is a function of the angle of rise χ.

We assume that the dielectric properties of our linear chiral STF are captured by a single–resonance Lorentz model [15], and thus

$$\begin{aligned}\epsilon_{a,b,c}(t) &= \delta(t) + \chi_{a,b,c}(t)\\ &= \delta(t) + p_{a,b,c}\left(\frac{2\pi c_0}{\lambda_{a,b,c}}\right)\sin\left(\frac{2\pi c_0}{\lambda_{a,b,c}}t\right)\exp\left(-\frac{c_0 t}{N_{a,b,c}\lambda_{a,b,c}}\right)\mathcal{U}(t)\,, \quad (6)\end{aligned}$$

where $\mathcal{U}(t)$ is the Heaviside step function, $c_0 = (\epsilon_0\mu_0)^{-1/2}$, and the oscillator strengths are $p_{a,b,c}$. The absorption bands, which occur around the resonance wavelengths $\lambda_{a,b,c}\left(1 + N_{a,b,c}^{-2}\right)^{-1/2}$, become narrower as $N_{a,b,c}$ increase. Additional terms may be added to (6) to model nonlinear dielectric responses, as illustrated elsewhere [16].

2.2. Derivation of Matrix Partial Differential Equation

Upon abbreviating partial differentiation $\partial/\partial_x = \partial_x$, $\partial/\partial_t = \partial_t$, and so forth, we insert (1) and (3) into the Maxwell curl equations

$$\nabla \times \underline{E}(\underline{r}, t) = -\partial_t \underline{B}(\underline{r}, t)\,, \quad (7)$$

$$\nabla \times \underline{H}(\underline{r}, t) = \partial_t \underline{D}(\underline{r}, t)\,, \quad (8)$$

with no source current densities present. For propagation in the z direction, $\partial_x \equiv \partial_y \equiv 0$ and $\nabla \equiv \underline{u}_z\,\partial_z$, and we obtain the following six differential equations:

$$\partial_z E_x(z,t) = -\mu_0\partial_t H_y(z,t)\,, \quad (9)$$

$$\partial_z E_y(z,t) = \mu_0\partial_t H_x(z,t)\,, \quad (10)$$

$$\partial_z H_x(z,t) = \epsilon_0\partial_t\left[\underline{u}_y\cdot\left(\underline{\underline{\epsilon}}_r * \underline{E}\right)(z,t)\right], \quad (11)$$

$$\partial_z H_y(z,t) = -\epsilon_0\partial_t\left[\underline{u}_x\cdot\left(\underline{\underline{\epsilon}}_r * \underline{E}\right)(z,t)\right], \quad (12)$$

$$0 = \partial_t\left[\underline{u}_z\cdot\left(\underline{\underline{\epsilon}}_r * \underline{E}\right)(z,t)\right], \quad (13)$$

$$0 = \partial_t H_z(z,t)\,. \quad (14)$$

Equation (14) implies $H_z(z,t) \equiv H_z(z)$, and the initial conditions to be prescribed later ensure that $H_z(z,t) = 0$. Therefore, we define the column 5–vector

$$[\underline{F}](z,t) = [E_x(z,t)\,,\ E_y(z,t)\,,\ H_x(z,t)\,,\ H_y(z,t)\,,\ E_z(z,t)]^T \quad (15)$$

to contain the five nonzero components of the electromagnetic field, with superscript ^{T} denoting the transpose. Then we write (9)–(13) compactly in matrix form as

$$\left[\underline{\underline{J}}\right] \partial_z \left[\underline{F}\right](z,t) = \left[\underline{\underline{Q}}\right] \partial_t \left[\underline{F}\right](z,t) + \epsilon_0 \partial_t \left\{ \left(\left[\underline{\underline{A}}\right] * \left[\underline{F}\right] \right)(z,t) \right\} . \quad (16)$$

In (16), the 5×5 matrix $\left[\underline{\underline{A}}\right](z,t)$ contains the constitutive properties. It is identically null–valued for $z \notin [z_0, z_1]$, while

$$\left[\underline{\underline{A}}\right](z,t) = \begin{bmatrix} 0 & 0 & 0 & 0 & 0 \\ 0 & 0 & 0 & 0 & 0 \\ A_{31}(z,t) & A_{32}(z,t) & 0 & 0 & A_{35}(z,t) \\ -A_{41}(z,t) & -A_{31}(z,t) & 0 & 0 & -A_{45}(z,t) \\ A_{45}(z,t) & A_{35}(z,t) & 0 & 0 & A_{55}(z,t) \end{bmatrix}, \quad z \in [z_0, z_1] , \quad (17)$$

with

$$A_{31} = \left(\epsilon_a \sin^2\chi + \epsilon_b \cos^2\chi - \epsilon_c\right) \sin\frac{\pi(z-z_0)}{\Omega} \cos\frac{\pi(z-z_0)}{\Omega} , \quad (18)$$

$$A_{32} = \left(\epsilon_a \sin^2\chi + \epsilon_b \cos^2\chi\right) \sin^2\frac{\pi(z-z_0)}{\Omega} + \epsilon_c \cos^2\frac{\pi(z-z_0)}{\Omega} - \delta(t) , \quad (19)$$

$$A_{35} = (\epsilon_b - \epsilon_a) \sin\chi \cos\chi \sin\frac{\pi(z-z_0)}{\Omega} , \quad (20)$$

$$A_{41} = \left(\epsilon_a \sin^2\chi + \epsilon_b \cos^2\chi\right) \cos^2\frac{\pi(z-z_0)}{\Omega} + \epsilon_c \sin^2\frac{\pi(z-z_0)}{\Omega} - \delta(t) , \quad (21)$$

$$A_{45} = (\epsilon_b - \epsilon_a) \sin\chi \cos\chi \cos\frac{\pi(z-z_0)}{\Omega} , \quad (22)$$

$$A_{55} = \epsilon_a \cos^2\chi + \epsilon_b \sin^2\chi - \delta(t) ; \quad (23)$$

the other two 5×5 matrixes in (16) are

$$\left[\underline{\underline{J}}\right] = \mathrm{diag}\,[1,1,1,1,0] , \quad (24)$$

$$\left[\underline{\underline{Q}}\right] = \begin{bmatrix} 0 & 0 & 0 & -\mu_0 & 0 \\ 0 & 0 & \mu_0 & 0 & 0 \\ 0 & \epsilon_0 & 0 & 0 & 0 \\ -\epsilon_0 & 0 & 0 & 0 & 0 \\ 0 & 0 & 0 & 0 & \epsilon_0 \end{bmatrix} . \quad (25)$$

Note that for calculations involving chiral STFs exhibiting magnetic properties, a 6–vector formalism is needed.

2.3. Finite–Difference Algorithm

An analytical solution for (16) is not known, and thus we employed a FD scheme. Upon discretization of both space and time as $z_i = i\Delta z$, $(i = 0,1,2,3,\ldots)$, and $t_n = n\Delta t$, $(n = 0,1,2,3,\ldots)$, replacing derivatives with central differences, and using the leapfrog method, (16) transforms into the matrix difference equation

$$\begin{aligned}\underline{[J]}\left(\frac{\underline{[F]}_{i+1}^{n}-\underline{[F]}_{i-1}^{n}}{2\Delta z}\right) &= \underline{\underline{[Q]}}\cdot\left(\frac{\underline{[F]}_{i}^{n+1}-\underline{[F]}_{i}^{n-1}}{2\Delta t}\right)\\ &+ \epsilon_0\sum_{m=1}^{n-1}\underline{\underline{[A]}}_{i}^{m}\cdot\left(\frac{\underline{[F]}_{i}^{n-m+1}-\underline{[F]}_{i}^{n-m-1}}{2}\right)\\ &+ \frac{\epsilon_0}{2}\underline{\underline{[A]}}_{i}^{n}\cdot\left(\underline{[F]}_{i}^{1}-\underline{[F]}_{i}^{0}\right).\end{aligned} \tag{26}$$

In (26),

$$\underline{[F]}_{i}^{n}=\underline{[F]}\left(z_i,t_n\right),\quad \underline{\underline{[A]}}_{i}^{n}=\underline{\underline{[A]}}\left(z_i,t_n\right) \tag{27}$$

and $\beta = c_0\Delta t/\Delta z < 1$ for stability [17]. On solving (26) for $\underline{[F]}_{i}^{n+1}$, we obtain the formula

$$\begin{aligned}\underline{[F]}_{i}^{n+1} &= \underline{[F]}_{i}^{n-1}+\beta\,\underline{\underline{[V]}}\cdot\left(\underline{[F]}_{i+1}^{n}-\underline{[F]}_{i-1}^{n}\right)\\ &- \underline{[C]}_{i}^{n}-\underline{\underline{[W]}}_{i}^{n}\cdot\left(\underline{[F]}_{i}^{1}-\underline{[F]}_{i}^{0}\right)\Delta t,\end{aligned} \tag{28}$$

whose last two terms must be calculated only if $z_i \in [z_0, z_1]$. In (28), the matrixes $\underline{\underline{[W]}}(z,t) = \epsilon_0\left[\underline{\underline{Q}}\right]^{-1}\underline{\underline{[A]}}(z,t)$ and $\underline{\underline{[V]}} = c_0^{-1}\left[\underline{\underline{Q}}\right]^{-1}\underline{\underline{[J]}}$. Due to the initial conditions we impose later, the final terms in (26) and (28) are identically null–valued and thus need not be computed.

The bulk of the convolution is contained in the term

$$\underline{[C]}_{i}^{n}=\sum_{m=1}^{n-1}\underline{\underline{[W]}}_{i}^{m}\cdot\left(\underline{[F]}_{i}^{n-m+1}-\underline{[F]}_{i}^{n-m-1}\right)\Delta t, \tag{29}$$

and direct calculation of it accounts for most of the computational time and memory requirements. A quicker algorithm is needed to handle the largest simulations. Such an algorithm, which exploits the chosen Lorentzian properties, and inspired by methods devised for simpler constitutive relations [18, 19], is described in detail elsewhere [20].

2.4. Initial and Boundary Conditions

For a numerical solution to be obtained, (28) requires initial and boundary conditions. The initial condition

$$\underline{[F]}_{i}^{0}=\underline{[0]}\ \forall i\geq 1 \tag{30}$$

ensures that the electromagnetic field is null–valued everywhere at time $t = 0$ except possibly at $z = 0$. The boundary condition

$$\underline{[F]}_{0}^{n}=g\left(t_n\right)\underline{[\varphi]}\left(t_n\right) \tag{31}$$

contains a pulse envelope $g(t)$ that modulates the amplitude of a carrier plane wave $\underline{[\varphi]}(t)$. The column vector $\underline{[\varphi]}(t)$ is defined as

$$\left[\underline{\varphi_\pm}\right](t)=\left[\cos\left(\omega_{\text{car}}t+\phi\right),\pm\sin\left(\omega_{\text{car}}t+\phi\right),\mp\eta_0^{-1}\sin\left(\omega_{\text{car}}t+\phi\right),\eta_0^{-1}\cos\left(\omega_{\text{car}}t+\phi\right),0\right]^T, \tag{32}$$

where $\left[\underline{\varphi}_{+}\right](t)$ represents a LCP plane wave, $\left[\underline{\varphi}_{-}\right](t)$ represents a RCP plane wave, both with carrier wavelength λ_{car} and carrier angular frequency $\omega_{\text{car}} = 2\pi c_0/\lambda_{\text{car}}$. The pulse envelope

$$g(t) = A_{\text{p}} \frac{t}{\tau_{\text{p}}} \exp(\frac{-2t}{\tau_{\text{p}}}) . \tag{33}$$

indicates a pulse with a rounded shape whose duration depends on the parameter τ_{p}; ϕ denotes the carrier phase.

2.5. Parameter Values

For the calculations reported in this chapter, we chose $h = +1$, i.e. the chosen chiral STFs are structurally right–handed. The oscillator strengths $p_a = 0.40, p_b = 0.52, p_c = 0.42$, angle of rise $\chi = 20°$, and structural half–period $\Omega = 200$ nm; we chose resonance wavelengths $\lambda_a = \lambda_c = 280$ nm, $\lambda_b = 290$ in the ultraviolet regime [15]. These choices fix the center wavelength of the Bragg regime λ_{Br} at approximately 516 nm [21]. Some variables, including $N_{a,b,c}$, A_{p}, τ_{p}, and ϕ, were picked to suit individual cases; their values are listed in the discussions of those cases. The variables Δz and β, and hence Δt, were also chosen based on the computational needs of each case.

3. Half–Space Problems

To examine the CBP in the time domain, we first consider problems in which an ultrashort optical pulsed plane wave falls normally upon a chiral STF half–space—i.e., we increase z_1 so that it is beyond the domain of computation.

3.1. Pulse Bleeding and the Light Pipe

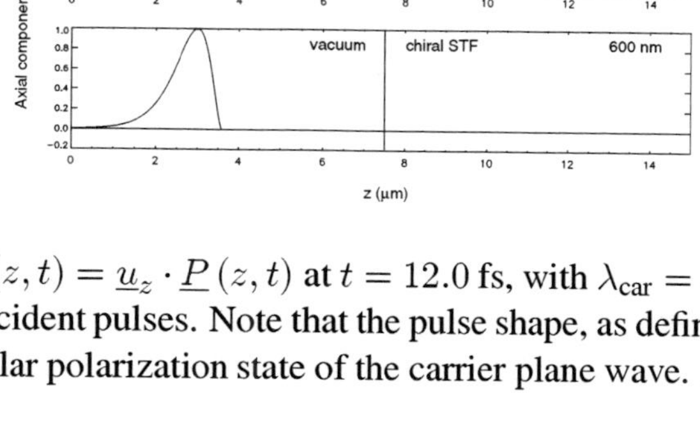

Figure 2. Plots of $P_z(z,t) = \underline{u}_z \cdot \underline{P}(z,t)$ at $t = 12.0$ fs, with $\lambda_{\text{car}} = 430, 516, 600$ nm (top to bottom) showing incident pulses. Note that the pulse shape, as defined by P_z, is the same regardless of the circular polarization state of the carrier plane wave. *Adapted from Geddes and Lakhtakia [22].*

For these calculations [22], $N_{a,b,c} = 100$, $A_{\rm p} = 1$ V/m, $\tau_{\rm p} = 2\lambda_{\rm car}/c_0$, and $\phi = 0$. The vacuum / chiral STF interface was positioned at $z_0 = 7.5$ μm, and we selected $\Delta z =$ 5 nm and $\beta = 0.9$ so that $\Delta t = 0.015$ fs. We examined pulsed plane waves with carrier wavelengths of $\lambda_{\rm car} = 430, 516$, and 600 nm, i.e., below, within, and above the Bragg regime. Note that even the spectrum of a pulse whose carrier wavelength lies outside the Bragg regime may overlap with it. Snapshots of the axial component of the Poynting vector, a measure of power per unit area, or

$$P_z(z,t) = \underline{u}_z \cdot \left[\underline{E}(z,t) \times \underline{H}(z,t)\right], \tag{34}$$

are plotted at two different instants of time in Figures 2–5. Note that when $P_z > 0$, energy flows in the $+z$ direction; and when $P_z < 0$, energy flows in the $-z$ direction. In Figure 2, at time $t = 12.0$ fs, only the incident pulses are captured in the snapshots, and the pulse shapes, which are independent of the polarization state of the carrier plane wave, are clearly visible. Figures 3–5 show P_z at time $t = 49.5$ fs, after the incident pulse has reached the vacuum / chiral STF interface. In those plots, part of each incident pulse has entered the chiral STF and become a refracted pulse, and part of each incident pulse has turned back from the film and become a reflected pulse. In Figure 3, the reflected pulses are of short duration and contain little energy because the incident pulses had LCP carrier plane waves. Thus, the CBP was not manifested by the structurally right–handed chiral STF. However, in Figure 4, and when $\lambda_{\rm car} = 516$ nm, the reflected pulse is of much longer duration and hence contains more energy; the CBP is manifested in this case. Furthermore, inspection of the magnified views of P_z near the vacuum / chiral STF interface at $t = 49.5$ fs, as shown in Figure 5, reveals that there is a continuous flow of energy across the interface when the conditions for the CBP are present, but not if otherwise. This flow of energy through a so–called light pipe, and termed pulse bleeding, is the spatiotemporal fingerprint of the CBP in the time domain. Fourier transformation of the time–domain results recovers the more familiar frequency domain results [22], and the pulse bleeding phenomenon is apparent for post–resonant Bragg regimes too [23].

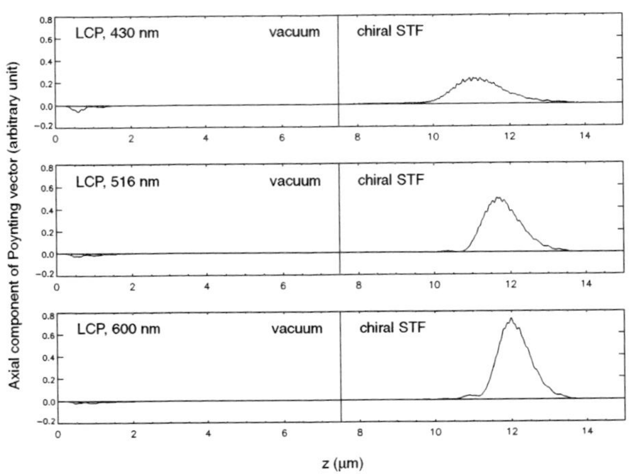

Figure 3. Plots of P_z at $t = 49.5$ fs for $\lambda_{\rm car} = 430, 516$, and 600 nm resulting from incident pulses whose carrier plane waves were LCP. *Adapted from Geddes and Lakhtakia [22].*

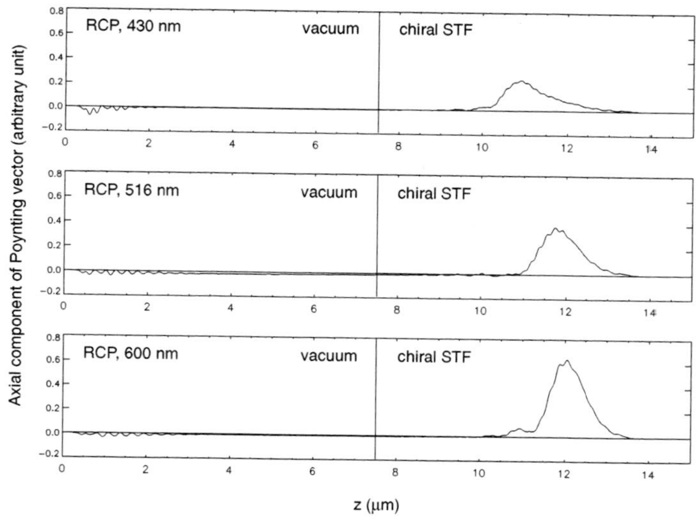

Figure 4. Same as Figure 3, except resulting from incident pulses whose carrier plane waves were RCP. *Adapted from Geddes and Lakhtakia [22].*

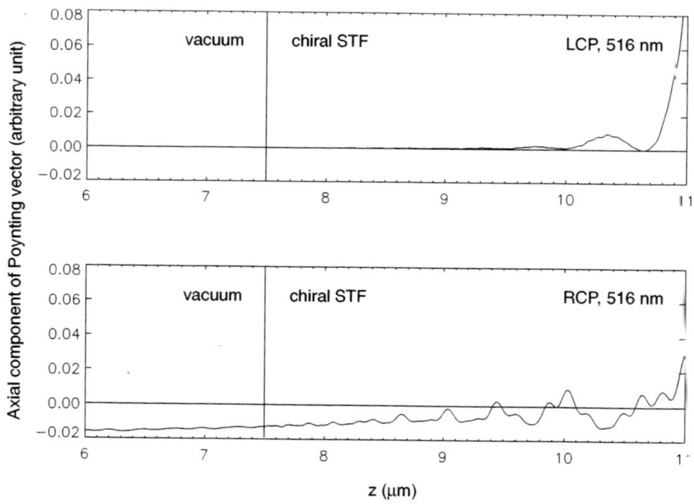

Figure 5. Magnified views of P_z in the region containing the vacuum / chiral STF interface from Figures 3 and 4, when $\lambda_{\rm car} = 516$ nm. Note that the light pipe is clearly absent when the incident pulses had LCP carrier plane waves (top), but present if otherwise (bottom). *Adapted from Geddes and Lakhtakia [22].*

3.2. Effects of Carrier Phase

Having clarified the role of $\lambda_{\rm car}$ and carrier plane wave polarization in the manifestation of the time domain CBP, we now turn to the effects of carrier phase. For these calculations [16], $N_{a,b,c} = 100$, $A_{\rm p} = 10^6$ V/m, and $\tau_{\rm p} = 3.44$ fs; the boundary between vacuum

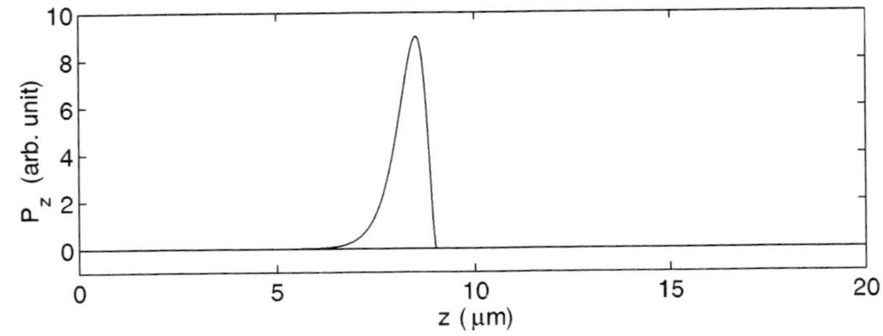

Figure 6. Plot of P_z showing the incident pulse shape at $t = 30.1$ fs. The shape is independent of both the polarization of the carrier plane wave and the carrier phase. The vacuum / chiral STF interface lies at $z = 20\ \mu$m, and hence is not shown. *Adapted from Geddes and Lakhtakia [16].*

and film lies at $z_0 = 20\ \mu$m. A snapshot of the incident pulse [1], whose shape is independent of carrier phase, wavelength, and polarization, is shown in Figure 6. Plots of P_z for reflected pulses at $t = 132$ fs for three carrier wavelengths ($\lambda_{\rm car} = 430, 516,$ and 600 nm), two carrier phases ($\phi = 0$ and $\pi/2$), and two carrier polarizations (LCP and RCP) are given in Figures 7 and 8. In the former, P_z at the leading edges of the reflected pulses increases with decreasing $\lambda_{\rm car}$, presumably because of the greater impedance mismatch between vacuum and film at lower wavelengths. Pulse bleeding is apparent in all reflected pulses when the incident carrier plane wave is RCP (Figure 8), because the spectrums of the incident pulses are wide enough to overlap the Bragg regime even when $\lambda_{\rm car}$ lies outside it. The rapid oscillations—which depend on the carrier phase ϕ—in P_z in the leading portions of those pulses are due to interference between LCP and RCP components in each reflected pulse—i.e., even though the carrier plane wave of an incident pulse is pure RCP, the reflected pulse contains both LCP and RCP components. The LCP component is created—at least largely—by reflection due to impedance mismatch near the surface of the film. The reflected pulses' long tails contain mostly RCP light due to pulse bleeding but little LCP light, and hence large oscillations in P_z are not observed there. The refracted pulse shapes in Figures 9 and 10 differ little with carrier phase, but notice that the refracted pulses carry more energy with increasing $\lambda_{\rm car}$ due to the presence of absorption resonances at ultraviolet wavelengths. These phenomenons, and reflection and refraction of pulses from nonlinear chiral STFs, were examined in more detail elsewhere [16], along with their connections to the interrelationship between phase, length, and time in optical nanotechnology [24].

[1]Note that though the scaling of the arbitrary units may be different from subsection to subsection in this chapter, comparisons of power densities are appropriate within each subsection.

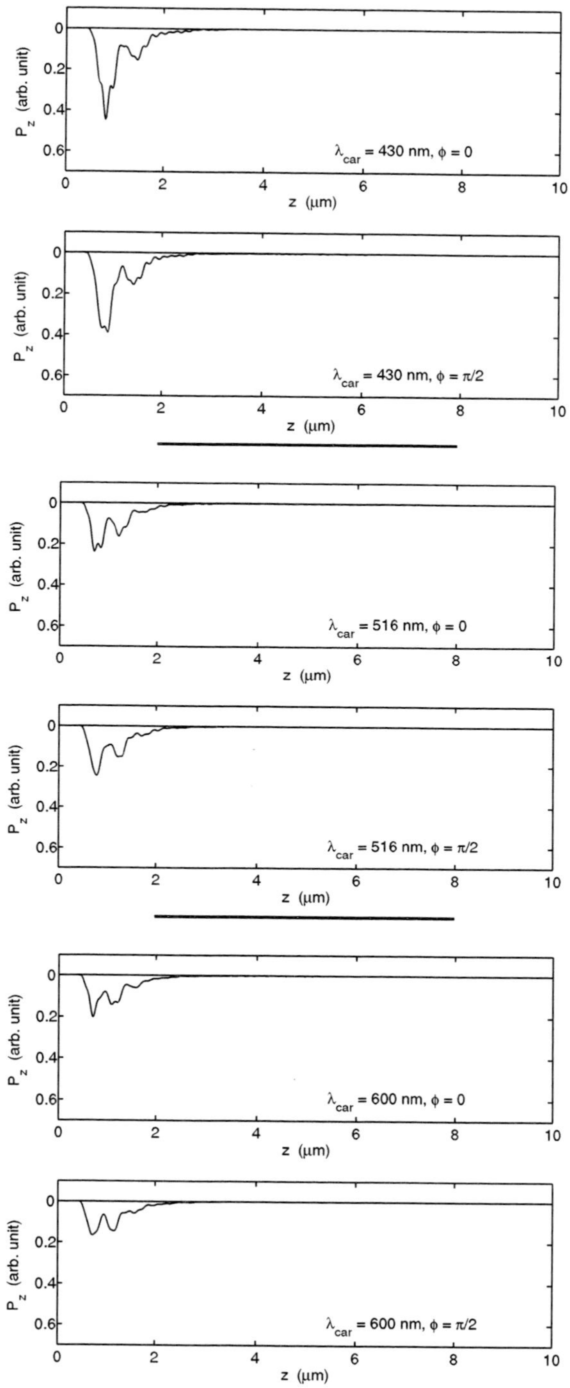

Figure 7. Plots of P_z at $t = 132$ fs, which show pulses reflected by a chiral STF. The carrier plane wave of the incident pulse was LCP; the carrier phase is either $\phi = 0$ or $\phi = \pi/2$; the carrier wavelength is $\lambda_{\text{car}} = 430$ nm (top), $\lambda_{\text{car}} = 516$ nm (middle), or $\lambda_{\text{car}} = 600$ nm (bottom). The vacuum / chiral STF interface at $z = 20$ μm is not depicted. *Adapted from Geddes and Lakhtakia [16].*

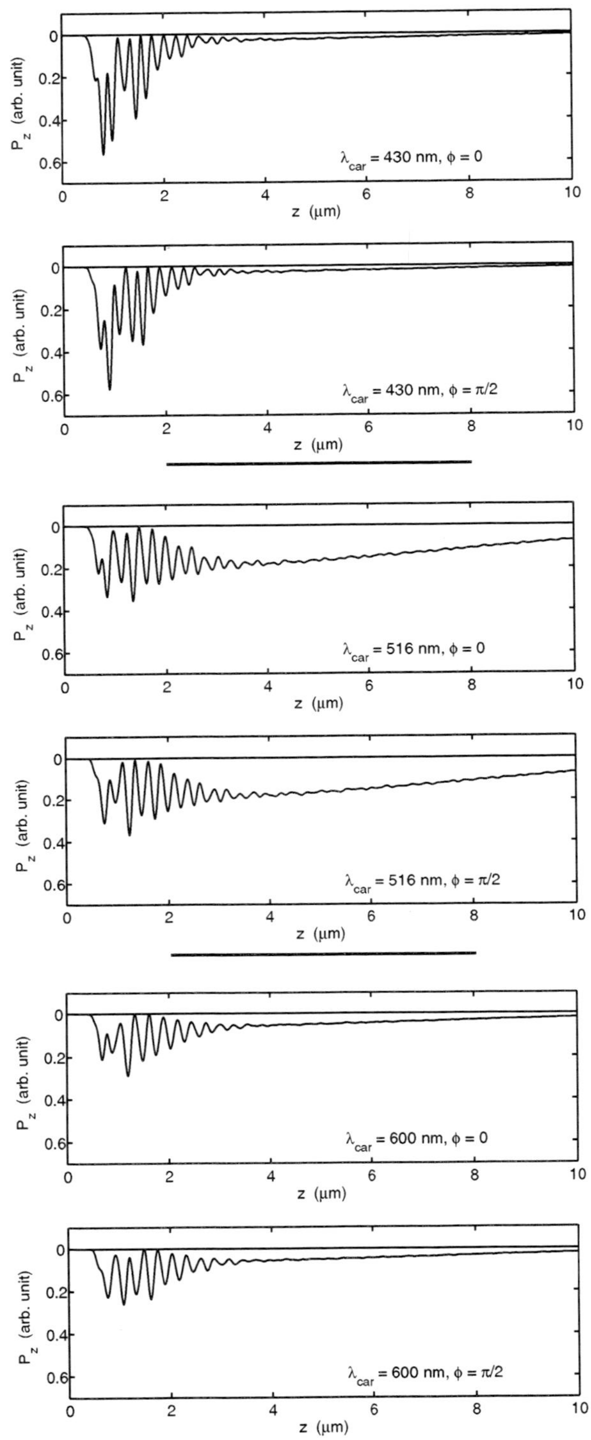

Figure 8. Same as Figure 7, except that the carrier plane wave of the incident pulse was RCP. *Adapted from Geddes and Lakhtakia [16].*

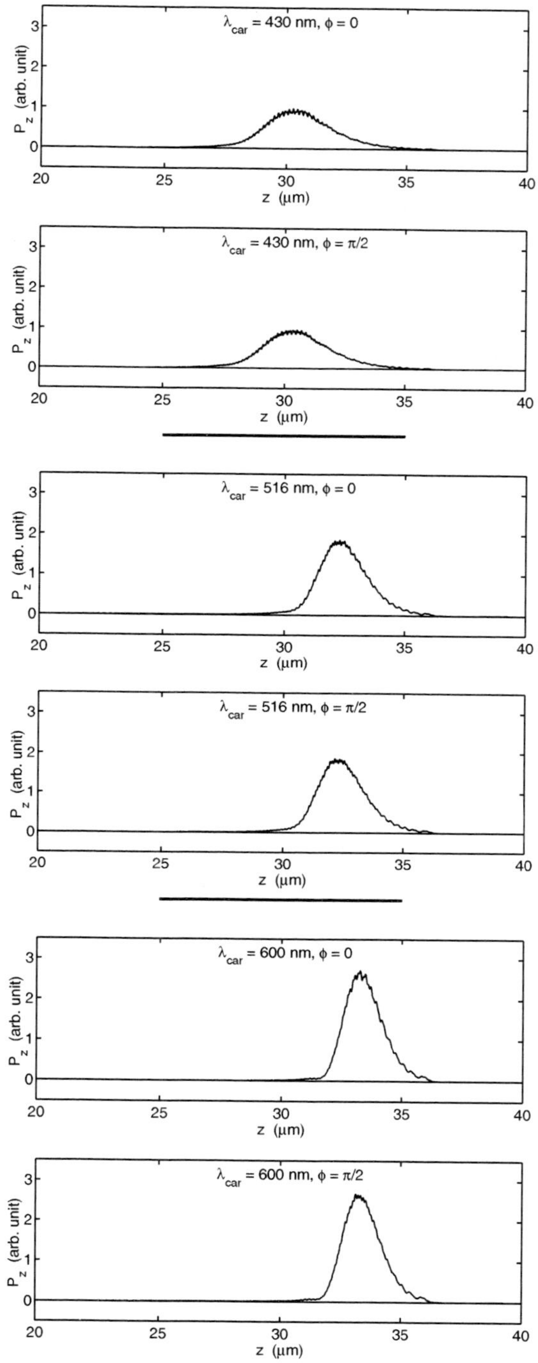

Figure 9. Plots of P_z at $t = 132$ fs, which show pulses refracted into a chiral STF. The carrier plane wave of the incident pulse was LCP; the carrier phase is either $\phi = 0$ or $\phi = \pi/2$; the carrier wavelength is $\lambda_{car} = 430$ nm (top), $\lambda_{car} = 516$ nm (middle), or $\lambda_{car} = 600$ nm (bottom). The vacuum / chiral STF interface at $z = 20\ \mu$m is not depicted. *Adapted from Geddes and Lakhtakia [16].*

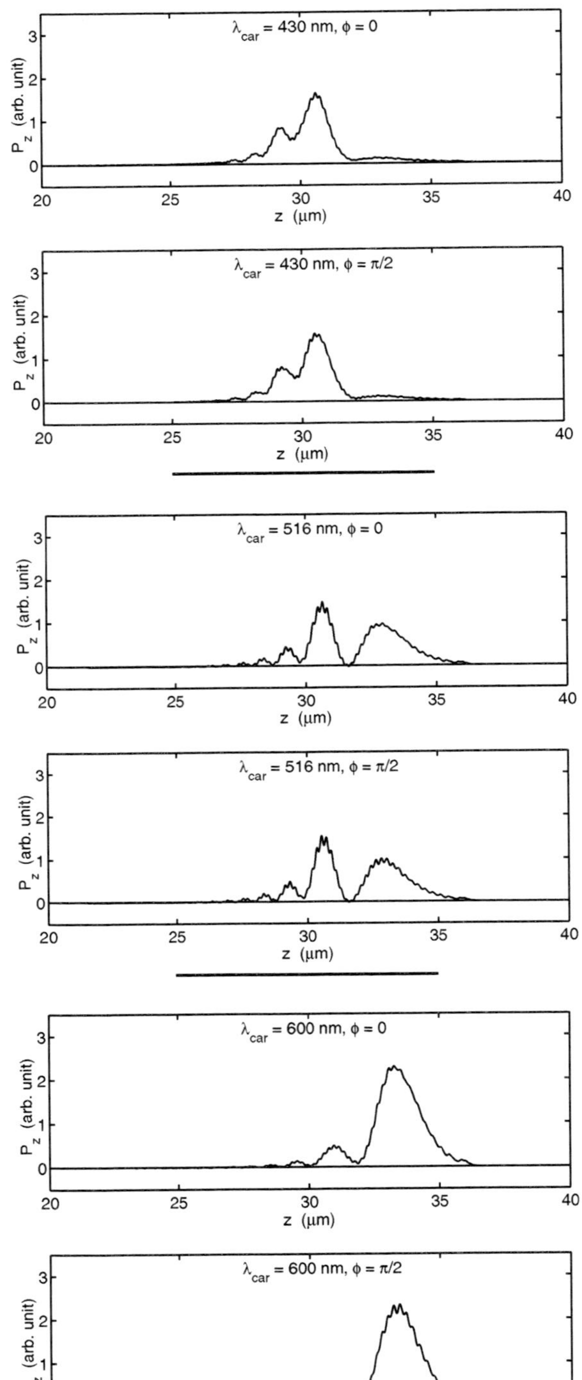

Figure 10. Same as Figure 9, except that the carrier plane wave of the incident pulse was RCP. *Adapted from Geddes and Lakhtakia [16].*

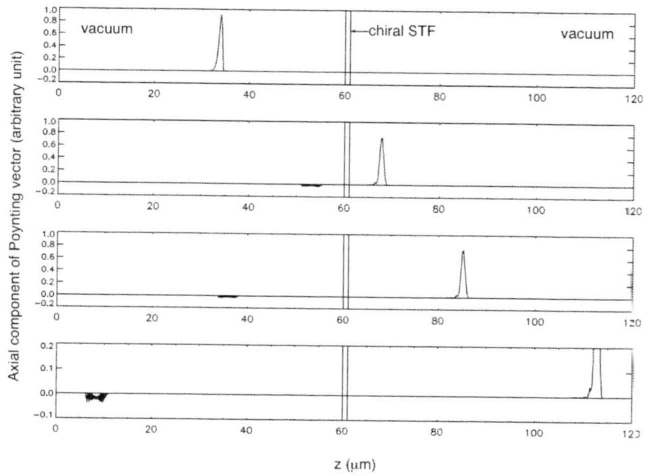

Figure 11. Snapshots of P_z at $t = 115.3, 230.6, 288.2$, and 380.4 fs—showing the spatiotemporal evolution of a narrow–extent pulse modulating a RCP carrier plane wave which impinges on a structurally right–handed chiral STF of thickness $z_1 - z_0 = 5\Omega$. The incident pulse strikes the film from the left, and gives rise to a reflected pulse (left of the film) and a transmitted pulse (right of the film). *Adapted from Geddes and Lakhtakia [20].*

4. Finite–Thickness Problems

In Section 3., FD computations elucidated the time–domain mechanism responsible for the CBP in chiral STFs—pulse bleeding through a light pipe. In this section, the phenomenon of pulse bleeding is used to explain the distortion of optical pulses as they traverse a chiral STF slab.

4.1. Ripening of the CBP

We now examine the ripening of the CBP, familiar from frequency domain research [25], as the thickness of a chiral STF slab increases. For these calculations [20], $N_{a.b,c} = 10^5$, $A_p = 1$ V/m, $\tau_p = 2\lambda_{car}/c_0$, and $\phi = 0$; the value $z_0 = 60\ \mu$m is fixed, while the slab thickness $z_1 - z_0$ was varied. We chose the grid size $\Delta z = 4$ nm and stability parameter $\beta = 0.9$ to fix $\Delta t = 0.012$ fs. Two cases are shown: in one $z_1 - z_0 = 5\Omega$; in the other $z_1 - z_0 = 100\Omega$. Plots of P_z are given at $t = 115.3, 230.6, 288.2$, and 380.4 fs to show the spatiotemporal evolution of each pulse. Conditions are right for the CBP to occur in both Figures 11 and 12, but the transmitted pulse at $t = 380.4$ fs in the former is distorted little in comparison to the incident pulse at $t = 115.3$ fs, while much distortion occurs in the latter case. This is because in the latter case, there are many more structural periods for the refracted pulse to pass through as it traverses the film, and hence pulse bleeding and distortion of the refracted pulse can occur for a longer time. Thus the transmitted pulse in Figure 12 possesses little energy compared to the transmitted pulse in Figure 11 in addition to its shape being more distorted. Note by examining Figure 13 that an incident pulse whose carrier plane wave is LCP passes through a 100Ω thick chiral STF having undergone some absorption, but

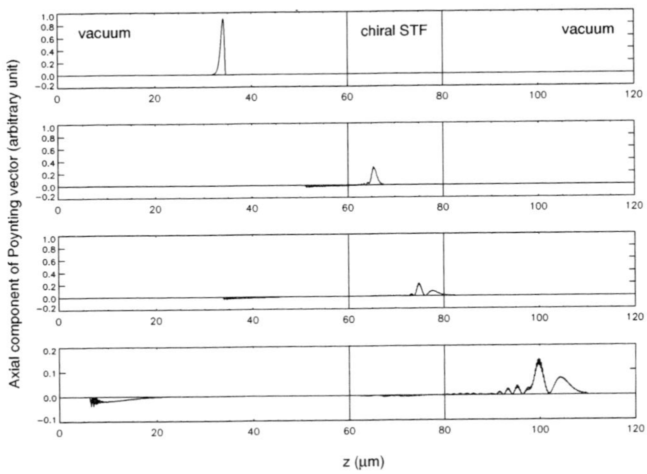

Figure 12. Same as Figure 11 except that the chiral STF is 100Ω thick. *Adapted from Geddes and Lakhtakia [20].*

not nearly the distortion experienced by an incident pulse whose carrier plane wave is RCP. Thus the CBP is at least partly responsible for the distortion of ultrashort pulses that traverse chiral STF slabs. More details about this process are available elsewhere [20].

4.2. Videopulse Propagation

Finally in this section, we examine the propagation of videopulses through chiral STF slabs. Videopulses are optical pulses of such short duration that they may encompass a few or even less than one optical cycle. For these calculations, $N_{a,b,c} = 10^5$, $A_p = 0.20$ V/m, $\tau_p = 0.688$ fs, and $\phi = 0$; the boundaries of the chiral STF were delineated by $z_0 = 8$ μm and $z_1 = 12$ μm. Figures 14–16 show the results. Even for incident videopulses of such short duration as these—about $1\frac{1}{2}$ optical cycles with $\lambda_{car} = 516$ nm—the phenomenon of pulse bleeding is apparent. Notice however, that both incident pulses with LCP and RCP carrier plane waves are quite distorted when they exit the chiral STF into the half–space $z > z_1$, an observation that indicates that while the CBP is important for shaping optical pulses, it is not the only effect which does so. More details about videopulse propagation through chiral STF slabs is given elsewhere [26].

5. Conclusion

Based on the results presented in Sections 3.–4., we can draw the following conclusions regarding the propagation of ultrashort optical pulsed plane waves through chiral STFs:

- the phenomenon of pulse bleeding, a polarization–dependent transfer of energy from a refracted pulse to a reflected pulse across a light pipe, is the spatiotemporal fingerprint of the CBP in the time domain (even for videopulses), and

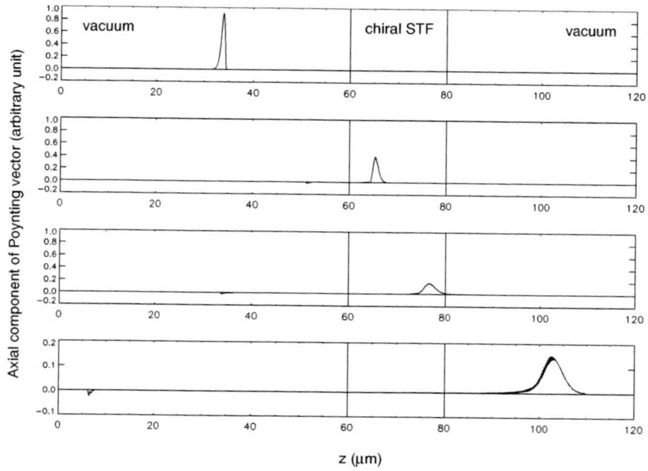

Figure 13. Same as Figure 12 except that the incident carrier plane wave is LCP. *Adapted from Geddes and Lakhtakia [20].*

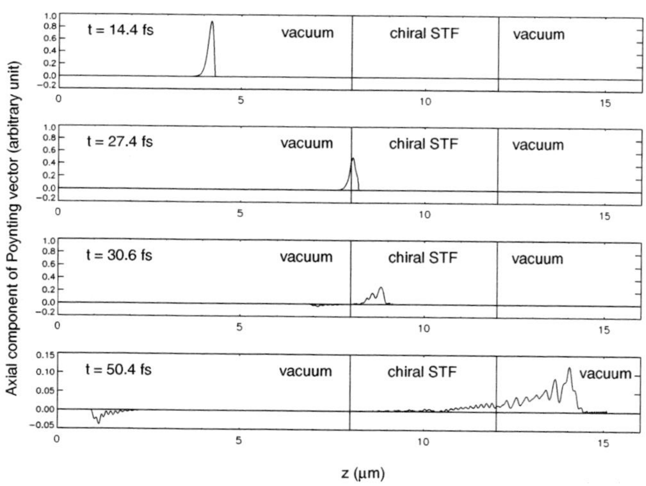

Figure 14. Plots of P_z at $t = 14.4, 27.4, 30.6,$ and 50.4 fs showing the tranformation of an incident pulse into reflected and transmitted pulses. The incident pulse, shown at left in the first panel, has a LCP carrier plane wave. *Adapted from Geddes and Lakhtakia [26].*

- pulse bleeding is responsible for at least some of the pulse shaping effects observed when ultrashort optical pulses traverse chiral STFs, but not all.

The mathematical analogy between the constitutive relations of chiral STFs and cholesteric liquid crystals (CLCs) implies that many of the conclusions of this chapter apply also to CLCs [27, 28, 29].

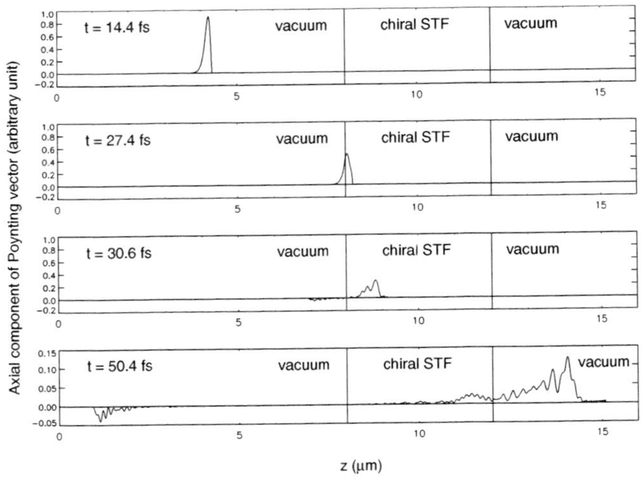

Figure 15. Same as Figure 15, except that the incident pulse possesses a RCP carrier plane wave. *Adapted from Geddes and Lakhtakia [26].*

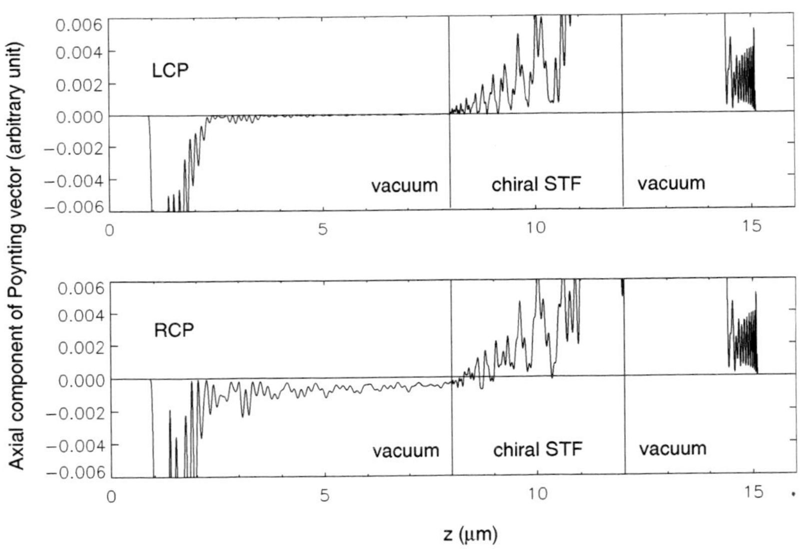

Figure 16. Magnified views of the bottom panels ($t = 50.4$ fs) of Figures 14 and 15 illustrating the pulse bleeding phenomenon for videopulses. *Adapted from Geddes and Lakhtakia [26].*

Many lines of research in time–domain pulse propagation through STFs and other complex materials, e.g. isotropic chiral materials [30], remain unexplored. To advance our understanding of such pulse propagation and use our knowledge to design pulse shaping devices, the phenomenology described in this chapter must be extended to include quantitative techniques for measuring pulse distortion.

Acknowledgments

I thank Akhlesh Lakhtakia for his patient guidance in the conduct of the research reviewed in this chapter. Computer facilities were supplied, in part, by the Pittsburgh Supercomputing Center. I am grateful for the support of an National Science Foundation Graduate Research Fellowship, a SPIE Educational Scholarship, and a Proctor & Gamble Summer Undergraduate Research Fellowship.

References

[1] M. W. Horn, M. D. Pickett, R. Messier, and A. Lakhtakia. Blending of nanoscale and microscale in uniform large–area sculptured thin film architectures. *Nanotechnology*, **15**:303–310, 2004.

[2] M. W. Horn, M. D. Pickett, R. Messier, and A. Lakhtakia. Selective growth of sculptured nanowires on microlithographic substrates. *J. Vac. Sci. Technol. B*, **22**:3426–3430, 2004.

[3] A. Lakhtakia and R. Messier. *Sculptured Thin Films: Nanoengineered Morphology and Optics*. SPIE Optical Engineering Press, Bellingham, WA, USA, 2005.

[4] I. J. Hodgkinson, A. Lakhtakia, and Q. h. Wu. Experimental realization of sculptured–thin–film polarization–discriminatory light–handedness inverters. *Opt. Eng.*, **39**:2831–2834, 2000.

[5] Q. Wu, I. J. Hodgkinson, and A. Lakhtakia. Circular polarization filters made of chiral sculptured thin films: Experimental and simulation results. *Opt. Eng.*, **39**:1863–1868, 2000.

[6] I. J. Hodgkinson, Q. h. Wu, A. Lakhtakia, and M. W. McCall. Spectral–hole filter fabricated using sculptured thin–film technology. *Opt. Commun.*, **177**:79–84, 2000.

[7] I. Hodgkinson and Q. h. Wu. Serial bideposition of anisotropic thin films with enhanced linear birefringence. *Appl. Opt.*, **38**:3621–3625, 1999.

[8] I. Hodgkinson, Q. h. Wu, B. Knight, A. Lakhtakia, and K. Robbie. Vacuum deposition of chiral sculptured thin films with high optical activity. *Appl. Opt.*, **39**:642–649, 2000.

[9] T. R. Gosnell and A. J. Taylor, editors. *Selected Papers on Ultrafast Laser Technology*. SPIE Optical Engineering Press, Bellingham, WA, USA, 1991.

[10] W. H. Knox. Ultrafast technology in telecommunications. *IEEE J. Sel. Topics Quantum Electron.*, **6**:1273–1278, 2000.

[11] N. Bloembergen. From nanosecond to femtosecond science. *Rev. Modern Phys.*, **71**:S283–S287, 1999.

[12] J. B. Geddes III. Traversal of optical pulses through dielectric thin–film helicoidal bianisotropic mediums. Master's thesis, The Pennsylvania State University, 2001.

[13] J. D. Jackson. *Classical Electrodynamics*. John Wiley & Sons, Inc., New York, NY, USA, 3rd edition, 1999.

[14] V. C. Venugopal and A. Lakhtakia. Sculptured thin films: Conception, optical properties, and applications. In O. N. Singh and A. Lakhtakia, editors, *Electromagnetic Fields in Unconventional Materials and Structures* , chapter 5, pages 151–216. Wiley Interscience, New York, NY, USA, 2000.

[15] C. Kittel. *Introduction to Solid State Physics*. Wiley Eastern, New Delhi, India, 1974.

[16] J. B. Geddes III and A. Lakhtakia. Effects of carrier phase on reflection of optical narrow–extent pulses from axially excited chiral sculptured thin films. *Opt. Commun.*, **225**:141–150, 2003.

[17] N. Gershenfeld. *The Nature of Mathematical Modeling*. Cambridge University Press, Cambridge, UK, 1999.

[18] R. Luebbers, F. P. Hunsberger, K. S. Kunz, R. B. Standler, and M. Schneider. A frequency–dependent finite–difference time–domain formulation for dispersive materials. *IEEE Trans. Electromagnetic Compat.*, **32**:222–227, 1990.

[19] R. J. Luebbers, F. Hunsberger, and K. S. Kunz. A frequency–dependent finite–difference time–domain formulation for transient propagation in plasma. *IEEE Trans. Antennas Propagat.*, **39**:29–34, 1991.

[20] J. B. Geddes III and A. Lakhtakia. Time–domain simulation of the circular Bragg phenomenon exhibited by chiral sculptured thin films. *Eur. Phys. J. Appl. Phys.*, **14**:97–105, 2001. *Erratum*: **16**:247, 2001.

[21] A. Lakhtakia. Spectral signatures of axially excited slabs of dielectric thin–film helicoidal bianisotropic mediums. *Eur. Phys. J. Appl. Phys.*, **8**:129–137, 1999.

[22] J. B. Geddes III and A. Lakhtakia. Reflection and transmission of narrow–extent pulses by axially excited chiral sculptured thin films. *Eur. Phys. J. Appl. Phys.*, **13**:3–14, 2001. Erratum: 16:247, 2001.

[23] J. Wang, A. Lakhtakia, and J. B. Geddes III. Multiple Bragg regimes exhibited by a chiral sculptured thin film half–space on axial excitation. *Optik*, **113**:213–222, 2002.

[24] A. Lakhtakia and J. B. Geddes III. Nanotechnology for optics is a phase–length–time sandwich. *Opt. Eng.*, **43**:2410–2417, 2004.

[25] V. C. Venugopal and A. Lakhtakia. On selective absorption in an axially excited slab of a dielectric thin–film helicoidal bianisotropic medium. *Opt. Commun.*, **145**:171–187, 1998. Erratum: **161**:370, 1999.

[26] J. B. Geddes III and A. Lakhtakia. Videopulse bleeding in axially excited chiral sculptured thin films in the Bragg regime. *Eur. Phys. J. Appl. Phys.*, **17**:21–24, 2002.

[27] J. B. Geddes III, M. W. Meredith, and A. Lakhtakia. Circular Bragg phenomenon and pulse bleeding in cholesteric liquid crystals. *Opt. Commun.*, **182**:45–57, 2000.

[28] M. W. Meredith and A. Lakhtakia. Time–domain signature of an axially excited cholesteric liquid crystal. Part I: Narrow–extent pulses. *Optik*, **111**:443–453, 2000.

[29] J. B. Geddes III and A. Lakhtakia. Time–domain signature of an axially excited cholesteric liquid crystal. Part II: Rectangular wide–extent pulses. *Optik*, **112**:62–66, 2001.

[30] A. Lakhtakia, editor. *Selected Papers on Natural Optical Activity*. SPIE Optical Engineering Press, Bellingham, WA, USA, 1990.

Joseph B. Geddes III graduated from the Pennsylvania State University with bachelors and masters degrees in 2001, and with a doctoral degree in 2006—all in Engineering Science and Mechanics. He won the Schreyer Honors College Dean's Research Award and the Xerox Research Award in 2001 for his BS and MS theses, respectively, and he was awarded a NSF Graduate Fellowship and a SPIE Educational Scholarship. His work centers on the electromagnetics of complex materials, and he has published twelve journal papers in that field. To date, his research has focussed on the time–domain responses of sculptured thin films to ultrashort optical pulses.

In: Frontiers in Optical Technology: Materials and Devices ISBN 1-60021-084-8
Editors: P.K. Choudhury and O.N. Singh, pp. 23-72 © 2007 Nova Science Publishers, Inc.

Chapter 2

ONE-DIMENSIONAL BIGYROTROPIC MAGNETIC PHOTONIC CRYSTALS: NON-DEFECT AND DEFECT STRUCTURES

N.N. Dadoenkova, I.L. Lyubchanskii, M.I. Lyubchanskii, E.A. Shapovalov[1] *and Y.P. Lee*[2]
[1]Donetsk Institute for Physics and Technology of the National Academy of Sciences of Ukraine, 72, R. Luxemburg str., 83114, Donetsk, Ukraine
[2]Quantum Photonic Science Research Center and Department of Physics, Hanyang University, 17 Haendoang-Dong, Sundong-Ku, Seoul, 133-791, Korea

Abstract

This chapter is devoted to a new class of photonic structures, namely, magnetic photonic crystals (MPCs), which are artificial periodic multilayer structures with either magnetic-nonmagnetic or magnetic-magnetic unit cells, with periods comparable with the wavelength of electromagnetic radiation. Special attention is paid to MPCs containing bigyrotropic magnetic materials which are characterized by the off-diagonal components of both permittivity and permeability tensors. The 4×4 transfer matrix method is applied for polar, longitudinal and transversal magneto-optical configurations, and the reflection/transmission spectra for normal and oblique incidence of electromagnetic radiation are presented. Linear optical (transmittivity) and magneto-optical (Faraday rotation) properties of MPCs are discussed. Recently published results of theoretical and experimental investigations of MPCs as well as possible applications of such structures are reviewed.

1. Introduction

Photonic crystals (PCs) are artificial one- (1D), two- (2D), or three-dimensional (3D) periodic structures made from materials with different refractive indexes (RIs) (1; 2), which show remarkable properties in the electromagnetic wave (EMW) spectra, namely existence of forbidden regions in the transmission and reflection spectra, called as photonic band gaps (PBGs). These PCs are very promising materials for application in modern photonics (3; 4). Magnetic photonic crystals (MPCs), in contrast to conventional PCs, are made from magnetic-magnetic, or magnetic-nonmagnetic components. Optical properties of MPCs are very sensitive to the direction and amplitude of the external magnetic field applied to MPC (5). Up to now a lot of papers devoted to the investigation of MPCs have been published (see, for example, review paper (5)).

In this Chapter we shall pay our attention to 1D MPCs, constructed of alternating magnetic and nonmagnetic layers. Particularly, we shall consider structures on the base of the magnetic yttrium-iron garnet (YIG) family and nonmagnetic gadolinium-gallium garnet (GGG), which are widely used in modern magneto-optics (6). One important property of YIG in the near infrared region is its bigyrotropy (7), i.e., dependence of YIG magneto-optical properties on off-diagonal components of both permittivity and permeability tensors. Optical properties of the bigyrotropic uniform and layered media were investigated in numerous publications (7; 8; 9; 10; 11; 12; 13; 14). The Chapter is organized as follows. In the first section we outline the solution of Maxwell equations for the uniform bigyrotropic medium. The second section is devoted to 4 x 4 transfer matrix method with application to the bigyrotropic magneto-optical layered structures. In the third section, we consider the ideal bigyrotropic MPCs with different orientations of the magnetization vector inside the magnetic layers. In the fourth section, we present the results of numerical simulations for non-defect MPCs. The fifth section is devoted to MPCs, containing a single defect (magnetic or nonmagnetic one), as well as to MPCs with the multiple defects. In conclusion, we outline a perspectives for a future investigations of 1D MPCs.

2. Uniform Bigyrotropic Media

Maxwell equations governing the electromagnetic field in the medium are

$$rot\boldsymbol{E} + \frac{1}{c}\frac{\partial \boldsymbol{B}}{\partial t} = 0, \qquad div\boldsymbol{B} = 0,$$
$$rot\boldsymbol{H} - \frac{1}{c}\frac{\partial \boldsymbol{D}}{\partial t} = \frac{4\pi}{c}\boldsymbol{j}, \qquad div\boldsymbol{D} = 4\pi\rho. \tag{1}$$

Here $\boldsymbol{E}$ and $\boldsymbol{D}$ are the macroscopic electric field and electric displacement vectors, $\boldsymbol{H}$ and $\boldsymbol{B}$ are the macroscopic magnetic field and induction vectors; ρ and $\boldsymbol{j}$ denote the densities of macroscopic charge and current, respectively. The vectors $\boldsymbol{D}$ and $\boldsymbol{B}$ are defined as

$$\boldsymbol{D} = \boldsymbol{E} + 4\pi\boldsymbol{P}, \qquad \boldsymbol{B} = \boldsymbol{H} - 4\pi\boldsymbol{M}. \tag{2}$$

In eq. (2) $\boldsymbol{P}$ and $\boldsymbol{M}$ are the polarization and magnetization vectors, respectively.

The boundary conditions for the field components at the surfaces and interfaces can be derived from eq. (1), and written in the form:

$$E_{1t} = E_{2t}, \qquad H_{1t} = H_{2t}, \tag{3}$$

$$D_{1n} = D_{2n}, \qquad B_{1n} = B_{2n}. \tag{4}$$

Here the indices t and n mean the tangential and normal components, respectively.

Further, we are going to introduce an external *dc* magnetic field, static magnetization, and *ac* fields of the EMW. We assume that *ac* magnetic field and magnetization are small as compared with *dc* ones. Than magnetization distribution can be found by solving Maxwell equations without taking into account the magnetic field of the EMW (Maxwell equations are reduced to the magnetostatic ones). The *ac* fields can be found as the solution of Maxwell equations with determined static magnetization. We are going to concentrate on the above problem. So, all the fields below are supposed to be *ac* ones and belong to the EMW. We assume the absence of macroscopic charges and *ac* currents.

Let us suppose that the electric displacement $\boldsymbol{D}$ and polarization $\boldsymbol{P}$ vectors depend on the electric field $\boldsymbol{E}$ of the EMW linearly; and similarly, the linear dependence on the magnetic field of the EMW is assumed for the magnetic induction $\boldsymbol{B}$ and magnetization $\boldsymbol{M}$ vectors:

$$\boldsymbol{D} = \hat{\epsilon}\boldsymbol{E}, \qquad \boldsymbol{B} = \hat{\mu}\boldsymbol{H}. \tag{5}$$

Here we neglected the spacial dispersion as well as vanishingly small magneto-electric effect (7).

For the cubic medium with the magnetization vector directed along the axis $\mathbf{Z}$, the electric permittivity $\hat{\epsilon}$ and magnetic permeability $\hat{\mu}$ tensors have the following form:

$$\hat{\epsilon} = \begin{pmatrix} \epsilon_\perp & i\epsilon' & 0 \\ -i\epsilon' & \epsilon_\perp & 0 \\ 0 & 0 & \epsilon \end{pmatrix}, \qquad \hat{\mu} = \begin{pmatrix} \mu_\perp & i\mu' & 0 \\ -i\mu' & \mu_\perp & 0 \\ 0 & 0 & \mu \end{pmatrix}. \tag{6}$$

For the ferrites, at the optical frequencies, the tensor $\hat{\mu}$ can be approximated by the unity tensor; at the far-infrared frequencies, $\hat{\epsilon}$ can be regarded as a scalar. Nevertheless, for YIG in the near infrared (IR) region, it is necessary to take into account the gyrotropic (off-diagonal) components in both $\hat{\epsilon}$ and $\hat{\mu}$ tensors, because at the IR frequencies they can give the comparable contribution to the magneto-optical effects.

For simplicity we shall use the coordinate system, connected with the wave vector of the EMW, that is convenient for the consideration of the oblique EMW propagation through the layered structure. We assume the axis $\mathbf{Z}$ to be normal to the layers, which are oriented in the $\mathbf{XY}$ plane.

Let us suppose that the plane EMW propagates in the $\mathbf{YZ}$ plane, and the wave vector is determined as

$$\boldsymbol{k} = \{0, k_y, k_z\}. \tag{7}$$

For the normal incidence ($k_y = 0$), the direction of the axes $\mathbf{X}$ and $\mathbf{Y}$ of the mobile coordinate system can be chosen arbitrary, but for some cases, the proper choice of the axes directions leads to considerable simplification in the formulas (see subsec. 3.1.1).

Let us consider the magnetization vector with an arbitrary direction in the mobile coordinate system: $\boldsymbol{M} = \{M\cos\phi\sin\theta, M\sin\phi\sin\theta, M\cos\theta\}$, where ϕ and θ are the azimuthal and polar angles, respectively. Using eq. (6), the dielectric susceptibility and magnetic permeability tensors can be easily rewritten relative to the mobile coordinate system via the transformation matrix $\hat{X}$

$$\hat{\hat{\epsilon}} = \hat{X}\hat{\epsilon}\hat{X}^{-1}, \qquad \hat{\hat{\mu}} = \hat{X}\hat{\mu}\hat{X}^{-1}, \tag{8}$$

where

$$\hat{X} = \begin{bmatrix} \cos\theta & 0 & \sin\theta \\ 0 & 1 & 0 \\ -\sin\theta & 0 & \cos\theta \end{bmatrix} \begin{bmatrix} \cos\phi & -\sin\phi & 0 \\ \sin\phi & \cos\phi & 0 \\ 0 & 0 & 1 \end{bmatrix}. \tag{9}$$

As one can see, the components of the tensors in eq. (8) become rather cumbersome, and further, we can use the general form for $\hat{\hat{\epsilon}}$ and $\hat{\hat{\mu}}$:

$$\hat{\hat{\epsilon}} = \begin{pmatrix} \epsilon_{xx} & \epsilon_{xy} & \epsilon_{xz} \\ \epsilon_{yx} & \epsilon_{yy} & \epsilon_{yz} \\ \epsilon_{zx} & \epsilon_{zy} & \epsilon_{zz} \end{pmatrix}, \qquad \hat{\hat{\mu}} = \begin{pmatrix} \mu_{xx} & \mu_{xy} & \mu_{xz} \\ \mu_{yx} & \mu_{yy} & \mu_{yz} \\ \mu_{zx} & \mu_{zy} & \mu_{zz} \end{pmatrix}. \tag{10}$$

In the further consideration no assumptions are made about $\hat{\epsilon}$ and $\hat{\mu}$, so, the formulas obtained in this section can also be applied for the magnetic media with non-cubic symmetry or for the absorbing media.

The electric $\boldsymbol{E}$ and magnetic $\boldsymbol{H}$ fields of the plane EMW can be represented in a complex form as

$$\boldsymbol{E}(y,z,t) = \boldsymbol{E}_0 \exp[i(\boldsymbol{kr} - \omega t), \qquad \boldsymbol{H}(y,z,t) = \boldsymbol{H}_0 \exp[i(\boldsymbol{kr} - \omega t) \tag{11}$$

with the wave vector $\boldsymbol{k}$ from eq. (7).

Substituting eq. (5) and eq. (11) in Maxwell eqs. (1), we obtain

$$\boldsymbol{k} \times \boldsymbol{E} - \Omega\hat{\mu}\boldsymbol{H} = 0, \qquad \boldsymbol{k} \times \boldsymbol{H} + \Omega\hat{\epsilon}\boldsymbol{E} = 0, \tag{12}$$

where we denoted

$$\Omega = \frac{\omega}{c}. \tag{13}$$

Using eqs. (7) and (10), the system of eqs. (12) can be rewritten in the matrix form:

$$\begin{bmatrix} \epsilon_{xx}\Omega & \epsilon_{xy}\Omega & \epsilon_{xz}\Omega & 0 & -k_z & k_y \\ \epsilon_{yx}\Omega & \epsilon_{yy}\Omega & \epsilon_{yz}\Omega & k_z & 0 & 0 \\ \epsilon_{zx}\Omega & \epsilon_{zy}\Omega & \epsilon_{zz}\Omega & -k_y & 0 & 0 \\ 0 & k_z & -k_y & \mu_{xx}\Omega & \mu_{xy}\Omega & \mu_{xz}\Omega \\ -k_z & 0 & 0 & \mu_{yx}\Omega & \mu_{yy}\Omega & \mu_{yz}\Omega \\ k_y & 0 & 0 & \mu_{zx}\Omega & \mu_{zy}\Omega & \mu_{zz}\Omega \end{bmatrix} \begin{bmatrix} E_x \\ E_y \\ E_z \\ H_x \\ H_y \\ H_z \end{bmatrix} = 0. \tag{14}$$

The system of eqs. (14) has a non-trivial solution, if its determinant is equal to zero, that leads to the following dispersion equation:

$$Ak_z^4 + Bk_z^3 + Ck_z^2 + Dk_z + E = 0. \tag{15}$$

The explicit expressions of the coefficients from eq. (15) are given in Appendix A.

In general, eq. (15) has four different solutions $k_z^{(j)}$, $(j = 1,2,3,4)$ (so called *normal modes*), just as in the anisotropic dielectric medium (15). But, unlike the later case, we cannot distinguish the ordinary mode (for which refractive index does not depend on propagation direction, i.e. on k_y) and the extraordinary mode (for which it depends).

Note that eq. (15) becomes biquadratic on k_z when $B = D = 0$. As it is shown in Appendix A, this happens when

$$\epsilon_{xy} = -\epsilon_{yx} \qquad \text{and} \qquad \mu_{xy} = -\mu_{yx}, \tag{16a}$$

$$\epsilon_{xy} = \epsilon_{yx} = \mu_{xy} = \mu_{yx} = 0, \quad or \quad \epsilon_{xz} = \epsilon_{zx} = \mu_{xz} = \mu_{zx} = 0 \tag{16b}$$

or when

$$k_y = 0. \tag{17}$$

In this case, z-components of the wave vectors satisfy the relations:

$$k_z^{(1)} = -k_z^{(2)} = k_+, \quad k_z^{(3)} = -k_z^{(4)} = k_-, \tag{18}$$

where the values $k_\pm$ are given by the expressions:

$$k_\pm^2 = (2A)^{-1}[-C \pm \sqrt{C^2 - 4AE}]. \tag{19}$$

In the case, when the condition determined by eq. (16b) takes place, the coefficients in eq. (15) do not change under the substitution $k_y \to -k_y$, and the medium is reciprocal, i.e.

$$\omega(\boldsymbol{k}) = \omega(-\boldsymbol{k}). \tag{20}$$

Let us discuss the structure of the normal modes, i.e. eigenmodes of eqs. (14). As usual, in the homogeneous linear system, for the normal mode j we must choose one basic amplitude, the field components of $\boldsymbol{e}^{(j)}$ or $\boldsymbol{h}^{(j)}$, and express five other field components of the normal mode j through it. We shall use the small letters for the fields of the normal modes and the capital letters for the *total* fields, which are the sum of the formers. Describing the EMW propagation through the layered structure, actually we need field components which enter the boundary conditions; they are $\{e_x^{(j)}, e_y^{(j)}, h_x^{(j)}, h_y^{(j)}\}$. Under the certain conditions (described further) the EMW in the layer splits into TM and TE modes, with two non-zero amplitudes from TM mode $\{e_y^{(j)}, h_x^{(j)}\}$, and two others $\{e_x^{(j)}, h_y^{(j)}\}$ for TE. So, in order to obtain the solution of the problem, we must take two basic amplitudes from the first set and two from the second one. We choose the amplitudes $h_y^{(1,2)}$ and $e_y^{(3,4)}$ (as we can see below, this gives us a certain symmetry in some coefficients). Then, the complete description of the electric $\boldsymbol{E}$ and magnetic $\boldsymbol{H}$ fields in an arbitrary point of the media can be given by the 4-component vector $\boldsymbol{\Psi}(z)$:

$$\boldsymbol{\Psi}(z) = \begin{bmatrix} h_y^{(1)}(z) \\ h_y^{(2)}(z) \\ e_y^{(3)}(z) \\ e_y^{(4)}(z) \end{bmatrix}. \tag{21}$$

The values of $\boldsymbol{\Psi}$ at two different points of media z_1 and z_2 are related via the diagonal matrix $\hat{E}(z)$:

$$\boldsymbol{\Psi}(z_2) = \hat{E}(z_2 - z_1)\boldsymbol{\Psi}(z_1), \tag{22}$$

where

$$\hat{E}(z) = \begin{bmatrix} \exp(ik_z^{(1)}z) & 0 & 0 & 0 \\ 0 & \exp(ik_z^{(2)}z) & 0 & 0 \\ 0 & 0 & \exp(ik_z^{(3)}z) & 0 \\ 0 & 0 & 0 & \exp(ik_z^{(4)}z) \end{bmatrix}. \tag{23}$$

The total fields $E_{x,y}(z)$, $H_{x,y}(z)$ are the sums of the fields $e_{x,y}^{(j)}(z)$, $h_{x,y}^{(j)}(z)$, $j = 1...4$ of the normal modes. The fields of the normal modes relate to basic amplitudes which enter $\mathbf{\Psi}(z)$ via the linear combinations with the coefficients obtained from eq. (14). So, the total fields are related to the basic amplitudes $\mathbf{\Psi}(z)$ via the matrix $\hat{A}$:

$$\begin{bmatrix} E_x(z) \\ E_y(z) \\ H_x(z) \\ H_y(z) \end{bmatrix} = \hat{A}\mathbf{\Psi}(z), \tag{24}$$

where the matrix $\hat{A}$ has the following form:

$$\hat{A} = \begin{bmatrix} V^{\epsilon(1)} & V^{\epsilon(2)} & -G^{\mu(3)} & -G^{\mu(4)} \\ G^{\epsilon(1)} & G^{\epsilon(2)} & 1 & 1 \\ F^{\epsilon(1)} & F^{\epsilon(2)} & -V^{\mu(3)} & -V^{\mu(4)} \\ 1 & 1 & F^{\mu(3)} & F^{\mu(4)} \end{bmatrix}. \tag{25}$$

The explicit form of elements of $\hat{A}$ is given in Appendix B.

In the next section we shall also use the inverse matrix $\hat{A}^{-1}$. The explicit expression for $\hat{A}^{-1}$ is given in Appendix C.

It should be emphasized, that when the following off-diagonal elements of the tensors $\hat{\epsilon}$ and $\hat{\mu}$ are vanishing

$$\epsilon_{xy} = \epsilon_{yx} = \mu_{xy} = \mu_{yx} = \epsilon_{xz} = \epsilon_{zx} = \mu_{xz} = \mu_{zx} = 0, \tag{26}$$

the eqs. (14) split into two independent subsystems, one of them containing variables $\{E_y, E_z, H_x\}$ and the other $\{E_x, H_y, H_z\}$. In this case, the normal waves in that layer can be represented as a superposition of TM and TE modes, and the wavenumbers k_z are

$$\begin{aligned} k_{TE}^{(\pm)} &= \frac{1}{2\mu_{zz}}\left[-(\mu_{yz}+\mu_{zy})k_y \pm \sqrt{(\mu_{yz}+\mu_{zy})^2 k_y^2 + 4\mu_{zz}\left(\epsilon_{xx}\Delta\mu_{yz}\Omega^2 - \mu_{yy}k_y^2\right)}\right], \\ k_{TM}^{(\pm)} &= \frac{1}{2\epsilon_{zz}}\left[-(\epsilon_{yz}+\epsilon_{zy})k_y \pm \sqrt{(\epsilon_{yz}+\epsilon_{zy})^2 k_y^2 + 4\epsilon_{zz}\left(\mu_{xx}\Delta\epsilon_{yz}\Omega^2 - \epsilon_{yy}k_y^2\right)}\right], \end{aligned} \tag{27}$$

were $\Delta\epsilon_{ij}$ and $\Delta\mu_{ij}$ are determined in Appendix A, eq. (116).

Then the matrix $\hat{A}$ reduces to the form

$$\hat{A} = \begin{bmatrix} V^{\epsilon(+)} & V^{\epsilon(-)} & 0 & 0 \\ 0 & 0 & 1 & 1 \\ 0 & 0 & -V^{\mu(+)} & -V^{\mu(-)} \\ 1 & 1 & 0 & 0 \end{bmatrix}, \tag{28}$$

where we introduced the notations

$$V^{\epsilon(\pm)} = \frac{(\mu_{zz}k_{TE}^{(\pm)} + \mu_{yz}k_y)\Omega}{\epsilon_{xx}\mu_{zz}\Omega^2 - k_y^2}, \qquad V^{\mu(\pm)} = \frac{(\epsilon_{zz}k_{TM}^{(\pm)} + \epsilon_{yz}k_y)\Omega}{\mu_{xx}\epsilon_{zz}\Omega^2 - k_y^2}. \tag{29}$$

and the inverse matrix $\hat{A}^{-1}$ is also simplified:

$$\hat{A} = \begin{bmatrix} (\Delta V^{\epsilon})^{-1} & 0 & 0 & -V^{\epsilon(-)}(\Delta V^{\epsilon})^{-1} \\ -(\Delta V^{\epsilon})^{-1} & 0 & 0 & V^{\epsilon(+)}(\Delta V^{\epsilon})^{-1} \\ 0 & -V^{\mu(-)}(\Delta V^{\mu})^{-1} & -(\Delta V^{\mu})^{-1} & 0 \\ 0 & V^{\mu(+)}(\Delta V^{\mu})^{-1} & (\Delta V^{\mu})^{-1} & 0 \end{bmatrix}, \tag{30}$$

where we introduced the notations

$$\Delta V^{\epsilon} = V^{\epsilon(+)} - V^{\epsilon(-)}, \qquad \Delta V^{\mu} = V^{\mu(+)} - V^{\mu(-)}. \tag{31}$$

Below we consider several examples of the media and magneto-optical configurations (MOCs).

2.1. Non-magnetic Medium with a Cubic Symmetry or an Isotropic Medium

The non-magnetic crystal with a cubic symmetry (or an isotropic medium) is characterized by the diagonal tensors $\hat{\epsilon}$ and $\hat{\mu}$ are diagonal:

$$\hat{\epsilon} = \epsilon\delta_{ij}, \qquad \hat{\mu} = \delta_{ij}, \tag{32}$$

where δ_{ij} is Kroneker delta symbol, k_z is

$$k_z^2 = n^2\Omega^2 - k_y^2, \tag{33}$$

and the RI is

$$n = \sqrt{\epsilon}. \tag{34}$$

Thus, the condition eq. (26) is satisfied by definition, and eq. (31) reduces to

$$V^{\epsilon(1)} = -V^{\epsilon(2)} = \frac{1}{n}, \qquad V^{\mu(3)} = -V^{\mu(4)} = n. \tag{35}$$

2.2. Magnetic Medium, Transversal MOC

In the case of the transversal MOC, the magnetization vector M is directed perpendicularly to the propagation direction of the EMW. The magnetic vector components are $M = \{M, 0, 0\}$. Then $\hat{\epsilon}$ and $\hat{\mu}$ have the form:

$$\hat{\epsilon} = \begin{pmatrix} \epsilon & 0 & 0 \\ 0 & \epsilon_{\perp} & i\epsilon' \\ 0 & -i\epsilon' & \epsilon_{\perp} \end{pmatrix}, \qquad \hat{\mu} = \begin{pmatrix} \mu & 0 & 0 \\ 0 & \mu_{\perp} & i\mu' \\ 0 & -i\mu' & \mu_{\perp} \end{pmatrix} \tag{36}$$

In this case, the condition eq. (26) is satisfied as well as eq. (16a). As it was mentioned above, the EMW splits into TM and TE modes, and eqs. (27) reduces to

$$k_{TM}^2 = \frac{\Delta\epsilon}{\epsilon_\perp}\mu\Omega^2 - k_y^2, \qquad k_{TE}^2 = \frac{\Delta\mu}{\mu_\perp}\epsilon\,\Omega^2 - k_y^2, \tag{37}$$

with

$$\Delta\epsilon = \epsilon_\perp^2 - (\epsilon')^2, \qquad \Delta\mu = \mu_\perp^2 - (\mu')^2. \tag{38}$$

Eq. (29) reduces to

$$V^{\epsilon(\pm)} = \frac{(\pm\mu_\perp k_{TE} - i\mu' k_y)\Omega}{\epsilon\mu_\perp\Omega^2 - k_y^2}, \qquad V^{\mu(\pm)} = \frac{(\pm\epsilon_\perp k_{TM} - i\epsilon' k_y)\Omega}{\mu\epsilon_\perp\Omega^2 - k_y^2}. \tag{39}$$

2.3. Magnetic Medium, Polar MOC

In this MOC the magnetization vector, M is directed along Z axis, perpendicularly to to the surface, and the tensors $\hat{\epsilon}$ and $\hat{\mu}$ have the form:

$$\hat{\epsilon} = \begin{pmatrix} \epsilon_\perp & i\epsilon' & 0 \\ -i\epsilon' & \epsilon_\perp & 0 \\ 0 & 0 & \epsilon \end{pmatrix}, \qquad \hat{\mu} = \begin{pmatrix} \mu_\perp & i\mu' & 0 \\ -i\mu' & \mu_\perp & 0 \\ 0 & 0 & \mu \end{pmatrix}. \tag{40}$$

Then the condition eq. (16b)is satisfied and eq. (19) simplifies to

$$k_\pm^2 = (\epsilon_\perp\mu_\perp + \epsilon'\mu')\Omega^2 - \frac{1}{2}(\frac{\epsilon_\perp}{\epsilon} + \frac{\mu_\perp}{\mu})k_y^2 \pm \sqrt{D_p}, \tag{41}$$

where

$$D_p = (\epsilon_\perp\mu' + \mu_\perp\epsilon')^2\Omega^4 + \frac{1}{4}\left(\frac{\epsilon_\perp}{\epsilon} - \frac{\mu_\perp}{\mu}\right)^2 k_y^4 - (\epsilon_\perp\mu' + \mu_\perp\epsilon')\left(\frac{\epsilon'}{\epsilon} + \frac{\mu'}{\mu}\right)\Omega^2 k_y^2. \tag{42}$$

For the polar MOC, the following symmetry relations between the elements of $\hat{A}$ in eq. (25) take place:

$$V^{\epsilon(1)} = -V^{\epsilon(2)} \equiv V^+, \qquad V^{\mu(3)} = -V^{\mu(4)} \equiv V^-, \tag{43a}$$

$$G^{\epsilon(1)} = -G^{\epsilon(2)} \equiv G^+, \qquad G^{\mu(3)} = -G^{\mu(4)} \equiv G^-, \tag{43b}$$

$$F^{\epsilon(1)} = F^{\epsilon(2)} \equiv F^+, \qquad F^{\mu(3)} = F^{\mu(4)} \equiv F^-. \tag{43c}$$

The explicit form of the elements of the matrix $\hat{A}$ can be written as

$$V^\pm = \frac{v^\pm}{z^\pm}, \qquad G^\pm = \frac{g^\pm}{z^\pm}, \qquad F^\pm = \frac{f^\pm}{z^\pm}, \tag{44}$$

where the following designations are made:

$$z^+ = \left[\epsilon(\epsilon_\perp\mu_\perp + \epsilon'\mu')\Omega^2 - \epsilon_\perp k_y^2 - \epsilon k_+^2\right] k_+, \tag{45a}$$

$$v^+ = \left[\epsilon_\perp \epsilon \Delta\mu \Omega^2 - \mu_\perp \left(\epsilon_\perp k_y^2 + \epsilon k_+^2\right)\right] \Omega, \tag{45b}$$

$$g^+ = i \left(\epsilon\, \epsilon' \Delta\mu \Omega^2 - \epsilon' \mu_\perp k_y^2 + \epsilon \mu' k_+^2\right) \Omega, \tag{45c}$$

$$f^+ = -i\epsilon(\epsilon_\perp \mu' + \mu_\perp \epsilon')\Omega^2 k_+. \tag{45d}$$

The values z^-, v^-, g^- and f^- can be easily obtained from z^+, v^+, g^+ and f^+, respectively, by the replacement $(\epsilon, k_+ \leftrightarrow \mu, k_-)$ (in accordance with the relations from Appendix B).

2.4. Magnetic Medium, Longitudinal MOC

In the case of the longitudinal MOC, M is directed parallel to the surface, and its components are $M = \{0, M, 0\}$. Thus, $\hat{\epsilon}$ and $\hat{\mu}$ have the following form:

$$\hat{\epsilon} = \begin{pmatrix} \epsilon_\perp & 0 & -i\epsilon' \\ 0 & \epsilon & 0 \\ i\epsilon' & 0 & \epsilon_\perp \end{pmatrix}, \qquad \hat{\mu} = \begin{pmatrix} \mu_\perp & 0 & -i\mu' \\ 0 & \mu & 0 \\ i\mu' & 0 & \mu_\perp \end{pmatrix}. \tag{46}$$

In this geometry, eq. (16b) is valid again and eq. (19) turns to

$$k_\pm^2 = \frac{1}{2} \left[\left(\frac{\Delta\epsilon}{\epsilon_\perp} \mu + \frac{\Delta\mu}{\mu_\perp} \epsilon \right) \Omega^2 - \left(\frac{\epsilon}{\epsilon_\perp} + \frac{\mu}{\mu_\perp} \right) k_y^2 \pm \sqrt{D_l} \right]. \tag{47}$$

In the equation above D_l is

$$D_l = \frac{1}{(\epsilon_\perp \mu_\perp)^2} \left[(\Delta\epsilon \mu_\perp \mu - \Delta\mu \epsilon_\perp \epsilon)^2 \Omega^4 + (\epsilon_\perp \mu - \mu_\perp \epsilon)^2 k_y^4 \right.$$
$$\left. + 2\epsilon_\perp \mu_\perp \left((\epsilon\mu' + \mu\epsilon')^2 - (\epsilon_\perp \mu - \mu_\perp \epsilon)^2 + 2\epsilon\mu(\epsilon_\perp \mu' + \mu_\perp \epsilon') \right) \Omega^2 k_y^2 \right]. \tag{48}$$

For the longitudinal MOC, the following relations between the components of $\hat{A}$ take place:

$$V^{\epsilon(1)} = -V^{\epsilon(2)} \equiv V^+, \qquad V^{\mu(3)} = -V^{\mu(4)} \equiv V^-, \tag{49a}$$

$$G^{\epsilon(1)} = G^{\epsilon(2)} \equiv G^+, \qquad G^{\mu(3)} = G^{\mu(4)} \equiv G^-, \tag{49b}$$

$$F^{\epsilon(1)} = -F^{\epsilon(2)} \equiv F^+, \qquad F^{\mu(3)} = -F^{\mu(4)} \equiv F^-. \tag{49c}$$

The explicit form of $\hat{A}$ elements are

$$V^+ = \frac{\mu\Omega}{k_+}, \qquad V^- = \frac{\epsilon\Omega}{k_-}, \tag{50}$$

where the values $F^\pm$, $G^\pm$ are defined as in eq.(44) with

$$z^+ = \left[\epsilon\, \epsilon_\perp \Delta\mu \Omega^2 - \mu_\perp \left(\epsilon k_y^2 + \epsilon_\perp k_+^2\right)\right] k_+, \tag{51a}$$

$$g^+ = i\mu(\epsilon_\perp \mu' + \mu_\perp \epsilon')\Omega k_y k_+, \tag{51b}$$

$$f^+ = -i\epsilon\mu(\epsilon_\perp \mu' + \mu_\perp \epsilon')\Omega^2 k_y. \tag{51c}$$

Again, z^-, g^- and f^- can be easily obtained from z^+, g^+ and f^+, respectively, by the substitution $(\epsilon, k_+ \leftrightarrow \mu, k_-)$.

3. 4×4 T-matrix Formalism

Let us consider the layered structure, introduced in the previous section. For description of the EMW propagation through these media, one of the most widely used and powerful methods is the T-matrix formalism. The EMW propagation through the layered structure, consisting of the isotropic (or cubic) layers only, is well investigated in the literature.

We define 4×4 *transfer matrix* $\hat{T}$, which relates the vectors $\Psi(z)$ between the arbitrary points z_1 and z_2:

$$vc\Psi(z_2) = \hat{T}(z_2, z_1)\mathbf{\Psi}(z_1) \tag{52}$$

The space between z_2 and z_1 can include an external boundary of the structure as well as the interfaces of the layers. One can see, that

$$\hat{T}(z_2, z_1) = \hat{T}(z_2, z)\hat{T}(z, z_1) = \hat{T}^{-1}(z_1, z_2). \tag{53}$$

This basic property of the transfer matrix is a direct consequence of its definition in eq. (52). If the points z_1 and z_2 belong to the same layer, then $\hat{T}$-matrix actually is the matrix $\hat{E}$ eq. (23).

In order to find $\hat{T}$-matrix, which relates vectors $\mathbf{\Psi}$ in at the points, belonging to the different layers, we must consider the boundary conditions determined by eq. (3). The boundary conditions between m-th and $(m+1)$-th layers have the following form:

$$H_{mx,y}(d_m) = H_{m+1x,y}(d_m), \quad E_{mx,y}(d_m) = E_{m+1x,y}(d_m), \tag{54}$$

where d_m is z-coordinate of the boundary.

With eq. (24), the boundary conditions eq. (54) becomes

$$\hat{A}_{m+1}\mathbf{\Psi}_{m+1}(d_m + 0) = \hat{A}_m\mathbf{\Psi}_m(d_m - 0), \tag{55}$$

from where one can obtain

$$\mathbf{\Psi}_{m+1}(d_m + 0) = \hat{A}^{-1}_{m+1}\hat{A}_m\mathbf{\Psi}_m(d_m - 0). \tag{56}$$

Passing from point z_1 to z_2, both of which belongs to the uniform layer of length d, we must multiply $\mathbf{\Psi}$-vector by $\hat{E}(z_2 - z_1)$, defined in eq. (23), and crossing each boundary between the mth and $(m+1)$th layers, one need to multiply $\mathbf{\Psi}$ by $\hat{A}^{-1}_{m+1}\hat{A}_m$.

In particular, $\hat{T}$-matrix which relates the vector $\mathbf{\Psi}_1(z)$ in the medium 1 near it's left boundary at $d_0 + 0$ and the vector $\mathbf{\Psi}_{k+1}(z)$ in medium $k+1$ near it's left boundary at $d_k + 0$ is

$$\hat{T}(d_k + 0, d_0 + 0) = (\hat{A}^{-1}_{k+1}) \prod_{m=2}^{k} [\hat{A}_m\hat{E}_m(d_m - d_{m-1}))(\hat{A}^{-1}_m)]\hat{A}_1\hat{E}_1(d_1 - d_0). \tag{57}$$

When the condition eq. (26) and, consequently, eqs. (27)-(31), are valid for each layer, then the TE- and TM-modes are eigenmodes for the whole structure, and the $\hat{T}$-matrix breaks down into two 2×2 non-zero blocks:

$$\hat{T} = \begin{bmatrix} \hat{T}_1 & 0 \\ 0 & \hat{T}_2 \end{bmatrix}. \tag{58}$$

In eq. (58), $\hat{T}_1$ and $\hat{T}_2$ describe the propagation of TE- and TM-modes, respectively.

2.5. Normal Incidence

For the normal incidence ($k_y = 0$), two particular cases are worth special consideration.

2.5.1. The Uniaxial Anisotropy Layers with the Symmetry Axis in *XY* Plane

We suppose that the layers have the uniaxial anisotropy with the symmetry axis laying in $\boldsymbol{XY}$ plane, i.e. $\hat{\epsilon}$ and $\hat{\mu}$ tensors in some coordinate system $X'Y'Z$, satisfy by eq. (26). If the anisotropy axes of the different layers coincide, one can use $\boldsymbol{X'Y'Z}$ coordinate system, and the simplified formulas of eqs. (27)-(31) for each of the layers. But if the system contains the layers with nonparallel symmetry axes, we cannot find any coordinate system to satisfy eq. (26) for all layer simultaneously. Further, the general formulas do not work properly in this case. As an illustration, we consider the normal incidence of the EMW on the structure consisting of two magnetic layers with the magnetization vectors directed perpendicularly one to other. We chose the direction of **X** axis along one the magnetization vectors and apply the formulas from subsection 2.2 to this layer. In the other layer, it seems that formulas from the subsection 2.4 can be used. Indeed, $g^{\pm}$ and $f^{\pm}$ are explicitly proportional to k_y, and it looks like $G^{\pm}$ and $F^{\pm}$ are equal to zero as they should. But one can note that k_+ becomes k_{TE} and, substituting eq. (37) to the equation for z^+ (eq. (51a)), one notes that it equal to zero. The same is true for z^-, so the values $G^{\pm}$ and $F^{\pm}$ become undefined. Such troubles happen also with the general formulas from Appendix B.

Thus, we must switch to the local coordinate system in the cases, similar to mentioned above. Suppose there is the layer m with the boundaries at d_{m-1} and d_m. We need to relate the total fields $E_{x,y}$ and $H_{x,y}$ at the points $d_{m-1} - 0$ and $d_m + 0$, and we write simply

$$\begin{bmatrix} E_x(d_m+0) \\ E_y(d_m+0) \\ H_x(d_m+0) \\ H_y(d_m+0) \end{bmatrix} = \hat{A}_m \hat{E}_m(d_m - d_{m-1}) \hat{A}_m^{-1} \begin{bmatrix} E_x(d_{m-1}-0) \\ E_y(d_{m-1}-0) \\ H_x(d_{m-1}-0) \\ H_y(d_{m-1}-0) \end{bmatrix}. \tag{59}$$

We suppose that the symmetry axis formes the angle ϕ with **X** axis, $\phi \neq 0$. Taking into account the lack of coincidence of the axis **X** with the anisotropy axis, we must rotate $E_{x,y}$ and $H_{x,y}$ before 'entering' layer and after 'exiting' it. Then, the relation we are interested in is

$$\begin{bmatrix} E_x(d_m+0) \\ E_y(d_m+0) \\ H_x(d_m+0) \\ H_y(d_m+0) \end{bmatrix} = \hat{R}(\phi) \hat{A}_m \hat{E}_m(d_m - d_{m-1}) \hat{A}_m^{-1} \hat{R}^{-1}(\phi) \begin{bmatrix} E_x(d_{m-1}-0) \\ E_y(d_{m-1}-0) \\ H_x(d_{m-1}-0) \\ H_y(d_{m-1}-0), \end{bmatrix} \tag{60}$$

where

$$\hat{R} = \begin{bmatrix} \hat{R}_1(\phi) & 0 \\ 0 & \hat{R}_1(\phi) \end{bmatrix} \tag{61}$$

with the rotation matrix

$$\hat{R}_1 = \begin{bmatrix} \cos(\phi) & \sin(\phi) \\ -\sin(\phi) & \cos(\phi). \end{bmatrix} \tag{62}$$

Of course, $\hat{E}_m$ and $\hat{A}_m$ in eq. (60) are defined by the eqs. (27)-(31). Note that if we deal with the structure consisting only of the layers with the symmetry axis directed along either $\mathbf{X}$ ($\phi = 0$) or $\mathbf{Y}$ axis ($\phi = \pi/2$) and layers with the cubic symmetry as in eq. (32), then eq. (58) is still valid as the EMWs are the eigenmodes for each layer and linearly polarized in $\boldsymbol{XZ}$ and $\boldsymbol{YZ}$ planes. So, for the whole structure, the 2×2 $\hat{T}$-matrices like those eq. (58) can be used. But, dealing with the layers, for which $\phi \neq 0$ and $\phi \neq \pi/2$, we need to use full 4×4 $\hat{T}$-matrix.

Certainly, if $k_y \neq 0$ or the symmetry axis does not lie in $\boldsymbol{XY}$ plane, then no rotation is needed, and one can use eq. (59) with general formulas for $\hat{E}_m$, $\hat{A}_m$.

2.5.2. Polar MOC

For the normal incidence in the polar MOC, eq. (41) becomes

$$k_{\pm}^2 = (\epsilon_{\perp} \pm \epsilon')(\mu_{\perp} \pm \mu')\Omega^2. \tag{63}$$

The matrix $\hat{A}$ reduces to

$$\hat{A} = \begin{bmatrix} 1/p_+ & -1/p_+ & -i & -i \\ -i/p_+ & i/p_+ & 1 & 1 \\ i & i & -p_- & p_- \\ 1 & 1 & -ip_- & ip_- \end{bmatrix} \tag{64}$$

with $p_{\pm}$ determined as

$$p_{\pm} = \sqrt{\frac{\mu_{\perp} \pm \mu'}{\epsilon_{\perp} \pm \epsilon'}}. \tag{65}$$

It is easy to note from eq. (64) that the positive sign in eq. (63) corresponds to the righthand (RH) circularly polarized wave (CPW), whereas the negative one refers to the lefthand (LH) CPW. By means of direct calculation one can show that the product $\hat{A}_2^{-1}\hat{A}_1$ (where $\hat{A}_1$ and $\hat{A}_2$ are taken from eq. (64)) and, consequently, the transfer matrix $\hat{T}$, have the block-diagonal form, like those in eq. (58). But now the blocks $\hat{T}_1$ and $\hat{T}_2$ correspond not to TE- and TM-modes but to forward and backward spreading RH, and LH circularly polarized waves (CPWs), which do not interact (7).

In the nonmagnetic media with the cubic symmetry for normal incidence there is no difference between TE and TM modes and any linear combination of two possible polarizations, including RH and LH CPW, are eigenmodes too. Taking RH- and LH-modes as eigenmodes in this case we can use eq. (64) for the matrix $\hat{A}$ with the only modification: $p_+ = p_- = n$, where n is given by eq. (34).

Thus, in the case of the normal incidence, for structures consisting of the layers, magnetized along the axis $\boldsymbol{Z}$, and of the nonmagnetic layers with cubic symmetry only, the 4×4 $\hat{T}$-matrix splits into two 2×2 $\hat{T}$-matrices as determined by eq. (58).

3. Periodic Structures

3.1. Bragg Reflections

To demonstrate the PBGs, appropriate to the layered structure, we consider the normal incidence of the EMW on the multilayer with the period D, consisting of different isotropic non-magnetic layers of lengthes d_i and permittivity ϵ_i (16). We suppose that axes are chosen as in sec. 2 and the wave vector is $\boldsymbol{k} = \{0, 0, q\}$. It is sufficient to consider the EMW with one non-zero electric field component, for example, $E_x = E$. The electric field can be easily obtained from eq. (12). One can expand the periodic function $\epsilon(x)$ on a Fourier series:

$$\epsilon(x) = \sum_{l=-\infty}^{+\infty} \epsilon_l e^{iglx} \tag{66}$$

with the reciprocal lattice wave number $g = 2\pi/D$. We search the solution of Maxwell equations in the form of Bloch waves:

$$E_q(z) = e^{-iqz} E_{qz} = e^{-iqz} \sum_l A_{q-lg} e^{il(2\pi/D)z}, \tag{67}$$

where E_{qz} is a function with the period D, and A_{q-lg} are the Fourier coefficients. The substitution of eq. (67) to the wave equation leads to the infinite system of linear equations. Using the plane wave expansion method, this system can be solved after truncation and the eigenfrequencies ω_n, as well as the eigenvectors $E_{q,n}$ can be obtained numerically.

One can use the approximate analytical treatment of these structures considering the wave vectors $\boldsymbol{q}$, which satisfy the following two conditions. The first is with $|q| \approx m\pi/D$, i.e.

$$|q - m\frac{2\pi}{D}| \approx q, \qquad m = \pm 1, \pm 2, \ldots., \tag{68}$$

and the second one is

$$q^2 \approx \left(\frac{\omega}{c}\right)^2 \epsilon_0, \tag{69}$$

where $\epsilon_0 = \left(\sum_i \epsilon_i d_i\right)/D$ is the dielectric function averaged over the period D.

After these assumptions, it turns out that only the coefficients A_q and A_{q-mg} in eq. (67) can be taken into account. Thus, the infinite system reduces to two equations. The dispersion equation can be easily obtained, and it can be shown that $\omega_+ < \omega < \omega_-$, where

$$\omega_\pm^2 = \frac{mq^2c^2}{\epsilon_0 \pm \epsilon_m}. \tag{70}$$

When the wave number q becomes imaginary, i.e. the wave is evanescent, that means appearance of the PBGs. The value $Im(q)$ achieves the maximum at the center of the PBGs and decreases to zero when ω turns to $\omega_\pm$. The wider gap correspond to the bigger $Im(q)$ and the larger absorbtion.

The frequency in the center of the band gap approximately satisfies Bragg's condition (16):

$$\frac{\omega}{c} \approx \frac{\pi m}{D\sqrt{\epsilon_0}}, \tag{71a}$$

or

$$D \approx \frac{m\lambda_0}{2\sqrt{\epsilon_0}}, \tag{71b}$$

where λ_0 is the free-space wavelength. When eq. (71a) is satisfied, the waves reflected from different layers, are in the same phase, and as a result the interference there exist the maximum in reflected and minimum in transmitted spectra.

The period D determines the PBGs positions. For the structure with two layers in the period (with RIs $n_{1,2}$) the ratio $n_1 d_1/n_2 d_2$ is responsible for the relative band gap width for different m. One can expect the maximal PBG width when the waves, reflected from both layers, are cophased and minimal width when they are anti-cophased. So, the m-th PBG has m maxima when

$$d_1 = \frac{(2j+1)\lambda_0}{4n_1}, \qquad d_2 = \frac{(2(m-j)-1)\lambda_0}{4n_2} \qquad j = 0, 1, \ldots m-1, \tag{72a}$$

or

$$\frac{d_1}{d_2} = \frac{(2j+1)n_2}{(2(m-j)-1)n_1}, \tag{72b}$$

and $(m-1)$ minima under the following condition:

$$d_1 = \frac{j\lambda_0}{2n_1}, \qquad d_2 = \frac{(m-j)\lambda_0}{2n_2} \qquad j = 1, 2, \ldots m-1. \tag{73}$$

When we investigate the oblique incidence, we expect the cut-off frequency

$$\omega_{cut} = \frac{k_y c}{\bar{n}}, \tag{74}$$

where $\bar{n}$ is the RI, averaged over the period. The EMWs with the frequencies, smaller than ω_{cut}, are evanescent ones as in the uniform medium. For the oblique incidence, instead of eq. (71a), the PBG frequency is approximated by:

$$\left(\frac{\omega}{c}\right)^2 \approx \left(\frac{2\pi m}{D\sqrt{\epsilon_0}}\right)^2 + \left(\frac{k_y}{\sqrt{\epsilon_0}}\right)^2. \tag{75}$$

Another important feature of PBG is a minimum of the band gap for TM mode at Brewster angle $\theta_B = arctan(n_2/n_1)$. At that angle of incidence Fresnel reflection coefficient for TM polarization turns to zero and the wave does not undergo Bragg reflection.

3.2. Dispersion Equation

Now we consider an infinite periodic stack forming 1D photonic crystal, with the period D. According to Bloch's theorem

$$\mathbf{\Psi}_j(z+D) = \hat{T}(z+D, z)\mathbf{\Psi}_j(z) = \lambda_j \mathbf{\Psi}_j(z) = \exp(iq_j D)\mathbf{\Psi}_j(z) \tag{76}$$

Here $\mathbf{\Psi}_j(z)$ $(j = 1, \ldots, 4)$ is the eigenfunction of the transfer matrix $\hat{T}(z+D, z)$ ($\hat{T}$ is given by eq. (57)); $\lambda_j = \exp(iq_j D)$ are the eigennumbers, and q_j plays the role of z-component of the wave vector.

Two values q_j are positive and two other are negative ones. If q_j is real, i.e. $|\exp(iq_jD)| = 1$, then the wave is propagating, otherwise there is a band gap.

In the case of the reciprocal propagation

$$\omega(k_y, q) = \omega(-k_y, -q) \tag{77}$$

the wavenumbers q_j satisfy the relations, analogous to eqs. (18):

$$q_1 = -q_2 = q_+, \quad q_3 = -q_4 = q_-. \tag{78}$$

Inversion symmetry of the structure is sufficient condition for reciprocity (27). Further on we shall use MPCs which period possess inversion symmetry and eq. (77)-(78) are supposed to be valid.

The characteristic equation for $\hat{T}$-matrix has the form:

$$\lambda^4 + I_3\lambda^3 + I_2\lambda^2 + I_1\lambda + I_0 = 0 \tag{79}$$

The invariant coefficients I_p, $(p = 0, 1, 2, 3)$ in eq. (79) are given in (9), (17)

$$I_0 = det(\hat{T}), \quad I_1 = -\frac{1}{6}[Tr^3(\hat{T}) - 3Tr(\hat{T})Tr(\hat{T}^2) + 2Tr(\hat{T}^3)],$$
$$I_2 = \frac{1}{2}[Tr^2(\hat{T}) - Tr(\hat{T}^2)], \quad I_3 = -Tr(\hat{T}) \tag{80}$$

Using the explicit form of λ_j in eq. (76), and eq. (78) one can rewrite the invariants of $\hat{T}$-matrix as:

$$I_0 = \lambda_1\lambda_2\lambda_3\lambda_4 = det(\hat{T}) = 1,$$
$$I_2 = \lambda_1\lambda_2 + \lambda_1\lambda_3 + \lambda_1\lambda_4 + \lambda_2\lambda_3 + \lambda_2\lambda_4 + \lambda_3\lambda_4 = 2[1 + 2\cos(q_1D)\cos(q_2D)],$$
$$I_3 = \lambda_1 + \lambda_2 + \lambda_3 + \lambda_4 = 2[\cos(q_1D) + \cos(q_2D)]. \tag{81}$$

Taking into account that the coefficients I_p are equal to the sum of p-th order principal minor, the value of I_1 can be rewritten as

$$I_1 = det(\hat{T})Tr(\hat{T}^{-1}). \tag{82}$$

Since $det(\hat{T}) = 1$ and $Tr(\hat{T}) = Tr(\hat{T}^{-1})$, from eq. (82) one can obtain that

$$I_1 = \lambda_1\lambda_2\lambda_3 + \lambda_1\lambda_2\lambda_4 + \lambda_2\lambda_3\lambda_4 = I_3. \tag{83}$$

Therefore, the set of dispersion equations for the photonic crystal has the form

$$\cos(q_1D) + \cos(q_2D) = \frac{1}{2}Tr(\hat{T}),$$
$$1 + 2\cos(q_1D)\cos(q_2D) = \frac{1}{4}[Tr^2(\hat{T}) - Tr(\hat{T}^2)]. \tag{84}$$

The solution of this set of equations is given by

$$\cos(q_{1,2}D) = \frac{1}{4}[Tr(\hat{T}) \pm \sqrt{2Tr^2(\hat{T}) - Tr(\hat{T}^2) + 8}]. \tag{85}$$

From eq. (58) the following dispersion equations for TM and TE modes can be obtained:

$$\cos(q_{TE}D) = \frac{1}{2}Tr(\hat{T}_1), \quad \cos(q_{TM}D) = \frac{1}{2}Tr(\hat{T}_2). \tag{86}$$

4. Numerical Calculations

As an example, we solved eq. (85) (or eq. (86))for the 1D MPC, composed of bigyrotropic YIG $Y_3Fe_5O_{12}$ layers of thickness d_1, alternating with nonmagnetic GGG $Gd_3Ga_5O_{12}$ of thickness d_2, which is transparent in the near-infrared regime. The unit cell with the period $D = d_1 + d_2$ contains one YIG and one GGG layers. We calculate PBG for three particular MOCs: polar (subsec. 2.3), transversal (subsec. 2.2) and longitudinal (subsec. 2.4). So, the permittivity and permeability tensors are given by eq. (40), eq. (36) or eq. (46), respectively. For the GGG eq. (32) is valid. At the free-space wavelength $\lambda = 1150$ nm YIG is characterized by $\epsilon_\perp = \epsilon_{||} = 4.64$ and $\mu_\perp = \mu_{||} = 1.0$; we set $\epsilon' = 0.02$ and $\mu' = 0.0065$. For GGG the dielectric permittivity is $\epsilon = 3.76$. We assume that $d_1 = d_2 = 0.5D$.

4.1. Transversal MOC

The MPC in transversal MOC (fig. 1) was studied in our paper (14). The solutions of the dispersion equation eq. (86) for TE-polarization are presented in fig. 2 (a) for the normal incidence ($k_y = 0$). The numbers $1, 2, 3$ denote the first three PBGs, $\omega^{(TE)}D/(2\pi c)$ and $k_zD/(2\pi)$ are the normalized frequency and normalized z-component of wave vector, respectively. Because the contrast between TE and TM modes is extremely small in this case, in fig.2 (b) we demonstrate the difference between spectra: $(\omega^{(TM)} - \omega^{(TE)})D/(2\pi)$ versus $k_zD/(2\pi)$. This small contrast is expected, as the RIs for TE and TM waves, approximated as $n_{TM} - n_{TE} \sim (\epsilon')^2 - (\mu')^2$, is very small. As the frequency $\omega(k_z)$ has minimums or maximums at the band gap edges (i.e. when $k_zD/(2\pi) = \pm 0.5, \pm 1.5, \ldots$ for odd PBGs and $k_zD/(2\pi) = 0, \pm 1, \ldots$ for even ones), then one can expect peculiarities in differences of ω for those k_z.

In fig. 3 the PBG spectra for $k_yD/(2\pi) = 0.5$ are presented. One can see the cutt-off frequency, shift of the PBG to higher frequencies, increase of the first PBG for TE-mode and decrease almost to zero for TM-mode, as mentioned in sec. 4.1.

The dependencies of the normalized PBG bandwidths $\Delta\omega^{(TE,TM)}D/(2\pi c)$ with respect to the normalized wavenumber $k_yD/2\pi$ are presented in fig. 4. Note that for TE-mode (fig. 4 (a)), all PBGs monotonically increase and the rate of their increase is larger for PBG with smaller numbers. For TM-mode, the first PBG has minimum near Brewster angle, which (as it is seen from the inset) do not reach zero; two other PBGs decrease monotonically.

4.2. Polar MOC

The polar MOC for the MPC is shown in fig. 5. This configuration is considered in our paper (13). As PBGs for this MOC look very likely to those for the transversal MOC, in Fig.6 we presented the differences between the spectra for the polar and transversal MOCs. For the normal incidence the RH and LH CPW are the eigenmodes for MPC with polar MOC, and, as both $(n_- - n_{TE})$ and $(n_+ - n_{TM})$ are approximately proportional to $\epsilon' + \mu'$. Thus, these differences are more pronounced, then those between $TE-$ and TM-modes for the transversal MOC (fig. 2(a)).

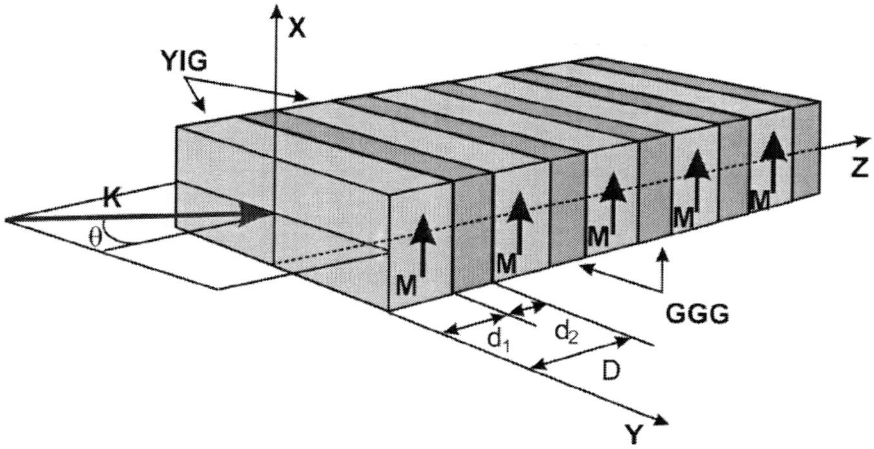

Figure 1. A fragment of the MPC containing a few periods of thickness D. The direction of the magnetization vector M in the YIG layers is shown by arrows. The EMW propagates in the YZ plane.

In fig. 7 the same differences are shown for $k_yD/(2\pi) = 0.5$. Now the RH- and LH-modes are no more eigenmodes and "+" and "-" signs mean signs of solution in eq. (85). One should be mentioned that in comparison with the normal incidence, the difference between the first PBGs becomes rather small; the difference between the second bands depends strongly on k_z, reaching zero at $k_zD/(2\pi) = \pm 0.5, 1.5, \ldots$ and sharp maximum at $k_zD/(2\pi) = 0, \pm 1, \ldots$. The difference between the third PBGs has the pronounced peculiarities at these points too. Note that for both cases $k_y = 0$ and $k_yD/(2\pi) = 0.5$, the differences $(\omega_p^{(-)} - \omega_t^{(TE)})D/(2\pi c)$ and $(\omega_p^{(+)} - \omega_t^{(TM)})D/(2\pi c)$ would be similar to each other if one of them is reflected against the horizontal axis $\Delta\omega = 0$. Also, for both cases the differences between the first bands are smallest ones. The dependencies of the normalized PBG bandwidths $\Delta\omega^{(\pm)}D/(2\pi c)$ with respect to the normalized wavenumber $k_yD/2\pi$ are presented in fig. 8. Note that "+" solution resembles TM-mode (first PBG has minimum) whereas "-" solution resembles TE-mode. That is why we estimate the values $(\omega^{(+)} - \omega^{(TM)})$ and $(\omega^{(-)} - \omega^{(TE)})$. As it is seen from the inset on fig. 8(b), the first PBG width turns to zero exactly, while the for TM-mode in the transversal MOC, the corresponding line does not touch the abscissa axis. Another distinction in the PBG bandwidths of the transversal and polar MOCs is the behaviour of that second PBGs. Comparing figs. 4(b) and fig. 8(b), one can observe the opposite behaviour of the bandwidths of the second modes: the "+" solution grows with the increase of k_y, while for the TM-mode it decays. In the case of polar MOC, for the "-" solution, at $k_y = 0$ the third PBG width exceeds the first one, while for TE-mode in the transversal MOC, the opposite is observed.

4.3. Longitudinal MOC

The longitudinal MOC for the MPC is shown on the fig. 9. The corresponding PBG spectra are presented in our paper (18). We demonstrate the differences between the longitudinal and transversal MOCs (fig.10). As for $k_y = 0$ the longitudinal and transversal MOCs

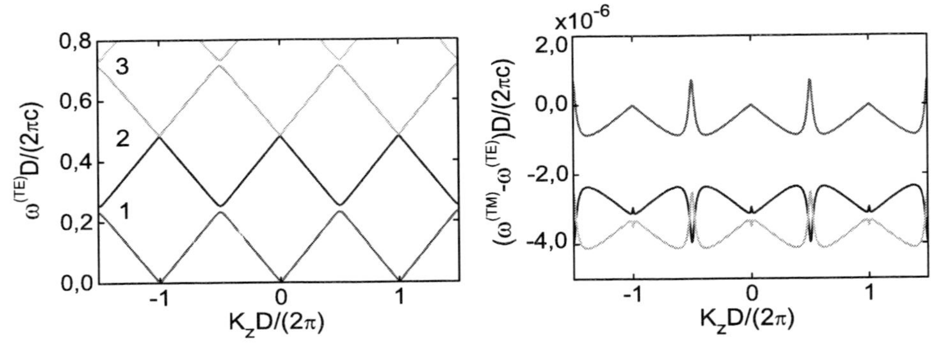

(a) (b)

Figure 2. (a): The normalized angular frequency $\omega^{(TE)}D/(2\pi c)$ of light *versus* normalized wave number $k_zD/(2\pi)$ of the eigenwave for the normal incidence ($k_y = 0$). (b): The difference of the normalized frequencies $(\omega^{(TM)} - \omega^{(TE)})D/(2\pi c)$ versus $k_zD/(2\pi)$; red - first band, blue - second, green - third.

can not be distinguished, we use the nonzero value $k_yD/(2\pi) = 0.1$. The differences $(\omega_l^{(-)} - \omega_t^{(TE)})D/(2\pi c)$ have the sharp minima at $k_zD/(2\pi) = 0, \pm1, \ldots$ and reach zero at $k_zD/(2\pi) = \pm0.5, \pm1.5, \ldots$. The differences for $k_yD/(2\pi) = 0.5$ are shown at fig. 11. They become larger, as distinction between MOCs increases with k_y. Now, the differences for the first bands have no minimum at $K_zD/(2\pi) = 0, \pm1, \ldots$; also the differences for the second and third bands have maxima of the absolute values at $k_zD/(2\pi) = 0, \pm1, \ldots$ instead of minima for $k_yD/(2\pi) = 0.1$. For both values of k_y, the same symmetry between $(\omega_l^{(-)} - \omega_t^{(TE)})D/(2\pi c)$ and $(\omega_l^{(+)} - \omega_t^{(TM)})D/(2\pi c)$ as for differences between the polar and transversal MOCs holds; and again, the differences between the first bands are the smallest ones.

The dependencies of the normalized PBG bandwidths $\Delta\omega^{(\pm)}D/(2\pi c)$ with respect to the normalized wave vector $k_yD/2\pi$ are presented in fig. 12. Again the "+" solution resembles TM-mode and the "-" solution - TE-mode. For small k_y, PBGs resembles those for the transversal MOC, whereas for larger k_y, they become similar to PBGs for the polar MOC. For example, the second PBG for the "-" solution decreases and for the "+" solution increases with k_y, in contrast to the second PBG for TE- and TM-modes. As well the first PBG for the "+" solution reaches zero (see the inset on fig. 12(b)), while those for TM-mode does not.

To conclude this section, we have considered the PBGs for the periodic MPC, the dependencies of their bandwidths on the angle of incidence (or k_y), as well as their change under the magnetization reorientation.

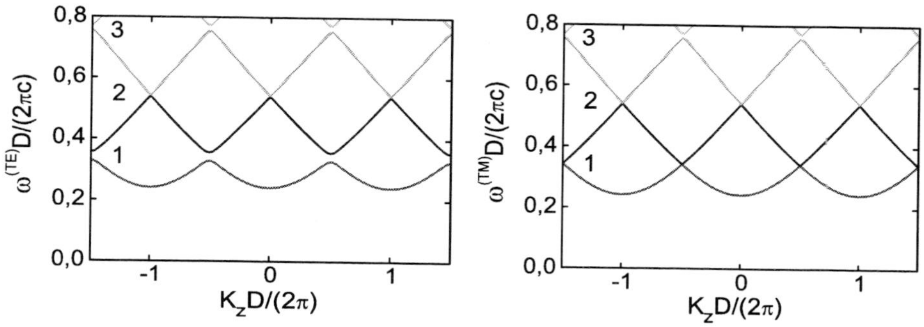

(a) (b)

Figure 3. The same as fig. 2 (a), but for $k_y D/(2\pi) = 0.5$; (a): TE-mode, (b): TM-mode.

5. Fresnel's Coefficients

If the transfer matrix $\hat{T}$ of the 1D system is known, then on can calculate the transmission and reflection coefficients of the system and the other output characteristics. Let us suppose, that the 1D system of flat slabs is embedded in the isotropic dielectric media i before and f behind it. If we want to calculate Fresnel coefficients, which are the ratio of the reflected or transmitted amplitudes of the electric field to incident ones, we must convert H-field amplitudes in $\mathbf{\Psi}_i$, $\mathbf{\Psi}_f$ to E-field amplitudes, using the fact that for the isotropic dielectric media with RI n, $H = nE$. Also, we must note that the y-components of the electric and magnetic fields are not equal to the full amplitude of modes, as the electric and magnetic fields have y- and z-components. For the full amplitudes of E- or H- fields, we obtain

$$(E, H)_{||} = \frac{1}{\cos\alpha}(E, H)_y, \tag{87}$$

where α is the angle between the wave vector and Z axis (the angle of incidence); and for the isotropic medium

$$\sin\alpha = \frac{k_y}{n\Omega}. \tag{88}$$

Thus, for conversion of the 4-component vectors $\mathbf{\Psi}_{i,f}$ to the vector, which components are the full electric amplitudes for the forward and backward waves of s- and p-polarizations, one should use the matrix $\hat{S}_{i,f}$

$$\begin{bmatrix} E_x^{s(+)} \\ E_x^{s(-)} \\ E_{||}^{p(+)} \\ E_{||}^{p(-)} \end{bmatrix}_{i,f} = \hat{S}_{i,f}\mathbf{\Psi}_{i,f} \equiv \frac{1}{\cos\alpha_{i,f}} \begin{bmatrix} n_{i,f}^{-1} & 0 & 0 & 0 \\ 0 & n_{i,f}^{-1} & 0 & 0 \\ 0 & 0 & 1 & 0 \\ 0 & 0 & 0 & 1 \end{bmatrix} \begin{bmatrix} h_y^{s(+)} \\ h_y^{s(-)} \\ e_y^{p(+)} \\ e_y^{p(-)} \end{bmatrix}_{i,f}. \tag{89}$$

In the eq. (88) we used the superscripts (s, p) and $(\pm)$ instead of numbers $1, 2, 3, 4$ in eq. (21), as for the isotropic dielectric media EMW is a superposition of s-polarized

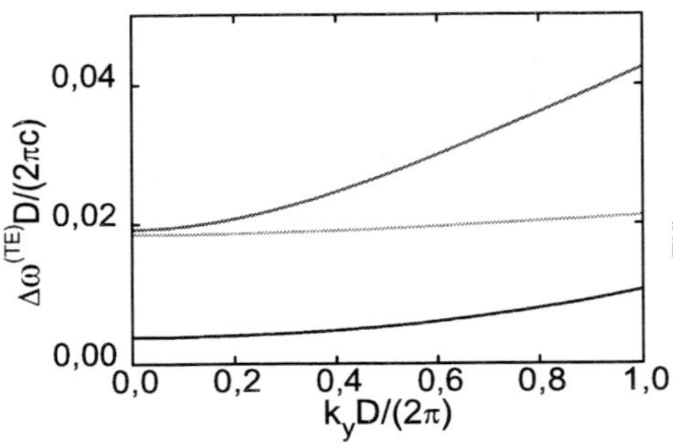

(a)

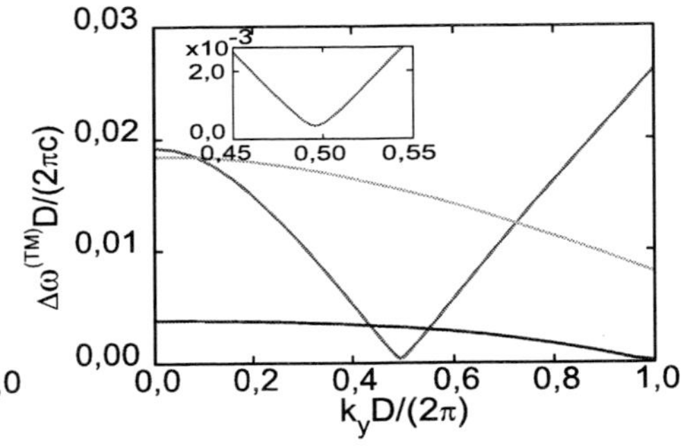

(b)

Figure 4. The PBG width *versus* $k_yD/(2\pi)$ for the transversal MOC; red line - first PBG, blue - second PBG, green - third PBG;(a): TE-mode, (b): TM-mode, on the inset the minimum of first PBG is shown.

(TE) and p-polarized (TM) modes; $(+)$ and $(-)$ refers to forward and backward waves, respectively.

Now one can write the system of equations for the transmitted t^s, t^p and reflected r^s, r^p amplitudes:

$$\begin{bmatrix} t^s \\ 0 \\ t^p \\ 0 \end{bmatrix} = \hat{M} \begin{bmatrix} a^s \\ r^s \\ a^p \\ r^p \end{bmatrix} = \hat{S}_f \hat{T}(d_f + 0, d_i - 0) \hat{S}_i^{-1} \begin{bmatrix} a^s \\ r^s \\ a^p \\ r^p \end{bmatrix}. \tag{90}$$

Here a^s, a^p are the incident amplitudes and r^s, r^p, t^s, t^p are related to them through matrices of Fresnel's coefficients:

$$\begin{bmatrix} r^s \\ r^p \end{bmatrix} = \begin{bmatrix} r^{s,s} & r^{s,p} \\ r^{p,s} & r^{p,p} \end{bmatrix} \begin{bmatrix} a^s \\ a^p \end{bmatrix}, \tag{91}$$

$$\begin{bmatrix} t^s \\ t^p \end{bmatrix} = \begin{bmatrix} t^{s,s} & t^{s,p} \\ t^{p,s} & t^{p,p} \end{bmatrix} \begin{bmatrix} a^s \\ a^p \end{bmatrix}, \tag{92}$$

where

$$r^{s,s} = \frac{1}{D_M}[M_{4,4}M_{2,1} - M_{4,1}M_{2,4}], \qquad r^{s,p} = \frac{1}{D_M}[M_{4,4}M_{2,3} - M_{4,3}M_{3,4}],$$
$$r^{p,s} = \frac{1}{D_M}[M_{2,2}M_{4,1} - M_{2,1}M_{4,2}], \qquad r^{p,p} = \frac{1}{D_M}[M_{2,2}M_{4,3} - M_{2,3}M_{4,2}] \tag{93}$$

and

$$D_M = M_{2,4}M_{4,2} - M_{2,2}M_{4,4}, \tag{94}$$

$$t^{s,s} = M_{1,1} + M_{1,2}r^{s,s} + M_{1,4}r^{p,s}, \qquad t^{s,p} = M_{1,3} + M_{1,2}r^{s,p} + M_{1,4}r^{p,p},$$
$$t^{p,s} = M_{3,1} + M_{3,2}r^{s,s} + M_{3,4}r^{p,s}, \qquad t^{p,p} = M_{3,3} + M_{3,2}r^{s,p} + M_{3,4}r^{p,p}. \tag{95}$$

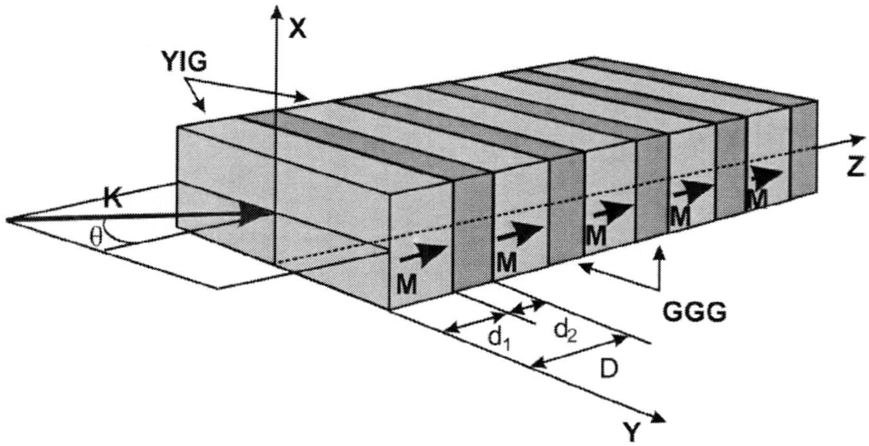

Figure 5. The same as fig. 1 for the polar MOC.

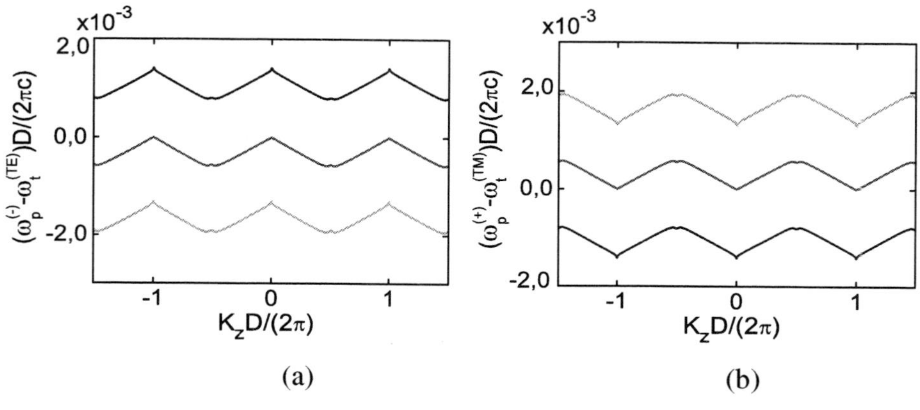

Figure 6. The differences of normalized frequencies of polar and transverse MOCs versus normalized wave number $k_zD/(2\pi)$ for the normal incidence ($k_y = 0$); (a): $(\omega_p^{(-)} - \omega_t^{(TE)})D/(2\pi c)$, (b): $(\omega_p^{(+)} - \omega_t^{(TM)})D/(2\pi c)$; red - the first band, blue - the second, green - the third.

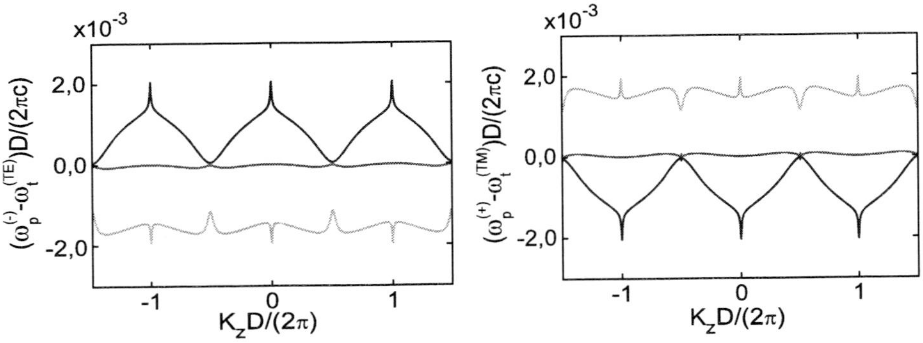

(a) (b)

Figure 7. The same as fig. 6, but for $k_y D/(2\pi) = 0.5$.

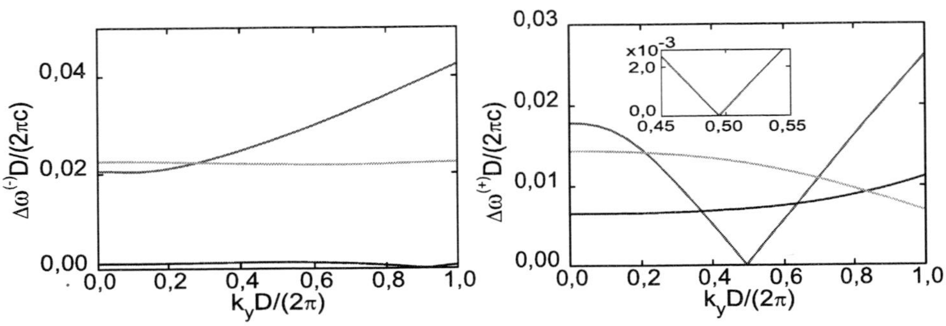

(a) (b)

Figure 8. The same as fig. 4, but for polar MOC; (a): the "-" solution, (b): the "+" solution.

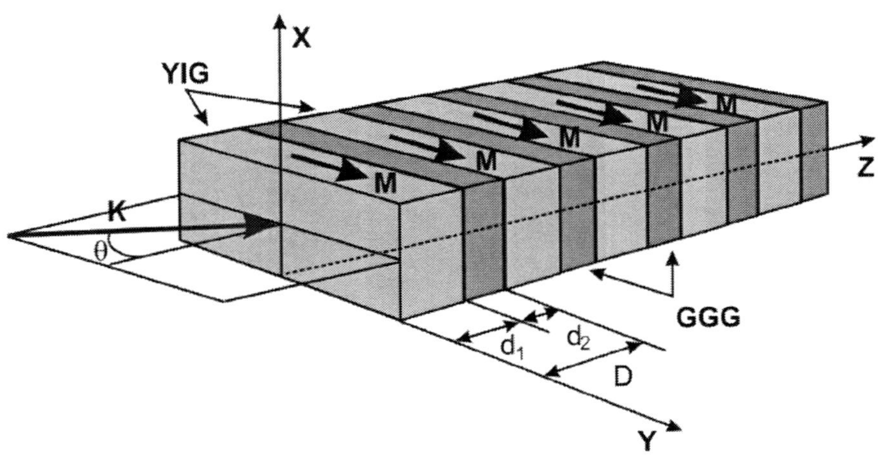

Figure 9. The same as fig. 1, but for the longitudinal MOC.

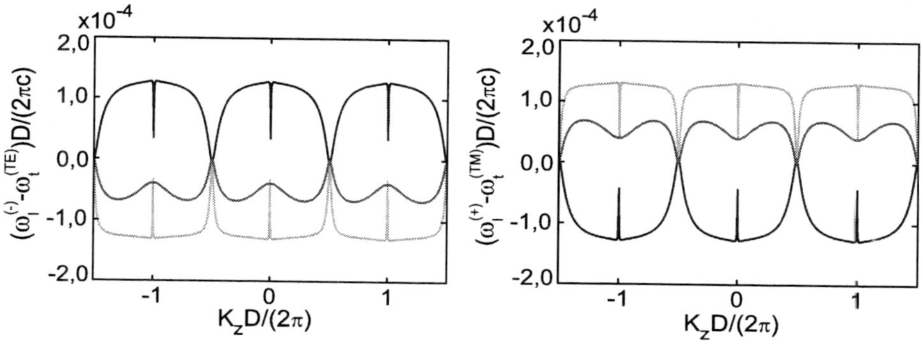

(a) (b)

Figure 10. The same as fig. 6, but for differences between the longitudinal and transversal MOCs and $k_yD/(2\pi) = 0.1$. (a): $(\omega_l^{(-)} - \omega_t^{(TE)})D/(2\pi c)$, (b): $(\omega_l^{(+)} - \omega_t^{(TM)})D/(2\pi c)$.

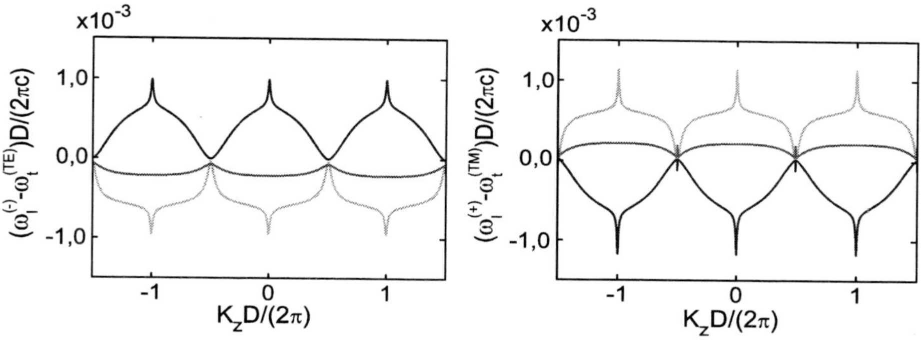

(a) (b)

Figure 11. The same as fig. 10, but for $k_yD/(2\pi) = 0.5$.

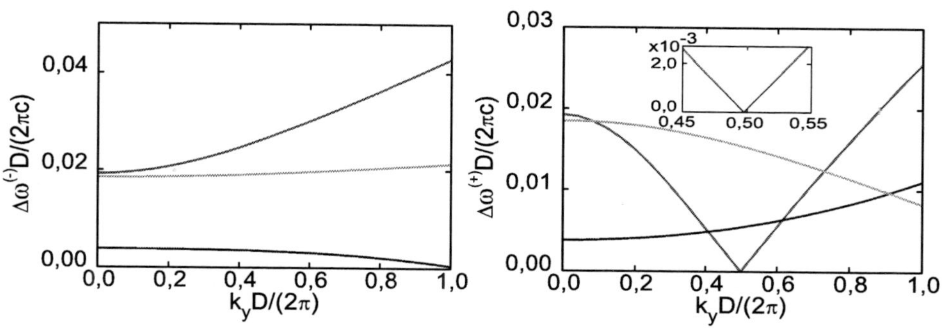

(a) (b)

Figure 12. The same as fig. 4, but for longitudinal MOC; (a): the "-" solution, (b): the "+" solution.

Intensities of the transmitted and reflected light are given by

$$T = |t^s|^2 + |t^p|^2, \tag{96a}$$

$$R = |r^s|^2 + |r^p|^2. \tag{96b}$$

In general, the transmitted and reflected waves are elliptically polarized even in the case of the incident linearly polarized wave (LPW). It can be described by the followings parameters: *i*) the sum of squared projections of $\boldsymbol{E}$ on the principle axes of the polarization ellipse ($E^2_{maj} + E^2_{min}$), which is proportional to the energy of the wave; *ii*) the angle θ between the major axis and the **X**-axis; *iii*) the ellipticity ψ. For the transmitted and reflected light the sums $E^2_{maj} + E^2_{min}$ are equal to T and R, respectively (see eqs. (96a) and (96b)). The ratio e between minor and major axes of polarization ellipse and the ellipticity ψ are related by the expressions

$$e = \tan(\psi), \qquad e \equiv E_{min}/E_{max}. \tag{97}$$

As $-1 < e < 1$, then $-45° < \psi < +45°$; the signs are defined so as $\psi = -45°$ and $\psi = +45°$ ($e = -1$ and $e = 1$) correspond to LH and RH CPW, respectively; and $\psi = 0°$ refers to the linear polarization. The angles θ and ψ can be given in terms of parameter $\chi \equiv E_{y1}/E_x$ (where $\mathbf{Y}_1$ is the axis perpendicular to **X** and the wave vector) (6):

$$\tan 2\theta = \frac{2Re\chi}{1 - |\chi|^2}, \tag{98}$$

$$\sin\psi = \frac{2Im\chi}{1 + |\chi|^2}. \tag{99}$$

The parameters χ for the transmitted and reflected waves are

$$\chi_t = \frac{t^p}{t^s}, \tag{100a}$$

$$\chi_r = \frac{r^p}{r^s}. \tag{100b}$$

Then, the absolute values of the major and minor axes of polarization ellipse are

$$E_{maj} = \sqrt{\frac{T}{1 + e^2}}, \tag{101a}$$

$$E_{min} = eE_{maj} = \sqrt{\frac{e^2 T}{1 + e^2}}. \tag{101b}$$

6. Numerical Results for Defect Modes

In the transmission spectra of the MPC, consisting of finite number of the periods, for the PBG region $\omega_+ < \omega < \omega_-$, (see, for example, eq. (70)) the vanishingly small transmission coefficient is observed. The PBG edges $\omega_\pm$ can be calculated exactly via $\hat{T}$-matrix from eqs. (85) or (86). If the strict periodicity is broken by a defect layer, the resonance peak (the defect mode) appears inside the PBG region. This effect cam be explained as follows. The PBG in spectra of the periodic structure appears because at some frequency the EMW is reflected coherently from each period and thus, the transmission is low. Due to the defect layer the waves, reflected from different parts of the MPC, can cancel each other and transmission becomes high. The defect in the MPC can be introduced by using an another material for some middle layer, or changing the middle layer length (defect of symmetry). In both methods the optical length (nd) of the defect layer is changed and for the frequency ω one can introduce the phase shift $\delta = \Delta(nd)\omega/c$. If $\delta = (2m+1)\pi/2$, the phase difference between the light reflected from the first part of PC (before the defect layer) and from the second one (behind the defect layer) is $(2m+1)\pi$ the corresponding reflected waves cancel each other. If the defect layer is localized not in the center of the MPC (asymmetric defect), the cancellation is not complete and the transmission resonance is lower than for the case of the symmetric defect. When we gradually increase length of middle layer (or it's refraction index) starting from it's value in strictly periodic PC then defect mode appear near upper edge of band gap to lower one. From (71a) obvious that for normal incidence defect mode reach center of band gap when

$$\Delta(nd) \approx \frac{D\sqrt{\epsilon_0}}{2m} \tag{102}$$

and disappears at the lower edge of the PBG when $\Delta(nd)$ is twice larger. The similar relations can be inferred (using eq. (75)) for the oblique incidence. By the same arguments, it can be shown that the defect mode moves from the lower edge of the PBG to the upper one when we decrease the central layer length (or it's RI). Further we shall introduce the defect mode by changing the central layer length, as it seems to be more practical.

6.1. One-Defect MPC

We have calculated the transmission coefficients at $\lambda_0 = 1.15\mu m$ for the MPC, constructed as $(NM)^a D_M (MN)^a$ and magnetized perpendicularly to the interfaces. Here a is a number of periods, N denotes the non-magnetic layer of thickness d_2 (we choose GGG with $n_2 = 1.94$ at $1.15\mu m$), M is the magnetic layer of thickness d_1 (in our calculations we used $Bi_{1.43}Y_{1.57}Fe_5O_{12}$ with $n_1 = 2.4$ at $1.15\mu m$ (19)), and D_M is magnetic $Bi:YIG$ layer of thickness d_M. Thus, in the center of the MPC we have one magnetic defect layer of thickness $2d_1 + d_M$. We suppose that MPC is surrounded by the medium with RI averaged over the MPC, so $\bar{n} = 2.15$. The electric and magnetic susceptibilities (20) contribute to the total Faraday rotation (FR), and the magnetic contribution can be calculated from the classical equation of motion for the magnetic moment (21). Then, subtracting the magnetic contribution from FR, measured in the experiment (22), one can obtain the electric contribution (see also (19)). Thus, for the chosen magnetic material, we obtained $\epsilon' = 0.005$, $\mu' = 1.65 \times 10^{-5}$. The period of the MPC is $D = d_1 + d_2$. We shall concentrate our attention on the first PBG. In order to ensure the maximal absorbtion and PBG width in it's center, the values $d_{1,2}$ should satisfy the relation $d_1 : d_2 = n_2 : n_1$ (eq. (72b) with $m = 1$ and $j = 0$), so $d_1 = 0.447D$ and $d_2 = 0.553D$. As follows from eq. (72a), each layer is quarter of the wavelength one for the frequency in the center of first PBG. As D_M is the defect offset (see fig. (13)); thus in the center of the MPC we have one magnetic

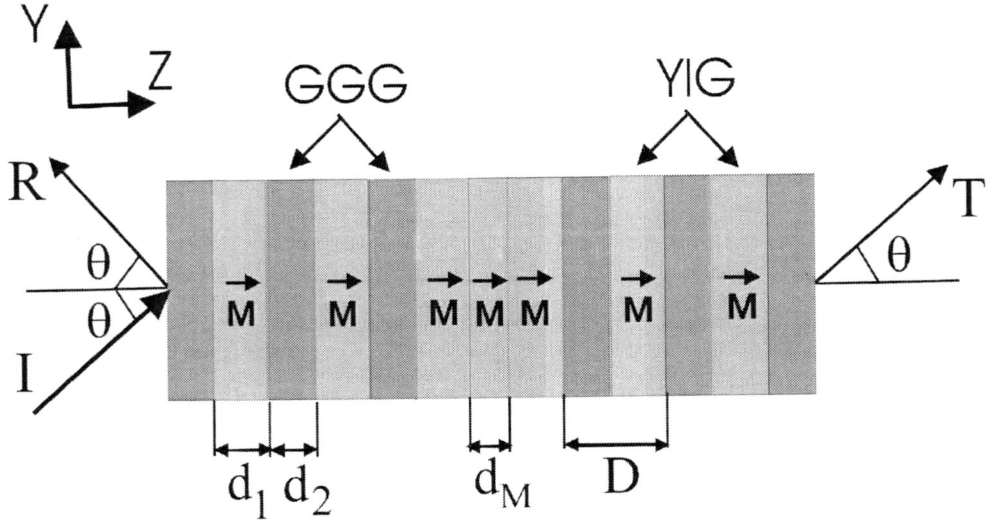

Figure 13. One defect MPC described by formula $(NM)^a D_M (MN)^a$ (M and N denote magnetic and nonmagnetic layers, respectively).

layer of thickness $2d_1 + d_M$. The value d_M can be taken to be negative, provided that $2d_1 + d_M > 0$. We present d_M as $d_M = 2md_1 + \Delta$, where $m = 0, 1, \ldots, -d_1 \le \Delta \le d_1$. Than, remembering that the phase shift produced by the layer M is $\delta = \pi/2$, we find that the additional phase shift due to the defect $MD_M M$ in comparison with magnetic layer M is $\delta = (2m+1)\pi/2 + \pi\Delta/(2d_1)$. For $\Delta = 0$ the shift is $\delta = (2m+1)\pi/2$ and the defect mode is located at the center of the first PBG. For $\Delta < 0$ ($\Delta > 0$) the defect mode is shifted from the center of the first PBG to higher (lower) frequencies. For $\Delta \approx -d_1$ ($\Delta \approx d_1$) the defect mode disappears at upper (lower) first PBG edge. Let us consider the case of normal incidence. As it was mentioned in subsec. 3.2, for the polar MOC the RH and LH CPWs are eigenmodes of the MPC and they do not mix with each other and, due to circular magnetic birefringence, produce two defect modes. In fig. 14, the transmission coefficients

$$T^{\pm} = |t^{\pm}|^2 = \left| \frac{E^{\pm}_{out}}{E^{\pm}_{in}} \right|^2, \tag{103}$$

are shown against the normalized frequency $(\omega D)/(2\pi c)$ for $a = 10$, where "+" and "-" refer to the RH- and LH-modes, respectively. As one can see from the fig. 14, $T^+ \approx T^-$, and the lines for T^+ and T^- are not distinguishable in this scale. One can clearly see the PBG and the defect mode located in its center (fig. 14 (a)) for $m = 0$, $\Delta = 0$ and near its edge (fig. 14 (b)) for $m = 0$, $\Delta = 0.22D$.

In fig. 15 these coefficients are given in more detail, as well as the phases $\phi_{\pm}$ of the RH- and LH- modes for $a = 10$ (fig. 15 (a)) and $a = 15$ (fig. 15 (b)). We also introduce two important dimensionless parameters: the magnetic splitting $2\alpha_0$ as the difference of the peak positions for $T^{\pm}$ (which is conditioned by the magnetically induced off-diagonal tensor components ϵ' and $\mu\prime$), and the peak width w at the point where $T^{\pm} = \frac{1}{2} T^{\pm}_{max}$ (see fig. 15 (b)). According to our estimations, for the polar MOC, $2\alpha_0 \simeq w = 1.63 \times 10^{-4}$. As one can see, the phase increases by π in the vicinity of the corresponding defect mode.

This figure represents two essentially different cases: when RH- and LH- peaks overlap, and when they do not.

The corresponding responses of MPC to the incident LPW are given in figs. 16 (a, b). In these figures we show the transmission coefficient, eq. (96a), the FR, eq. (98), and the ellipticity, eq. (99). These responses are better to understand using responses to incident RH and LH CPWs (23). Transmission coefficient T, FR and ellipticity of EMW are easily obtained from corresponding amplitudes and phases of RH and LH CPWs on which this EMW can be decomposed. Thus response of MPC to incident LPW can be obtained from $t^{\pm}$ using formulas (6):

$$T = \frac{1}{2}(T^+ + T^-) \tag{104}$$

$$\theta_F = -\frac{1}{2}(\phi_+ - \phi_-) \tag{105}$$

$$e = \frac{|t^+| - |t^-|}{|t^+| + |t^-|} \tag{106}$$

It is obvious that figs. 16 (a,b) are in accordance with figs. 15 , respectively, and formulas (104)-(106). Maximum of $\Delta\phi = \phi_- - \phi_+$ and, consequently, FR is observed in the middle of distance between the centers of the RH and LH defect modes in figs. 15. The ellipticity has the maximum and minimum in the centers of RH and LH defect modes and it turns to zero in the middle of the distance between the centers of RH and LH defect modes, where $t^+ = t^-$. In the form of transmission coefficient T one can see the difference between the cases of overlapping fig. 15 (a) and non-overlapping fig. 15 (b) RH and LH-modes. In the first case (corresponding to maximal value $\theta_F \approx 7°$), T has one peak with the maximum $T \approx 0.98$, and in the second one (minimum in $T \approx 0.47$ corresponds to maximum in $\theta_F \approx 47°$), two peaks with a minimum between them. So, we see that the large T corresponds to moderate FR (though larger than for the uniform material), and the large FR corresponds to small T. This behaviour seems to be typical for the one-defect structures (23).

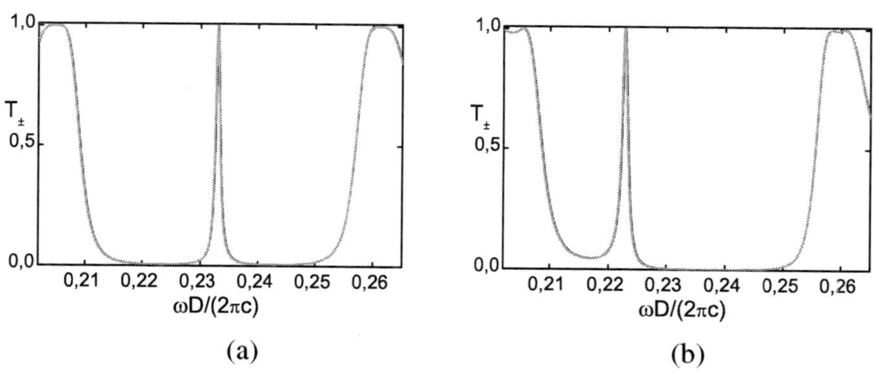

Figure 14. The transmission coefficients $T^{\pm}$ (a): the centered defect mode, $d_M = 0$; (b): the off-centered defect mode $d_M = 0.22D$. Red an green lines correspond to T^+ and T^-, respectively.

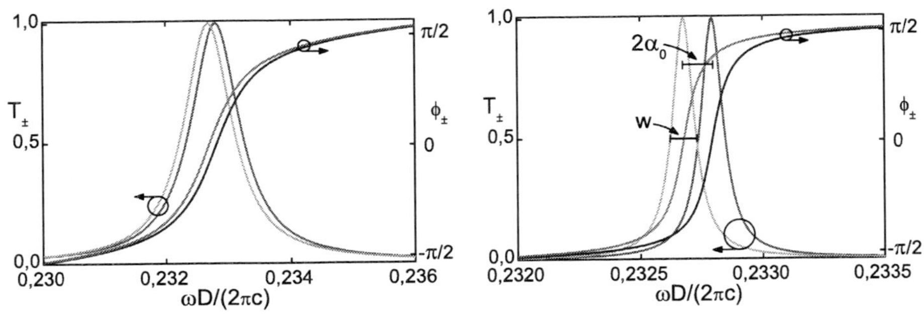

(a) (b)

Figure 15. The transmission coefficients for the MPC. The red line corresponds to T^+ (RH-mode) and green one refers T^- (LH-mode); the blue and magenta lines denote ϕ_+ and ϕ_-, respectively ; (a): a=10, (b): a=15.

In all calculations above we supposed that MPC is surrounded by the medium with RI, averaged over MPC, in order to avoid the effect related to Fresnel reflections at the surfaces of MPC. If MPC is surrounded by vacuum, its response strongly depends on the type of the surface layers. The case for which the surface layer has a small RI, is illustrated by fig. 17 (a): it is well-known that the quarter wavelength layer, placed before the optical system with RI, which is lower than the RI, averaged over the period of the system, increases the transmission. So the transmission for any frequency is larger (or the defect mode is wider) than that in fig. 15. If the surface layer has a large RI (fig. 17 (b)), the situation is reversed. Further we shall surround the MPC by the medium with an averaged RI.

The larger is the magnetic splitting, the better are the magneto-optical properties. The magnetic splitting depends on the difference of optical paths for the RH- and LH-modes, consequently, it depends on magnetic material parameters (ϵ' and μ') and thickness of the defect layer, but it is independent of the number of periods. The thickness of the defect layer determines also a position of the defect mode inside the PBG. The peak width decreases strongly with the number of the periods, and it increases with the shift of the defect modes from the center of PBG to its edges (because the imaginary part of the wave vector decreases towards the PBG edges). These dependencies are shown in figs. 18 and 19. We assume $d_M = 2md_1 + \Delta$, $m = 0, 1, 2 \ldots$. If $\Delta = 0$, then the defect mode is located in the center of the PBG at $\omega D/(2\pi c) \approx 0.233$. In fig. 18 (a), the ratios Δ/D for the MPCs with magnetic and nonmagnetic defects are plotted versus the corresponding defect mode frequency ($\omega_a \equiv (\omega_+ + \omega_-)/2)D/(2\pi c)$ in the cases of $m = 0$ and $m = 1$. The MPC with the non-magnetic defect has the structure $(MN)^a D_N (NM)^a$, where D_N is the non-magnetic GGG layer of thickness $d_N = 2d_2 + \Delta$. As one can see from fig. 18, the behaviour of Δ/D is similar. For the same shift in ω, the non-magnetic defect requires larger Δ than for magnetic, as $n_1 > n_2$. The magnetic splitting $2\alpha_0$ (in units of $D/(2\pi c)$) versus ω_a is shown in fig. 18 (b) for the magnetic and non-magnetic defects and $m = 0, 1, 2$. One can see, that the magnetic splitting is larger for the magnetic defects, then for non-magnetic ones for all ω_a. The magnetic splitting increases with m for the MPC with a magnetic

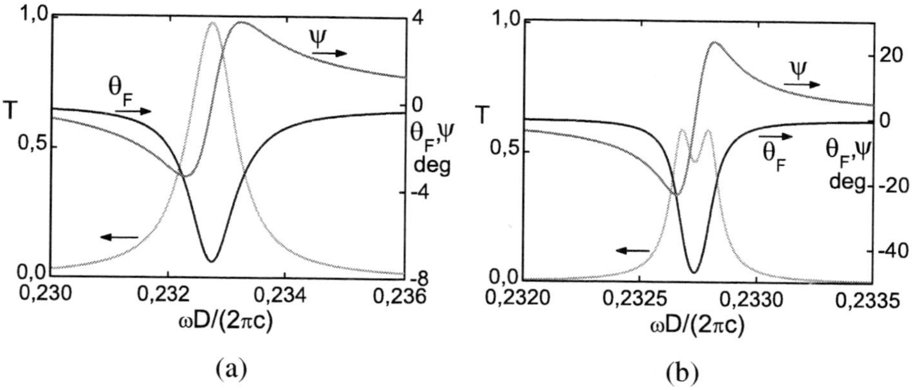

Figure 16. The transmission coefficient T (green line), FR (blue line), the ellipticity ψ (magenta line) versus the normalized frequency (a): m=10; (b): m=15.

defect and it decreases with m for the non-magnetic defect. That's why it is preferable to use MPCs with magnetic defects and $m > 0$. It should be mentioned that the magnetic splitting has the largest increment in the case of the magnetic defect, and $m = 0$. For both figures it is assumed that $a = 10$, for other values of a, the figures are almost the same. For the magnetic defect the peak widths w versus w_a are presented in fig. 19 (a) ($a = 10$) and fig. 19 (b) ($a = 15$). For the case of the non-magnetic defect the peak widths are similar. One can be mentioned that the ratio of peak widths at the PBG edge to those in the center increases with the number of periods from 2.7 for $a = 10$ to 8.3 for $a = 15$.

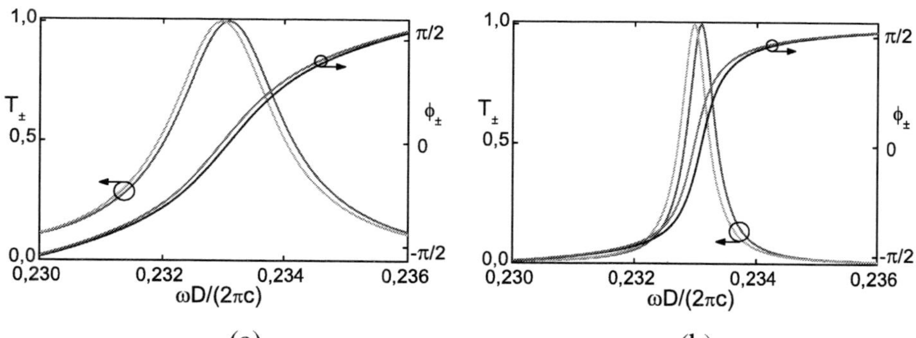

Figure 17. The same as in fig. 15, but the MPCs are surrounded by vacuum; (a): for $(NM)^{10}(MN)^{10}$; (b): for $M(NM)^9(MN)^9M$.

We also present some results for the oblique incidence and compare them with the previously obtained for the normal incidence for $a = 10$. First of all, the PBG and the defect mode shift to higher frequencies because the same value of $k_z^2 = |k|^2 - k_y^2$ corresponds to higher frequency in eq. (75). Also, there is a difference between the defect mode at the center of PBG and near it's edge (see fig. 14). As RH- and LH-modes are no more the

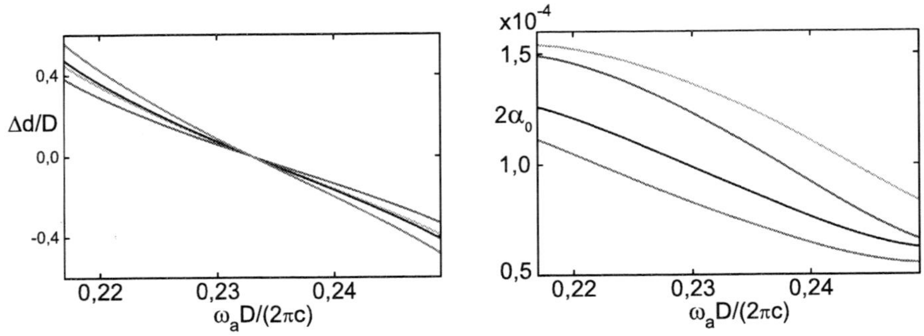

(a) (b)

Figure 18. The dependence of the parameter $\delta D/D$ (a) and magnetic splitting $2\alpha_0$ (b) on the defect mode frequency ω_a. The magnetic defect: red line for $m = 0$, green line for $m = 1$; nonmagnetic defect: blue line for $m = 0$, magenta line for $m = 1$.

eigenmodes in the case of the oblique incidence of the light, and the 'mixed' transmission coefficients appear as well:

$$T'^{\pm} = |t'^{\pm}| = \left| \frac{E^{\mp}_{out}}{E^{\pm}_{in}} \right|^2 . \tag{107}$$

The fig. 20 (a) demonstrates that for the centered defect mode, RH- and LH-modes are good approximations to the eigenmodes, and the mixed terms $t'^{+} \approx t'^{-}$ are small. That leads to the fact that the response to the linearly polarized input is quite similar to the case of the normal incidence, except for the defect mode broadening for the p-polarized input (see fig. 21). But, as it is seen from fig. 20 (b), for the off-centered defect mode, the mixed terms $t'^{+} \approx t'^{-}$ have comparable values with t^{+}, t^{-}. As a result, in the response to the linearly polarized input light, the minimum of FR and zero of ellipticity do not correspond to the maximum of the transmission (fig. 22). In both cases for p-polarized input, the PBG is smaller than for s-polarized one (see (16)) and, consequently, absorption at it's center is smaller too, leading to more wide defect mode for p-polarized input.

6.2. Two-Defect MPC

The correlation between high transmission and large FR, mentioned in the subsec. 7.1, can be avoided with two-defect MPC. The detailed explanation of the better performance of the two-defect MPC was given by Steel et al. in (23; 24). We shall demonstrate this fact using MPC made of $Bi : YIG$ and GGG, as in the previous subsection. Now the MPC contains two magnetic defects, and it is described by formula $(NM)^a D_M (MN)^b M D_M (MN)^a$ (see subsec. 7.1; again we introduce $d_M = 2md_1 + \Delta$). In contrast to our case, the magnetic and non-magnetic defects are characterized by formula $(MN)^a (NM)^b (MN)^a$. The response to RH and LH input in eq. (103) for $a = 13$, $b = 24$, $m = 0$ and $\Delta = 0$ is shown in fig. 23 (a). Now the RH and LH defect modes split into two peaks at the frequencies $\omega^{-}_{1,2}$ and $\omega^{+}_{1,2}$. As well, the figure shows the magnetic splitting $2\alpha_0$ and splitting of the defect mode σ, the mode width w, which is defined as difference between two points

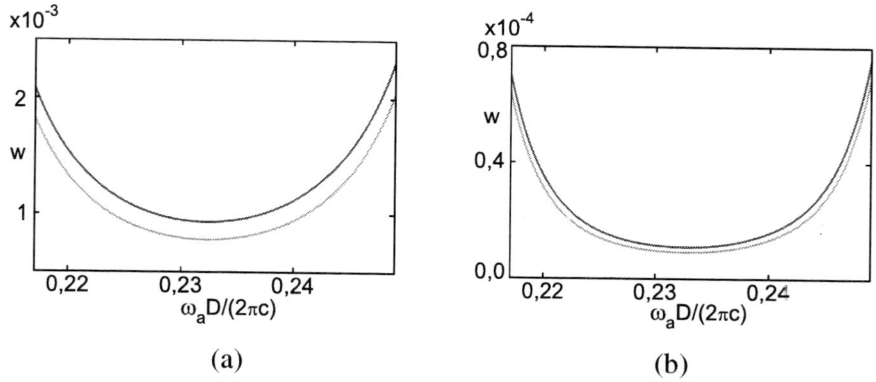

Figure 19. Peak width w versus defect mode frequency (a): $a = 10$, (b): $a = 15$, red line for $m = 0$, green line for $m = 1$, dotted line for $m = 2$.

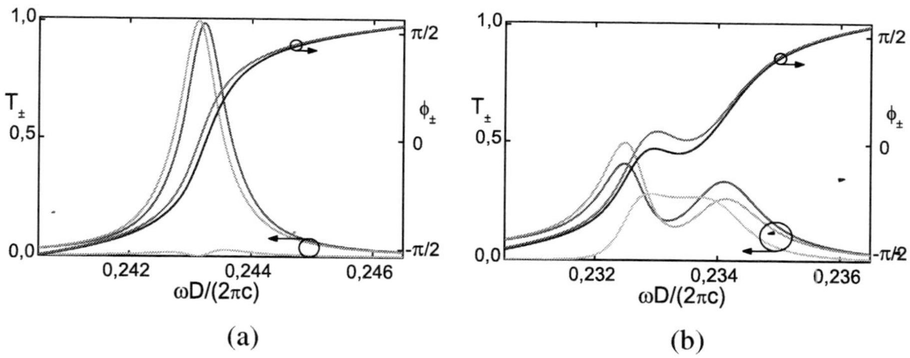

Figure 20. The same as in fig. 15, blue line for T'_+, $k_y D/(2\pi) = 0.15$; (a): centered defect mode ($d_M = 0$), (b): off-center one ($d_M = 0.22D$).

where $T^{\pm} = 0.98$. One can see, that the phase increases by π near each peak, so after traversing both peaks, the phase is advanced by 2π.

Using coupled mode theory, Steel et al. (23) analyzed dependence of those parameters on the total length of MPC $L \approx 2L_2 + L_1$ and parameter $f \equiv L_1/L_2$. They found that the magnetic splitting $2\alpha_0$ is constant, fixed by magneto-optical strength of material. For very short L, so that $w \gg 2\alpha_0$, the defect modes of each polarization overlap, and the phase difference $\phi_+ - \phi_-$ is very small (like in fig. 15 (a)). As L increases, the peak width w decreases (just as for the one-defect MPC from the previous subsection) and the peak separation σ decreases too (as the distance between two defects increases and their interaction becomes weaker). The relation of the peak separation σ to peak width w decreases with increase of f: for $f \lesssim 1.5$, two peaks for both RH and LH defect modes are well separated, as f increases they becomes less separated, as in fig. 23 (a) ($f = 1.88$), and finally, when $f = 2$, two peaks merge into a single broad resonance.

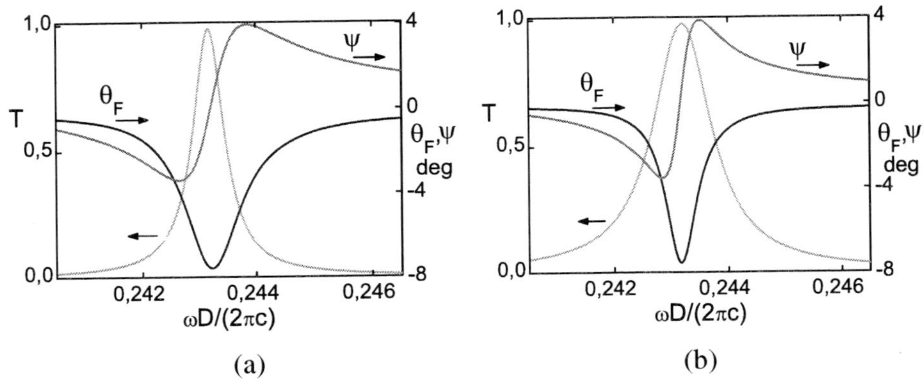

Figure 21. The same as fig. 16, the centered defect mode ($d_M = 0$), $k_y D/(2\pi) = 0.15$; (a): s-polarized input, (b): p-polarized input.

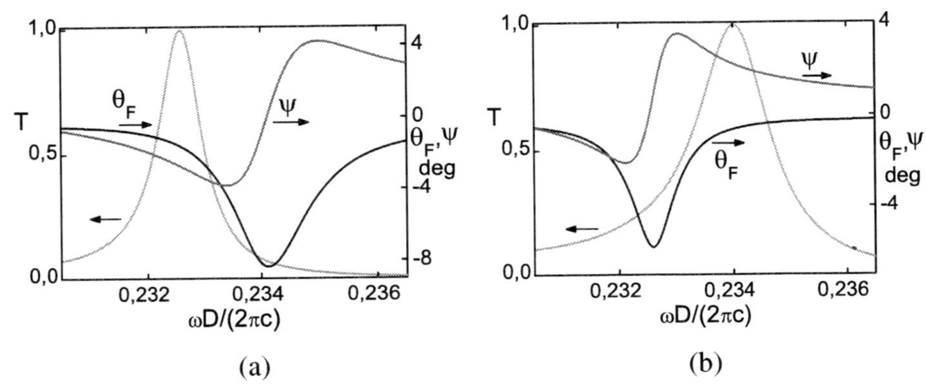

Figure 22. The same as fig. 21, the off-center defect mode ($d_M = 0.22D$); (a): s-polarized input, (b): p-polarized input.

The main idea of optimization of MPC characteristics is to find some parameters, which make close enough the rightmost peak for LH defect mode and the leftmost peak for RH defect mode (i.e. $\omega_2^- \approx \omega_1^+$) as in fig. 23 (a). Then we reach the phase difference $\phi_+ - \phi_-$ high enough (which leads to large FR) with still high transmission. The response to the linear polarized light is shown in fig. 23 (b). In the coupled mode approximation, L and f can take the arbitrary values, so varying them one can achieve desired FR at high transmission. But in the discrete system L and f are discrete values, as they can be changed only by the period D, so not all values of FR are achievable. If we try to get MPC with the practically important value of FR $\theta_F = -45°$ (then MPC in combination with two polarizers can be used as an optical isolator), we shall meet with difficulties, as for the MPC with $a = 13$, $b = 24$, $m = 0$ (fig. 23 (b)), the FR becomes $\theta_F = -41.4°$, whereas for MPC with $a = 13$, $b = 25$, $m = 0$ it is $\theta_F = -52°$ and $T = 0.93$. Steel et al. (23; 24) suggested to vary RI of the non-magnetic material, but there exists another possibility. As one can see from fig. 18 (b) and fig. 19, we can change the magnetic splitting $2\alpha_C$ and mode width by varying the offset Δ and mode position inside PBG, and we can do this *continiously*. The figure 24 (a) shows the width of the rightmost LH peak w_2^-, the width of the leftmost RH peak w_1^+, and the absolute value of difference between the peak frequencies $|\omega_1^+ - \omega_2^-| = |2\alpha_0 - \sigma|$ versus $\omega_a D/(2\pi c)$, where ω_a is the averaged frequency of $\omega_{1,2}^\pm$. The dependence of the offset Δ on the frequency shift ω_a is similar to those in fig. 18 (a), and is not presented here. We note that at the higher ω_a, the following inequality is satisfied $w_2^- > w_1^+$, while for the lower ω_a, vice versa. While $|\omega_1^+ - \omega_2^-| \leq w_2^-, w_1^+$, the response to the linear input is larger than 0.98. In fig. 24 (b) the plots of the transmission coefficient and FR versus $\omega_a D/(2\pi c)$ are shown. The magnetic splitting $2\alpha_0$ increases monotonically when ω_a decreases (fig. 18 (b)) and σ increases considerably when the defect modes (ω_a) approaches the PBG edges; because of this concurrence the value $\omega_1^+ - \omega_2^-$ has an extremum (maximum) near the PBG center, ($|\omega_1^+ - \omega_2^-|$ has a minimum). The increase of difference $\omega_1^+ - \omega_2^-$ leads to the increase of FR; that is why it is easy to reduce FR, but it is hard to increase it. Indeed, the maximum of FR in fig. 24 (b) (at $\omega_a D/(2\pi c) \approx 0.231$) corresponds to $-43.5°$, but we can get any FR up to $-35°$ with still high transmission. So, if we use the MPC with FR higher than $-45°$ for ω_a at the center of the PBG, we can be able to tune FR to $-45°$. This is illustrated in fig. 25. Again, $\omega_1^+ - \omega_2^-$ has maximum, but this time it changes sign so as $|\omega_1^+ - \omega_2^-|$ exhibits two minima which correspond to two transmission maxima. So, at $\omega_a D/(2\pi c) \approx 0.237$, the value θ_F reaches $-45°$ with $T = 0.999$. Another illustration for $a = 13, b = 24, m = 1$ is shown in fig. 26. This time, the value $\theta_F = -45°$ is reached near $\omega_a D/(2\pi c) \approx 0.2355$ with $T = 0.984$. Successfully, for $a = 13, b = 13$ and $m = 1$, we achieve $\theta_F = -45.7°$ with $T = 0.986$ without any offset from the center of the PBG.

Another way to decrease the magnetic splitting $2\alpha_0$ and FR is using of the MPC with the magnetization vector directed at the angle γ to the $\boldsymbol{Z}$-axis. In this case, the MOC is no more polar one and tensors $\hat{\epsilon}$ and $\hat{\mu}$ are given by (8)-(9) with $\theta = \gamma$ and $\phi = 0$. This time one can not use formulas from subsec. 2.3, but one has to employ the general formulas from Appendixes A, B. Surely, the RH- and LH-modes are no more the eigenmodes. But, even for the transversal MOC ($\gamma = 90°$), the difference between n_{TM} and n_{TE} is very small ($\sim (\epsilon')^2$), and for the numbers a and b used for the polar MOC, the TM and TE transmission peaks look very similar. Thus, any linear combinations of TM and TE (and RH-, LH-modes, too) for those values a and b are good approximations to the eigenmodes. Actually, we found that the 'mixed' terms in eq. (107) are smaller then 10^{-6}. Fig. 27 (a) demonstrates the magnetic splitting $2\alpha_0$ and the difference $|\Delta\omega| = |\omega_1^+ - \omega_2^-|$ in units $D/(2\pi c)$ versus $\cos(\gamma)$ for $a = 13, b = 24, m = 1$ and $a = 14, b = 25, m = 0$ ($\Delta = 0$ for both cases). As one can see from figs. 27 (a), all these dependencies are almost linear. The corresponding responses to the linear input are shown in fig. 27 (b). Both MPCs produce $\theta_F = -45°$ with almost the same transmission coefficient T as if tuned by $\Delta \neq 0$. But in the former case, tuning is limited by the fact that T decreases sharply for the defect mode near the PBG edges, and even now, for very low θ_F, the transmission is high enough. The drawback is that the $\boldsymbol{Z}$-axis is the preferable direction of the magnetization for $Bi : YIG$, and any deflection from it can require an external dc magnetic field.

The third interesting possibility of tuning the transmission and FR is using of the oblique incident light. Now we consider the oblique incidence of the light with the wavelength

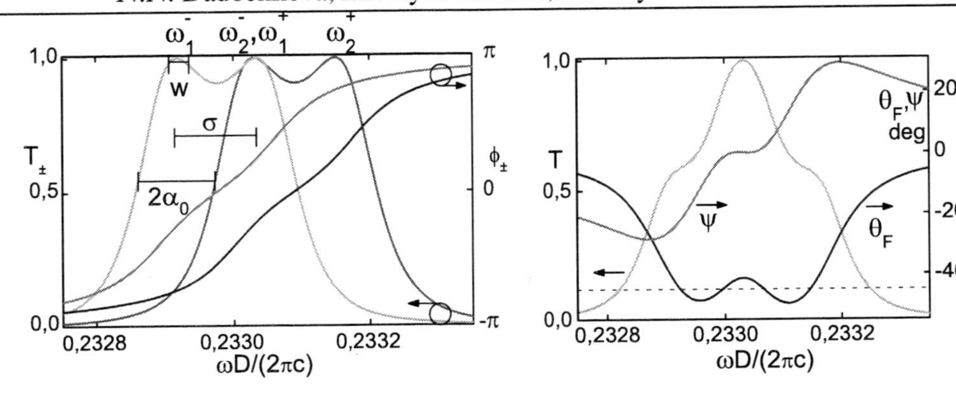

(a) (b)

Figure 23. The response from the two-defect MPC with a=13, b=24, m=0, $\Delta = 0$ versus the normalized frequency to: (a) $RH-$ and $LH-$ input light and (b) to the linearly polarized light. (a): The red an green lines correspond to the transmission coefficients T^+ and T^-, respectively; the blue and magenta lines are ϕ_+ and ϕ_-. (b): The green and blue lines correspond to the transmission coefficient T and FR, respectively. The ellipticity ψ is depicted with the magenta line. The black dotted line corresponds to $\theta_F = -45°$.

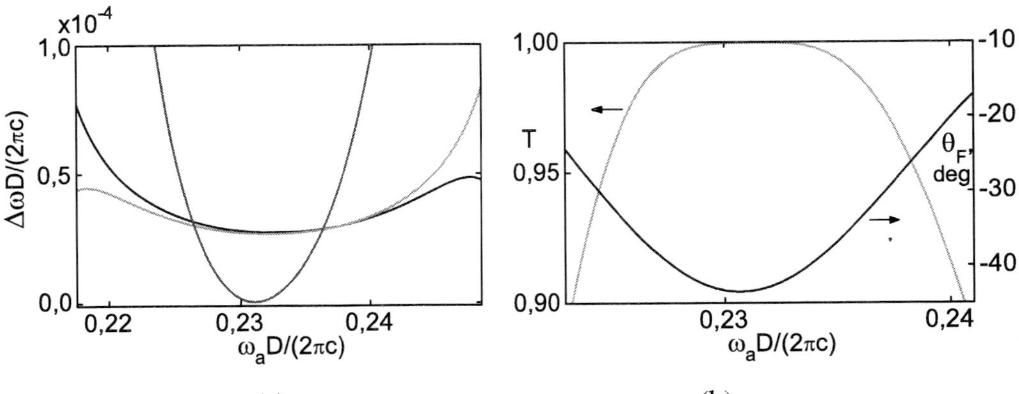

(a) (b)

Figure 24. (a) The width of the rightmost LH w_2^- (green line) and leftmost RH w_1^+ (blue line) peaks and the absolute value of difference between the peak frequencies $|\omega_1^+ - \omega_2^-| = |2\alpha_0 - \sigma|$ (red line) versus $\omega_a D/(2\pi c)$. Dash-dotted line denotes $w_1^+ + w_2^-$. (b) The transmission coefficient (green line) and FR (blue line) versus $\omega_a D/(2\pi c)$.

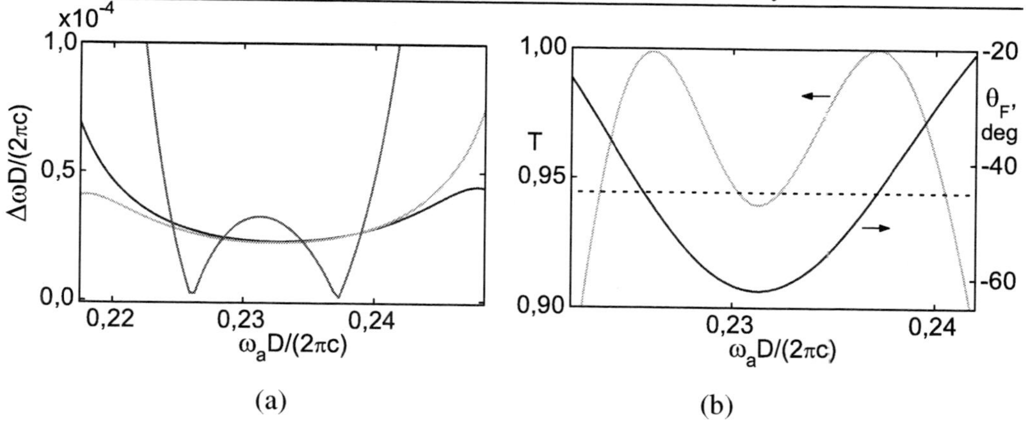

Figure 25. The same as fig. 24, but for $a = 13$, $b = 24$, $m = 1$; black dotted line in (b) corresponds to $\theta_F = -45°$.

$1.55\ \mu m$ on the MPC, created of Ce-doped YIG with $n_1 = 2.21$, $\epsilon' = 0.009$ at $1.55\ \mu m$. We assume $\mu' = 1.65 \times 10^{-5}$ as for $Bi : YIG$; the nonmagnetic material is GGG with RI $n_2 = 1.926$ at $1.55\ \mu m$. Now we choose the thicknesses of the layers to be $d_1 = 0.4657D$, $d_2 = 0.5343D$. We suppose that the MPC is surrounded by the medium with $\bar{n} = 2.06$ (see (25)). For the MPC composed of these materials and parameters $a = 16$, $b = 29$, we obtain $\theta_F = -40.8°$ at $k_y = 0$, and $\theta_F = -45.0°$ at $k_yD/(2\pi) = 0.195$. The response to the incident CPWs is given in fig. 28 (a). Nevertheless the 'mixed' terms $T'^+ \approx T'^-$ (eq. (107)) achieve quite large values, they tend to zero at frequency where $T^+ \approx T^- \approx 1$. This leads to the fact that in the response to the linearly polarized input the ellipticity is small enough at this frequency (fig. 29). Again, the response to the p-polarized input is wider because it has three maxima, whereas the response to the s-polarized input has only one maximum. Beside the growth of FR with increasing of k_y, the ellipticity at maximum of transmission also grows (or one can say that zero ellipticity do not coincide with the maximum of T). In fig. 28 (b) we show dependency of the FR θ_F and the ellipticity e on $k_yD/(2\pi)$ for $a = 16$, $b = 29$, $m = 0$ and $a = 16$, $b = 28$, $m = 1$ (transmission is about 0.99) for s-polarized input. The corresponding results for the p-polarized input look similarly, except that the ellipticity e has the opposite sign. In the former case at $k_yD/(2\pi) = 0.195$, we obtained $\theta_F = -45°$, $T = 0.995$ and $|e| = 0.075$. It means that 5% of the output light energy has undesirable perpendicular polarization; and in the later case $\theta_F = -45°$ at $k_yD/(2\pi) = 0.032$, $T = 0.989$ and $|e| = 0.003$. This growth of ellipticity puts limitations of $\sim 4°$ to FR increase.

6.3. Three and Four Defects: Flat-Top Defect Modes

Similarly to the two-defect MPC, the defect mode of the three-defect MPC has three peaks. The flat-top defect modes appears when these three peaks merge in a single broad peak. Flat-top response for non-magnetic structures were investigated by Li and Lit (26) and then for magnetic ones by Levy et al. (25). Let us consider the MPC described by the formula $(NM)^aD_M^{(1)}(MN)^bMD_M^{(2)}M(NM)^bD_M^{(1)}(MN)^a$, where M is the magnetic layer of Ce :

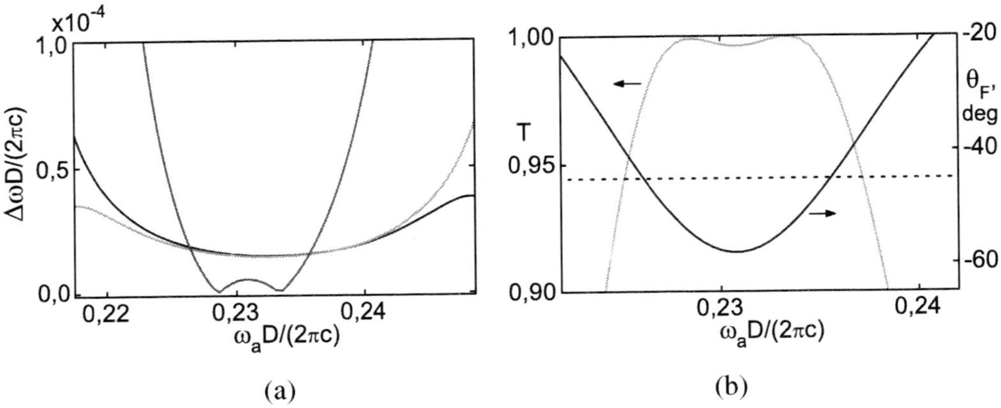

Figure 26. The same as fig. 25, but for $a = 14, b = 25, m = 0$.

YIG and N is GGG layer, $D_M^{1,2}$ are $Ce : YIG$ magnetic defect layers of the thickness $D_M^{(1,2)} = 2m_{1,2}d_1 + \Delta^{(1,2)}$ $(m = 0, 1, 2\ldots)$. All other parameters are the same as at the end of the previous subsection. The response of this MPC to the iincident CPWs is shown in fig. 30 (a). One can see the wide (flat-top) RH and LH defect modes, shifted one from another. Each mode consists of three combined peaks. The phases $\phi_\pm$ advance by 3π through the entire defect structure. The response of the tree-defect MPC to the incident LPW is shown in fig. 30 (b). The flat-top transmission resonance is observed at the intersection of the RH and LH defect modes, and again, the high FR corresponds to large difference $\phi_+ - \phi_-$.

Again, the increase (decrease) of the MPC length leads to the decrease (increase) of RH and LH defect mode widths. But for the flat-top defect resonances with $f > 2$, in contrast to the discrete peaks with $f < 2$ investigated in previous section, the peak frequency mismatch is not so essential: the transmission resonances with high T exist in wide range of L, and f parameters, but the defect mode width decreases and FR increases with the increase of L. The defect mode width can be calculated using the fact that the center of the defect modes $\omega D/(2\pi c) = D/\lambda_0 = 0.2429$ should correspond to $1.55\ \mu m$. This allows us to obtain the period D and then the length of MPC, and $\Delta\lambda = \lambda_0|\Delta\omega|/\omega$. Two definitions of $\Delta\lambda$ are possible - one ($\Delta\lambda_T$) is related to decrease of transmission coefficient T (we used $T = 0.95$ in order to determine $\Delta\lambda_T$), another ($\Delta\lambda_\theta$) is related to decrease of FR from it's value in the center of flat-top defect mode (we used $\Delta\theta_F = 5^\circ$ in order to determine $\Delta\lambda_\theta$). Both $\Delta\lambda_T$ and $\Delta\lambda_\theta$ are shown in fig. 30 (b). We found that in our case $\Delta\lambda_T > \Delta\lambda_\theta$ and so further on we use $\Delta\lambda_\theta$. The figures 31 (a) and (b) show the dependence of the mode width $\Delta\lambda_\theta$ and FR on the parameter a for the MPC, containing magnetic defects with $m_{1,2} = 0$ and $m_{1,2} = 1$. For this type of MPC, in order to obtain good flat-top defect modes (with f slightly over 2) one need to put $b = 2a$. We also calculated $\Delta\lambda_\theta$ and FR for MPC with non-magnetic defects $(MN)^a D_N^{(1)} (NM)^b N D_N^{(2)} N (MN)^b D_N^{(1)} (NM)^a$, and MPC with the defects both types $(NM)^a D_M^{(1)} (MN)^b D_N^{(2)} (NM)^b D_M^{(1)} (MN)^a$ (the later was investigated in (25)); here $D_N^{(1,2)}$ are non-magnetic defect offsets with thicknesses $d_N^{(1,2)} = \Delta^{(1,2)}$. In these figures, we consider the centered defect modes with $\Delta^{(1)} = \Delta^{(2)} = 0$. In the former case, we again put $b = 2a$ but in the later one $b = 2a + 1$.

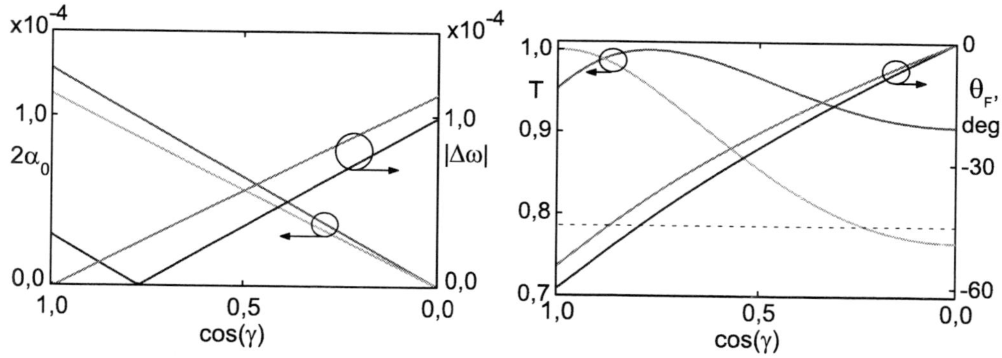

(a) (b)

Figure 27. (a): The magnetic splitting $2\alpha_0$ and the difference $|\Delta\omega| = |\omega_1^+ - \omega_2^-|$ versus $\cos(\gamma)$. The red line is $2\alpha_0$, blue line is $|\Delta\omega|$ for $a = 13$, $b = 24$, $m = 1$, green line denotes $2\alpha_0$, and magenta line is $|\Delta\omega|$ for $a = 14$, $b = 25$, $m = 0$. (b): The transmission coefficient T and FR versus $\cos(\gamma)$: for the parameters $a = 13$, $b = 24$, $m = 1$, (the red line is T, the blue line is FR. For the parameters $a = 14$, $b = 25$, $m = 0$: the green line is T, the magenta line is FR.

One can clearly observe that for all MPCs $\Delta\lambda_\theta$ decreases and FR increases with the period number a (i.e. MPC length L). This corresponds to decreasing RH and LH defect modes for larger a with the same magnetic splitting $2\alpha_0$; thus the region of intersection of the RH and LH defect modes becames smaller and the phase difference $\phi_+ - \phi_-$ grows. In fig. 18 (b) one can see that for the MPC with a magnetic defect, the magnetic splitting $2\alpha_0$ is higher than for the MPC with a non-magnetic defect and for the former defects, it increases with the parameter m. The lower $\Delta\lambda_\theta$ and higher FR correspond to larger $2\alpha_0$ (see fig. 31). But MPCs of different structures have different lengths for the same a. In fig. 32 (a) the lengths of MPCs are shown. As was mentioned in subsec. 7.1, the magnetic splitting can be changed by varying the length of defects; thus, FR can be tuned to the desired value. We found that the offset lengths $|\Delta^{(1)}|$ and $|\Delta^{(2)}|$ should be different, especially for large $|\Delta^{(1,2)}|$, otherwise the defect mode is distorted. The defect modes for equal $\Delta^{(1,2)}$ are shown in fig. 33 (a) and it is destroyed completely. If $\Delta^{(2)}$ is slightly tuned, then the off-centered defect mode looks like the centered one, as it seen from fig. 33 (b).

In Table 1 the parameters of MPCs with $\theta_F = -45°$ are given. Here M, N and MN denote MPC with magnetic, nonmagnetic and both defects. For the MPCs of the types M and N, for each value a corresponds $b = 2a$, in the case of MN MPC we choose $b = 2a + 1$. The quality of flatness of the defect modes is characterized by the ripple factor $R_T \equiv (T_{max} - T_{min})/(T_{max} + T_{min})$.

The bandwidth $\Delta\lambda_\theta$ the length L for different defects are shown in fig. 34 (a). From this figure, it is seen that the bandwidth increase and the MPC length decrease in the sequence N-MN-M(0,0)-M(1,1) (the numbers in brackets correspond to the value of m for both defects). This fact is in good agreement with the those that the magnetic splitting is larger for the magnetic defects then for nonmagnetic ones and in the former case it is larger for defects with higher thickness. The ripple factor R_T (fig. 34 (b)) for these MPCs does not have

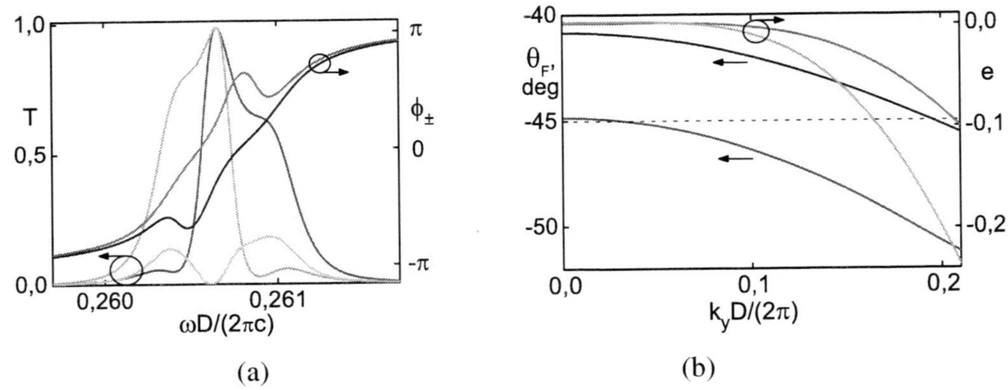

Figure 28. (a) The same as fig. 20, but for two-defect $Ce:YIG$ MPC with $a = 16$, $b = 29$, $m = 0$, $k_yD/(2\pi) = 0.195$; (b): The red line is FR, the blue line is the ellipticity e for $a = 16$, $b = 29$, $m = 0$; the green line is FR, the magenta line is the ellipticity for $a = 16$, $b = 28$, $m = 1$.

any tendency but remains small enough. As for M(0,1), it has the same bandwidth and the length slightly larger than M(0,0), but considerably lower ripple factor. The MPC M(1,0) has slightly larger bandwidth and length but greater ripple factor.

We also calculated the same characteristics for the four-defect MPCs described by the following formulas:

$(NM)^a D_M^{(1)} (NM)^b M D_M^{(2)} (MN)^c N D_M^{(2)} (MN)^b M D_M^{(1)} (MN)^a$ (magnetic defects)

$(MN)^a D_N^{(1)} (NM)^b N D_N^{(2)} (NM)^c N D_N^{(2)} (NM)^b N D_N^{(1)} (NM)^a$ (nonmagnetic ones)

$(NM)^a D_M^{(1)} M N^b D_N^{(2)} N M^c D_M^{(1)} M N^b D^{(2)} N M^a$ (both type of defects).

In fig. 35 (a), fig. 35 (b) and fig. 32 we present dependencies of the bandwidth, FR and MPC length, respectively, on parameter a. Those figures look very similarly to those for the three-defect MPCs. The parameters $b = 2a + 1$, $c = 2a + 3$ are used for the MPC with magnetic and nonmagnetic defects and $b = 2a + 2$, $c = 2a + 4$ for the MPC with both types of defects.

The parameters for the MPCs with $\theta_F = -45°$ are summarized in the Table 2 and shown in fig. 36. Again, the bandwidth increases and the MPC length decreases in the sequence N-MN-M(0,0). Now the MPCs M(1,1) falls out of this sequence. The MPC length is larger then expected because the required offsets $|\Delta^{(i)}|$ are larger in comparison with the previous MPCs. This leads to larger normalized frequency $\omega D/(2\pi c)$ and period $D = 1.55(\omega D/(2\pi c))$ and smaller bandwidth. As for the MPCs M(0,1) and M(1,0), the first has very low ripple factor again, while the later has a large R_T, without pronounced features in length and bandwidth.

We can see that for the three- and four-defect MPCs with flat-top defect modes the FR can be tuned to the desired value by variation of the defect thickness, but the first (last) defect in the MPCs formulas above and the other defects must have different thicknesses, even if they are made of the same material, otherwise the defect mode becomes distorted. Comparing the magnetic and nonmagnetic defect, we can see that the increase of the num-

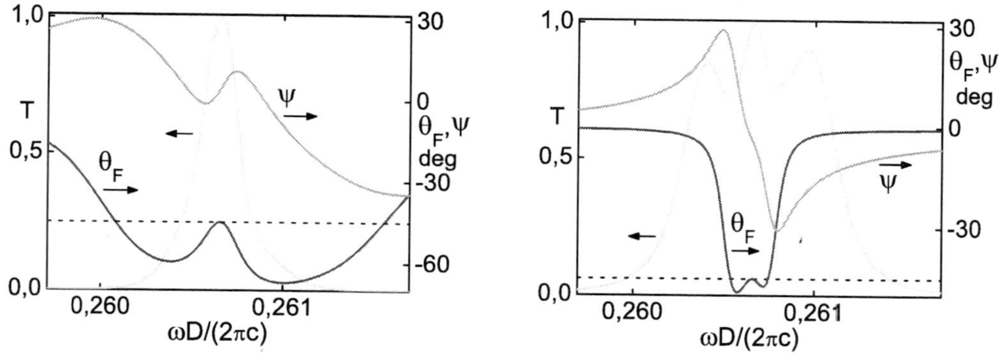

(a) (b)

Figure 29. The same as on fig. 16, but for the two-defect $Ce : YIG$ MPC with $a = 16$, $b = 29$, $m = 0$, $k_y D/(2\pi) = 0.195$; (a) s-polarized input, (b) p-polarized input.

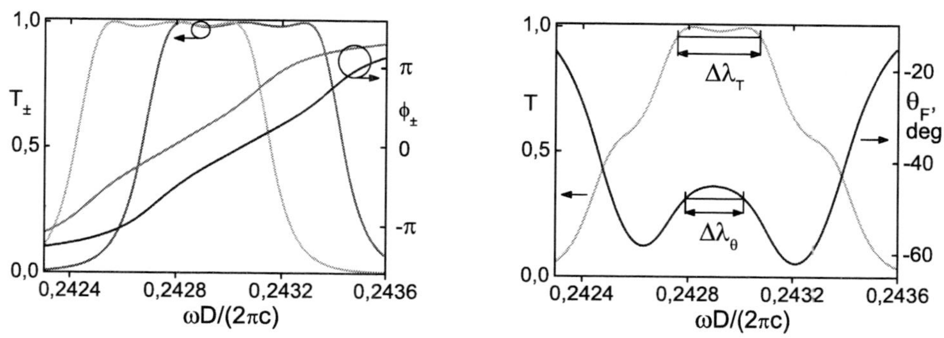

(a) (b)

Figure 30. (a) and (b): the same as on fig. 23 (a) and (b), but for the three-defect $Ce : YIG$ MPC with parameters $a = 13$, $b = 26$, $m_1 = m_2 = 1$, $\Delta^{(1)} = \Delta^{(2)} = 0$.

ber of defects (or the total length of the defects) leads to the smaller length of MPC with the fixed FR and larger the bandwidth of the defect mode, with the exception of cases when the required offsets are too large.

7. Conclusion

In this Chapter we have developed the phenomenological description of the non-defect bigyrotroppic MPCs as well as the MPCs with the single and multiple magnetic and non-magnetic defects. It was shown that the position and width of PBG is strongly dependent on the mutual orientation of the magnetization vector inside the magnetic layers and the wave vector of the electromagnetic wave. We have investigated the dependencies of the

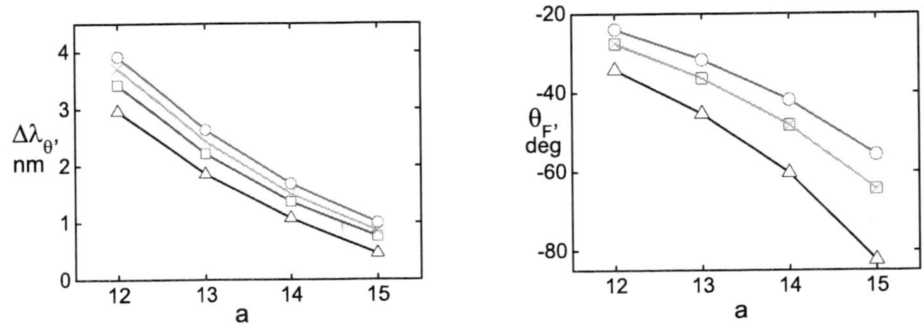

(a) (b)

Figure 31. The dependence of the mode width $\Delta\lambda_\theta$ (a) and FR (b) on the parameter a for the MPC, containing magnetic defects with $m_{1,2} = 0$ and $m_{1,2} = 1$.The magenta line with circles corresponds to MPC with non-magnetic defects, the green line with crosses corresponds to MPCs with magnetic and non-magnetic defects, the red line with squares corresponds to MPC with magnetic defects for $m_{1,2} = 0$, and the blue line with triangles corresponds to MPC with magnetic defects for $m_{1,2} = 1$.

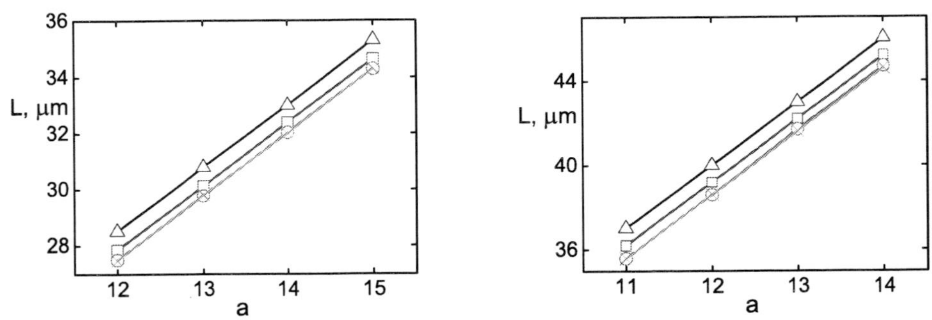

(a) (b)

Figure 32. The length of MPCs versus a. The notations are the same, as on fig. 31.

Table 1. Parameters of three-defect MPCs with FR of $-45°$

Def. type	a	m_1	m_2	$\Delta^{(1)}/D$	$\Delta^{(2)}/D$	$\omega D/(2\pi c)$	$\Delta\lambda_\theta$, nm	L, μm	T_{max}	R_T
N	15	0	0	-0.087	-0.0884	0.2456	1.31	34,6	0.999	0.0122
MN	14	0	0	-0.0305	-0.0357	0.2441	1.46	32.5	0.997	0.0078
M	14	0	0	-0.0305	-0.0311	0.2441	1.61	32.1	0.999	0.0089
	14	0	1	-0.0665	-0.0766	0.2441	1.61	32.4	1.000	0.0002
	14	1	0	-0.087	-0.0786	0.2457	1.70	33.0	0.999	0.0283
	13	1	1	-0.004	-0.0042	0.2432	1.86	30.8	0.999	0.0083

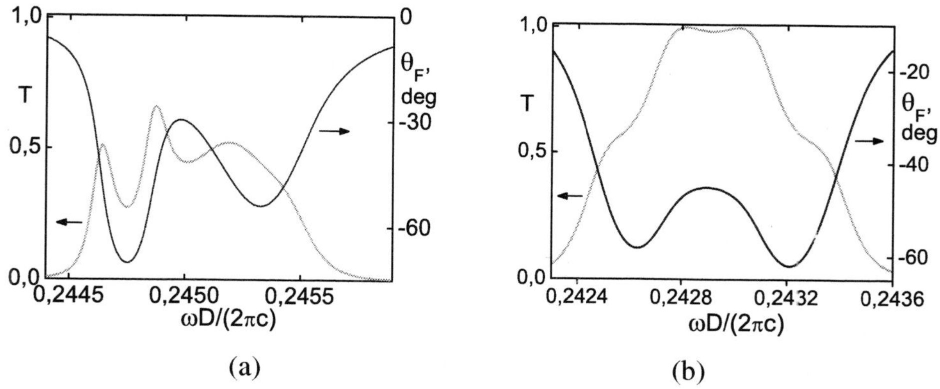

Figure 33. The same as in fig. 23 (b), but the parameters of the MPC are $a = 14$, $b = 28$, $m_1 = 0$, $m_2 = 1$; (a): $\Delta^{(1)} = \Delta^{(2)} = -0.0665D$; (b): $\Delta^{(1)} = -0.0665D$, $\Delta^{(2)} = -0.0766D$.

Table 2. Parameters of four-defect MPCs with FR of -45°

Def. type	a	m_1	m_2	$\Delta^{(1)}/D$	$\Delta^{(2)}/D$	$\omega D/(2\pi c)$	$\Delta\lambda_\theta$, nm	L, μm	T_{max}	R_T
N	13	0	0	-0.069	-0.071	0.2452	2.00	41.9	1.000	0.0031
MN	12	0	0	-0.0046	-0.0053	0.2432	2.31	39.2	1.000	0.0021
M	12	0	0	-0.004	-0.0042	0.2432	2.52	38.6	1.000	0.0021
	12	0	1	-0.065	-0.076	0.2454	2.29	39.6	1.000	0.0005
	12	1	0	-0.051	-0.0466	0.2447	2.72	39.5	1.000	0.0102
	12	1	1	-0.103	-0.1067	0.2463	2.45	40.3	0.999	0.0040

position and width of one-defect MPC and magnetic field-induced splitting as the function of the internal dc magnetic field as well of the incidence angle. The central and off-central defect modes were considered. We showed the possibility to optimize the transmission coefficients and Faraday rotation parameters. We also considered two-, three-, and four-defect MPC structures and showed the possibilities of optimization of transmittivity and Faraday rotation by changing of thicknesses of defect layers, as well as the magnetization vector direction inside the defect layers. In the future, we are planning to investigate the influence of optical nonlinearities on the optical and magneto-optical properties of MPCs. This study is very interesting and promising because cubic nonlinearity in the electric polarization leads to change of the refractive index of the MPC and can lead to the new effects and possibilities in the area of magneto-photonics.

Acknowledgments

We would like to thank Prof. P. Choudhury and Prof. A. Lakhtakia, who encourage us to prepare this Chapter. We also wish to express our gratitude to Prof. A. Lakhtakia and Prof. Th. Rasing for fruitful collaboration. Ukrainian team is supported by INTAS under Grant No.03-51-3784. This work is also supported by the KOSEF through Quantum Photonic Research Center, Seoul, Korea. The contribution of M.I.L. is done with support of the Grant of President of Ukraine for young scientists, and contribution of E.A.Sh. is supported by

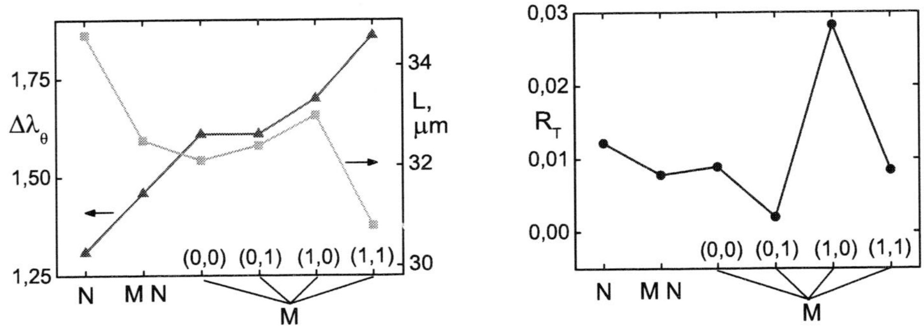

(a) (b)

Figure 34. The characteristics of the MPCs with $\theta_F = -45.0°$. The notations M, N and MN mean the MPCs with the magnetic, nonmagnetic and both defects, the numbers for MPCs with the magnetic defects mean (m_1,m_2); (a): the mode width $\Delta\lambda_\theta$ (red triangles), the MPC length (green squares); (b): the ripple factor R_T.

fellowship of the National Academy of Sciences of Ukraine for young researchers.

A Coefficients of Dispersion Eq. (14)

The coefficients which are given in eq. (14) are

$$A = \epsilon_{zz}\mu_{zz}, \tag{108}$$

$$B = [\epsilon_{zz}(\mu_{yz} + \mu_{zy}) + \mu_{zz}(\epsilon_{yz} + \epsilon_{zy})]k_y^3, \tag{109}$$

$$C = C_\Omega \Omega^2 + C_{k_y} k_y^2, \tag{110}$$

$$D = D_{\Omega k_y}\Omega^2 k_y + D_{k_y} k_y^3, \tag{111}$$

$$E = E_\Omega \Omega^4 + E_{\Omega k_y}\Omega^2 k_y^2 + E_{k_y} k_y^4. \tag{112}$$

In eq. (110) the following substitutions are used:

$$\begin{aligned} C_\Omega = {} & \epsilon_{zz}\mu_{zz}(\epsilon_{xy}\mu_{xy} + \epsilon_{yx}\mu_{yx}) - \epsilon_{zz}(\epsilon_{yx}\mu_{yz}\mu_{zx} + \epsilon_{xy}\mu_{zy}\mu_{xz}) \\ & - \mu_{zz}(\mu_{yx}\epsilon_{yz}\epsilon_{zx} + \mu_{xy}\epsilon_{zy}\epsilon_{xz}) + \epsilon_{xz}\epsilon_{zy}\mu_{xz}\mu_{zy} + \epsilon_{zx}\epsilon_{yz}\mu_{zx}\mu_{yz} \\ & - \Delta\epsilon_{xz}\Delta\mu_{yz} - \Delta\epsilon_{yz}\Delta\mu_{xz}, \end{aligned} \tag{113a}$$

$$C_{k_y} = \epsilon_{zz}\mu_{yy} + \epsilon_{yy}\mu_{zz} + (\epsilon_{zy} + \epsilon_{yz})(\mu_{zy} + \mu_{yz}). \tag{113b}$$

In eq. (111) the following substitutions are used:

$$\begin{aligned} D_{\Omega k_y} = {} & \epsilon_{yy}(\epsilon_{xz}\mu_{xz}\mu_{zy} + \epsilon_{zx}\mu_{zx}\mu_{yz}) + \mu_{yy}(\mu_{xz}\epsilon_{xz}\epsilon_{zy} + \mu_{zx}\epsilon_{zx}\epsilon_{yz}) \\ & + \epsilon_{zz}(\epsilon_{xy}\mu_{xy}\mu_{yz} + \epsilon_{yx}\mu_{yx}\mu_{zy}) + \Delta\epsilon_{yz}[\mu_{xy}\mu_{zx} + \mu_{yx}\mu_{xz} - \mu_{xx}(\mu_{yz} + \mu_{zy})] \\ & + \mu_{zz}(\mu_{xy}\epsilon_{xy}\epsilon_{yz} + \mu_{yx}\epsilon_{yx}\epsilon_{zy}) + \Delta\mu_{yz}[\epsilon_{xy}\epsilon_{zx} + \epsilon_{yx}\epsilon_{xz} - \epsilon_{xx}(\epsilon_{yz} + \epsilon_{zy})] \\ & - (\epsilon_{xy}\epsilon_{yz}\mu_{xz}\mu_{zy} + \epsilon_{yx}\epsilon_{zy}\mu_{zx}\mu_{yz}) - (\mu_{xy}\mu_{yz}\epsilon_{xz}\epsilon_{zy} + \mu_{yx}\mu_{zy}\epsilon_{zx}\epsilon_{yz}) \\ & - \epsilon_{yy}\mu_{zz}(\epsilon_{xz}\mu_{xy} + \epsilon_{zx}\mu_{yx}) - \mu_{yy}\epsilon_{zz}(\mu_{xz}\epsilon_{xy} + \mu_{zx}\epsilon_{yx}), \end{aligned} \tag{114a}$$

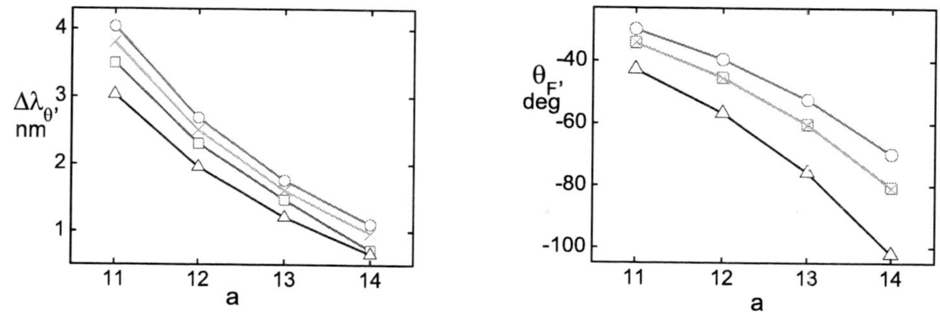

(a) (b)

Figure 35. The same as in fig. 31, but for the four-defect MPCs.

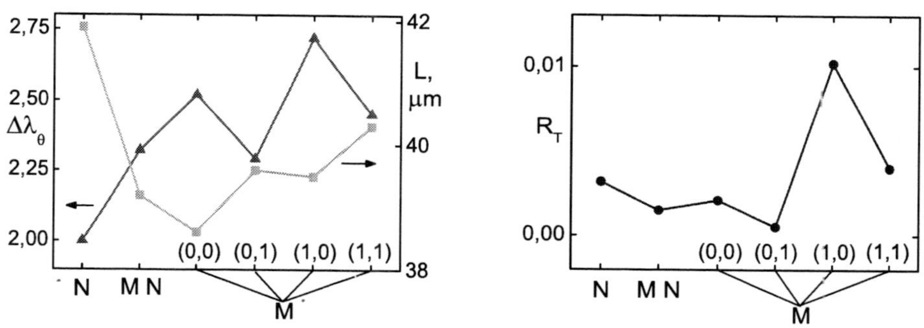

(a) (b)

Figure 36. The same as in fig. 34, but for the four-defect MPCs.

$$D_{k_y} = \epsilon_{yy}(\mu_{yz} + \mu_{zy}) + \mu_{yy}(\epsilon_{yz} + \epsilon_{zy}). \tag{114b}$$

In eq. (112) following substitutions are used:

$$E_{\Omega} = det(\hat{\epsilon})\, det(\hat{\mu}) \tag{115a}$$

$$\begin{aligned} E_{\Omega k_y} = {} & \epsilon_{yy}\mu_{yy}(\epsilon_{xz}\mu_{xz} + \epsilon_{zx}\mu_{zx}) + \epsilon_{xy}\epsilon_{yz}\mu_{xy}\mu_{yz} + \mu_{xy}\mu_{yz}\epsilon_{xy}\epsilon_{yz} \\ & - \epsilon_{yy}(\epsilon_{xz}\mu_{xy}\mu_{yz} + \epsilon_{zx}\mu_{yx}\mu_{zy}) - \mu_{yy}(\mu_{xz}\epsilon_{xy}\epsilon_{yz} + \mu_{zx}\epsilon_{yx}\epsilon_{zy}) \\ & - \Delta\epsilon_{xy}\Delta\mu_{yz} - \Delta\mu_{xy}\Delta\epsilon_{yz}, \end{aligned} \tag{115b}$$

$$E_{k_y} = \epsilon_{yy}\mu_{yy}. \tag{115c}$$

In eqs. (113a), (114a) and eq. (115b) the following substitution is used

$$\Delta\epsilon_{ij} = \epsilon_{ii}\epsilon_{jj} - \epsilon_{ij}\epsilon_{ji}, \quad \Delta\mu_{ij} = \mu_{ii}\mu_{jj} - \mu_{ij}\mu_{ji} \quad i,j = x,y,z. \tag{116}$$

B Elements of Matrix $\hat{A}$

We represent elements of matrix $\hat{A}$ eq. (25) in the form where the index m is dropped for simplicity and all parameters are assumed to belong to the mth layer:

$$V^{s(j)} = \frac{v^{s(j)}}{z^{s(j)}}, \quad G^{s(j)} = \frac{g^{s(j)}}{z^{s(j)}}, \quad F^{s(j)} = \frac{f^{s(j)}}{z^{s(j)}} \qquad s = \epsilon, \mu; \quad j = 1, 2, 3, 4, \tag{117}$$

$$v^{\epsilon(j)} = (v^{\epsilon}_{\Omega}\Omega^2 + v^{\epsilon}_{k_y}k_y^2 + v^{\epsilon}_{k_yk_z}k_yk_z + v^{\epsilon}_{k_z}k_z^2)\Omega, \tag{118}$$

$$f^{\epsilon(j)} = (f^{\epsilon}_{k_y}k_y + f^{\epsilon}_{k_z}k_z)\Omega^2, \tag{119}$$

$$g^{\epsilon(j)} = (g^{\epsilon}_{\Omega}\Omega^2 + g^{\epsilon}_{k_y}k_y^2 + g^{\epsilon}_{k_yk_z}k_yk_z + g^{\epsilon}_{k_z}k_z^2)\Omega, \tag{120}$$

$$z^{\epsilon(j)} = (z^{\epsilon}_{\Omega k_y}k_y + z^{\epsilon}_{\Omega k_z}k_z^{(j)})\Omega^2 + z^{\epsilon}_{k_z^3}(k_z^{(j)})^3 + z^{\epsilon}_{k_y^3}k_y^3 + z^{\epsilon}_{k_z^2k_y}(k_z^{(j)})^2k_y + z^{\epsilon}_{k_zk_y^2}k_z^{(j)}k_y^2. \tag{121}$$

In eq. (118) the following substitutions are used:

$$v^{\epsilon}_{\Omega} = det(\hat{\mu})\Delta\epsilon_{yz}, \tag{122a}$$

$$v^{\epsilon}_{k_y} = -\Delta\mu_{yz}\epsilon_{yy}, \tag{122b}$$

$$v^{\epsilon}_{k_yk_z} = -\Delta\mu_{yz}(\epsilon_{yz} + \epsilon_{zy}), \tag{122c}$$

$$v^{\epsilon}_{k_z} = -\Delta\mu_{yz}\epsilon_{zz}. \tag{122d}$$

In eq. (119) the following substitutions are used:

$$f^{\epsilon}_{k_y} = -[\Delta\epsilon_{yz}(\mu_{yy}\mu_{xz} - \mu_{xy}\mu_{yz}) + \Delta\mu_{yz}(\epsilon_{yy}\epsilon_{zx} - \epsilon_{yx}\epsilon_{zy})], \tag{123a}$$

$$f^{\epsilon}_{k_z} = -[\Delta\epsilon_{yz}(\mu_{zz}\mu_{xy} - \mu_{xz}\mu_{zy}) - \Delta\mu_{yz}(\epsilon_{zz}\epsilon_{yx} - \epsilon_{zx}\epsilon_{yz})]. \tag{123b}$$

In eq. (120) the following substitutions are used:

$$g^{\epsilon}_{\Omega} = det(\hat{\mu})(\epsilon_{yz}\epsilon_{zx} - \epsilon_{zz}\epsilon_{yx}), \tag{124a}$$

$$g^{\epsilon}_{k_y} = \epsilon_{yx}\Delta\mu_{yz} - \epsilon_{yz}(\mu_{yy}\mu_{xz} - \mu_{xy}\mu_{yz}), \tag{124b}$$

$$g^{\epsilon}_{k_yk_z} = \epsilon_{yz}(\mu_{zz}\mu_{xy} - \mu_{xz}\mu_{zy}) - \epsilon_{zz}(\mu_{yy}\mu_{xz} - \mu_{xy}\mu_{yz}) + \epsilon_{zx}\Delta\mu_{yz}, \tag{124c}$$

$$g^{\epsilon}_{k_z} = \epsilon_{zz}(\mu_{zz}\mu_{xy} - \mu_{xz}\mu_{zy}). \tag{124d}$$

In eq. (121) the following substitutions are used:

$$z^{\epsilon}_{\Omega k_y} = (\epsilon_{yy}\epsilon_{zx} - \epsilon_{yx}\epsilon_{zy})(\mu_{zz}\mu_{yx} - \mu_{zx}\mu_{yz}) + \Delta\epsilon_{yz}(\mu_{xx}\mu_{yz} - \mu_{yx}\mu_{zx}), \tag{125a}$$

$$z^{\epsilon}_{\Omega k_z} = \Delta\epsilon_{yz}\Delta\mu_{xz} - (\epsilon_{zz}\epsilon_{yx} - \epsilon_{zx}\epsilon_{yz})(\mu_{zz}\mu_{yx} - \mu_{zx}\mu_{yz}), \tag{125b}$$

$$z^{\epsilon}_{k_y^3} = -\epsilon_{yy}\mu_{yz}, \tag{125c}$$

$$z^{\epsilon}_{k_zk_y^2} = -[\epsilon_{yy}\mu_{zz} + \mu_{yz}(\epsilon_{yz} + \epsilon_{zy})], \tag{125d}$$

$$z^{\epsilon}_{k_z^2 k_y} = -[\mu_{zz}(\epsilon_{yz} + \epsilon_{zy}) + \epsilon_{zz}\mu_{yz}], \tag{125e}$$

$$z^{\epsilon}_{k_z^3} = \epsilon_{zz}\mu_{zz}. \tag{125f}$$

The values $V^{\mu(j)}, F^{\mu(j)}, G^{\mu(j)}$ can be easily obtained from $V^{\epsilon(j)}, F^{\epsilon(j)}, G^{\epsilon(j)}$ substituting $\epsilon \leftrightarrow \mu$. This fact and different signs before $V^{\epsilon(j)}$, $V^{\mu(j)}$ and $G^{\epsilon(j)}$, $G^{\mu(j)}$ in eq. (25) correspond to our choice of the basic amplitudes and certain symmetry of Maxwell's equations: in eq. (12) the left equation turns to the right one and vice versa with the substitution $(\epsilon, \boldsymbol{E}) \leftrightarrow (\mu, -\boldsymbol{H})$.

C Elements of Matrix $\hat{A}^{(-1)}$

The elements of matrix $\hat{A}^{-1}$ are

$$(\hat{A}^{-1})_{1,1} = \frac{1}{D_A}[-\Delta V^{\mu} - F^{\epsilon(2)}\Delta G^{\mu} + G^{\epsilon(2)}\Delta(VG)^{\mu}], \tag{126a}$$

$$(\hat{A}^{-1})_{1,2} = \frac{1}{D_A}[-\Delta(VF)^{\mu} + F^{\epsilon(2)}\Delta(FG)^{\mu} - V^{\epsilon(2)}\Delta(VG)^{\mu}], \tag{126b}$$

$$(\hat{A}^{-1})_{1,3} = \frac{1}{D_A}[\ \Delta F^{\mu} + V^{\epsilon(2)}\Delta G^{\mu} + G^{\epsilon(2)}\Delta(FG)^{\mu}], \tag{126c}$$

$$(\hat{A}^{-1})_{1,4} = \frac{1}{D_A}[-V^{\epsilon(2)}\Delta V^{\mu} - F^{\epsilon(2)}\Delta F^{\mu} + G^{\epsilon(2)}\Delta(VF)^{\mu}]. \tag{126d}$$

The elements $(\hat{A}^{-1})_{2,1}$, $(\hat{A}^{-1})_{2,2}$, $(\hat{A}^{-1})_{2,3}$, $(\hat{A}^{-1})_{2,4}$ can be easily obtained from $-(\hat{A}^{-1})_{1,1}$, $-(\hat{A}^{-1})_{1,2}$, $-(\hat{A}^{-1})_{1,3}$, $-(\hat{A}^{-1})_{1,4}$, respectively, substituting $(V^{\epsilon(2)}, F^{\epsilon(2)}, G^{\epsilon(2)}) \leftrightarrow (V^{\epsilon(1)}, F^{\epsilon(1)}, G^{\epsilon(1)})$.

Then the elements $(\hat{A}^{-1})_{3,3}$,$(\hat{A}^{-1})_{3,4}$, $(\hat{A}^{-1})_{4,3}$, $(\hat{A}^{-1})_{4,4}$, $(\hat{A}^{-1})_{3,1}$, $(\hat{A}^{-1})_{3,2}$, $(\hat{A}^{-1})_{4,1}$, $(\hat{A}^{-1})_{4,2}$ can be obtained from $(\hat{A}^{-1})_{1,1}$, $(\hat{A}^{-1})_{1,2}$, $(\hat{A}^{-1})_{2,1}$, $(\hat{A}^{-1})_{2,2}$, $(\hat{A}^{-1})_{1,3}$, $(\hat{A}^{-1})_{1,4}$, $(\hat{A}^{-1})_{2,3}$, $(\hat{A}^{-1})_{2,4}$, respectively, by substitution ($\epsilon \leftrightarrow \mu$).

In eq. (126a) the following substitutions are used:

$$D_A \equiv det(\hat{A}) = \Delta F^{\epsilon}\Delta F^{\mu} + \Delta V^{\epsilon}\Delta V^{\mu} - \Delta G^{\epsilon}\Delta(VF)^{\mu} - \Delta G^{\mu}\Delta(VF)^{\epsilon} + \\ + \Delta(VG)^{\epsilon}\Delta(VG)^{\mu} + \Delta(FG)^{\epsilon}\Delta(FG)^{\mu} \tag{127a}$$

$$\Delta K^s = K^{s(i)} - K^{s(j)}, \tag{127b}$$

$$\Delta(KL)^s = K^{s(i)}L^{s(j)} - K^{s(j)}L^{s(i)}, \tag{127c}$$

In eqs. (127b)-(127c) $K, L = V, F, G$, $s = \epsilon, \mu$; $i = 1, j = 2$ for $s = \epsilon$ and $i = 3, j = 4$ for $s = \mu$.

References

[1] Jannopoulos, J. D.; Meade, R. D.; Winn, J. N. *Photonic crystals*. Princeton University Press: Princeton, US, 1995 pp. 1-137.

[2] Sakoda, K. *Optical propeties of photonic crystals*. Springer-Verlag: Berlin, 2001.

[3] Busch, K.; Lölkes, S; Wehrspohn, R. W.; Föll, H., editor. *Photonic Crystals. Advances in Design, Fabrication and Characterization.* Wiley - VCH: Weinheim, DE, 2004, pp. 1-354.

[4] Inoue, K.; Ohtaka, K., editor. *Photonic Crystals. Physics, Fabrication and Application.* Springer Series in Optical Sciences. Springer: Berlin, DE, 2004, pp. 1-320.

[5] Lyubchanskii, I. L.; Dadoenkova, N. N.; Lyubchanskii, M. I.; Shapovalov, E. A.; Rasing, Th. *J. Phys. D: Appl. Phys.* 2003, 36, R277-287.

[6] Zvezdin, A. K.; Kotov, V. A. *Modern Magnetooptics and Magnetooptical Materials*. IOP: Bristol, 1997, pp. 1-386.

[7] Gurevich, A. G.; Melkov, G. A. *Magnetisation Oscilation and Waves*. CRC Press: New York, 1996, pp. 1-445.

[8] Borisov, S. B.; Lyubchanskii, I. L. *Opt. Spectrosk.* 1988,65, 219-222.

[9] Borisov, S. B.; Dadoenkova, N. N.; Lyubchanskii, I. L. *Opt. Spectrosk.* 1993, 74, 670-675.

[10] Ivanov, O. V.; Sementsov, D. I. *Pure Appl. Opt.* 1997, 6, 455-464.

[11] Makarov, D. G.; Davilov, V. V.; Kovalenko, V. F. *J. Phys. D: Appl. Phys.* 2004, 37, 530-535.

[12] Makarov, D. G.; Davilov, V. V.; Kovalenko, V. F. *Phys. Stat.Sol (a)* 2004, **201**, 130-138.

[13] Lyubchanskii, I. L.; Dadoenkova, N. N.; Lyubchanskii, M. I.; Shapovalov, E. A.; Lakhtakia, A.; Rasing, Th. *Appl. Phys. Lett.* 2004, 53, 214-217.

[14] Lyubchanskii, I. L.; Dadoenkova, N. N.; Lyubchanskii, M. I.; Shapovalov, E. A.; Lakhtakia, A.; Rasing, Th. *Phys. Stat. Sol. (a)* 2004, 201, 3338-3344.

[15] Yeh, P. *Surf. Sci.* 1980, 96, 41-53.

[16] Yariv, A.; Yeh, P. *Optical waves in crystals*. John Wiley & Sons: New York, 1984.

[17] Korn G. A.; Korn T. M. *Handbook on Mathematics for Scientists and Engineers*. McGraw-Hill: New-York, 1961.

[18] Lyubchanskii, I. L.; Dadoenkova, N. N.; Lyubchanskii, M. I.; Shapovalov, E. A.; Lakhtakia, A.; Rasing, Th. *Proc. SPIE* 2004, 5508, 184.

[19] Hansen, P.; Krumme, J.-P. *Thin Solid Films* 2004, 114, 69-107.

[20] Crossley, W. A.; Cooper, R. D.; Page, J. L.; van Stapele, R. P. *Phys. Rev.* 1969, 181, 896-904.

[21] Wangness, R. K. *Phys. Rev.* 1954, 95, 339-345.

[22] Krumme, J.-P.;Doormann, V.; Klages, C.-P. *Appl. Opt.* 1984, 23, 1184-1192.

[23] Steel, M. J.; Levy, M.; Osgood, R. M. *J. Lightwave Technol.* 2000, 18, 1297-1308.

[24] Steel, M. J.; Levy, M.; Osgood, R. M. *IEEE Photon. Technol. Lett.* 1999, 12, 1171-1173.

[25] Levy, M.; Yang, H. C.; Steel, J. M.; Fujita, J. *J. Lightwave Technol.* 2001, 19, 1964-1969.

[26] Li, W.; Lit, J. W. Y. *J. Lightwave Technol.* 1997, 15, 1405-1409.

[27] Figotin, A.; Vitebskiy, I. *Phys. Rev. B* 2003, 67, 165210.

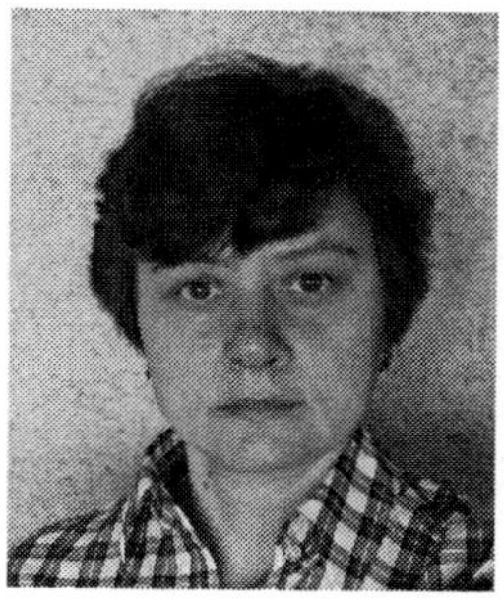

Nataliya N. Dadoenkova started her scientific activity in 1988 in Donetsk Institute for Physics and Technology of the Academy of Sciences of Ukrainian SSR (Donetsk, USSR), and received Ph.D. degree in Physics in 1994. During 1994 - 1998 she worked as a Junior Researcher, and during 1998 - 2000 as a Researcher in the Department for Theory of Electronic and Kinetic Properties of Nonlinear Systems in Donetsk Institute for Physics and Technology of the National Academy of Sciences of Ukraine. During 2001 - 2002 she was a Post Doctoral Fellow in the Institute of Microstructure Physics of Max Planck Society (Halle/Salle, Germany) where she worked on first principle calculation of the nonlinear magneto-optical properties of Ni/Cu multilayers. Presently she is working as a Senior Researcher in Donetsk Institute for Physics and Technology of the National Academy of Sciences of Ukraine. She is involved in joint research with West European Scientists in the framework of INTAS projects. Her current research interests lie in the theory of linear and nonlinear optics of homogeneous and layered magnetic mediums, ab-initio calculation of the nonlinear magneto-optical properties of magnetic superlattices, and magnetic photonic crystals. She is the co-author of about fifty papers in refereed journals.

Igor L. Lyubchanskii started his scientific activity in 1973 in Donetsk Institute for Physics and Technology of the Academy of Sciences of Ukrainian SSR (Donetsk, USSR) as Ph.D. student. He received a Ph.D. degree in Theoretical and Mathematical Physics in 1983, and D.Sc. degree in Solid State Physics in 1996. During 1976 2000 he worked as Junior Researcher, Researcher, and Senior Researcher in the Department for Theory of Electronic and Kinetic Properties of Nonlinear Systems in Donetsk Institute for Physics and Technology of the National Academy of Sciences of Ukraine (Donetsk, Ukraine). Since 2000 till present he has been working as a Leading Researcher at the same Department. He is also an Adjunct Professor in the Department of Radiophysics at the Faculty of Physics of Donetsk National University. During 1998 1999 he worked as Visiting Scientist in the Center for Nanospinics of Spintronic Materials and the Department of Physics of the Advanced Institute of Science and Technology (Taejon, Republic of Korea). Since 1994 till present he has been working together with the researchers from the University of Nijmegen (The Netherlands) in the framework of joint INTAS projects supported by the European Union. Since 2004 he has been collaborating with the scientists from the Quantum Photonic Science Research Center of Hanyang University (Seoul, Republic of Korea). His current research interests lie in the theory of linear and nonlinear optical effects in uniform and layered magnetic materials, and magnetic photonic crystals. He is an author and co-author of about one hundred research papers in refereed journals. He is a Member of SPIE. In 2004 he was awarded with the State Prize of Ukraine on Science and Technology together with the group of Ukrainian Scientists for the selected publications on New Optical and Magneto-Optical Phenomena in Antiferromagnets.

Maxim I. Lyubchanskii started his scientivic activity in 1995 in Donetsk Institute for Physics and Technology of the National Academy of Sciences of Ukraine (Donetsk, Ukraine), and received Ph.D. degree in Physics of Magnetism in 2001. During 1998-2002 he worked as Research Engineer, and since 2002 till now he has been working as a Junior

Researcher in the Department for Theory of Electronic and Kinetic Properties of Nonlinear Systems in Donetsk Institute for Physics and Technology of the National Academy of Sciences of Ukraine. He is involved in joint research with West European Scientists in the framework of INTAS projects. His research interests lie in the theory of linear and nonlinear optics of uniform and layered magnetic mediums, optical properties of magnetic interfaces, and magnetic photonic crystals. He is a co-author of about twenty research papers in refereed journals.

Eugene A. Shapovalov started his scientific activity in 2003 in Donetsk Institute for Physics and Technology of the National Academy of Sciences of Ukraine (Donetsk, Ukraine) as a Ph.D. student. His current research interests lie in the theory of magnetic photonic crystals. He is involved in joint research with West European Scientists in the framework of INTAS projects. He is a co-author of about ten papers in referred journals.

Young Pak Lee started his research career in 1975 from the Agency for Defense Development, Seoul (Korea), and received Ph.D. degree in Physics at Iowa State University, Ames (U.S.A.) in 1987. From 1987 to 1988, he worked as a Research Associate at USDOE Ames Laboratory in Ames, Iowa. In 1989, he joined Research Institute of Industrial Science and Technology, Pohang (Korea) as a Senior Researcher, where he was until 1994. Also, during 1989 1990, he was a Researcher at the Institute of Physical and Chemical Research, Wako (Japan). In 1994, he became an Associate Professor and a Professor, later, at the Department of Physics, Sunmoon University, Asan (Korea), where he worked until 1999. Since 2000, he is a Professor at the Department of Physics, Hanyang University (Seoul). Presently he is the Director of Quantum Photonic Science Research Center (Seoul) designated by the Korean Ministry of Science and Technology. His current research interests lie in magneto-optical transport properties, electronic structures of thin films and nanostructures, analysis of magnetic surfaces and films by using synchrotron radiation, nanoscopic investigation of morphology and magnetic domain, nonlinear and time-resolved magneto-optical spectroscopy, and characterization of the surfaces and the interfaces of thin films including multilayers. He has authored and co-authored over four-hundred and fifty research publications in these areas. He is the Secretary General and Chair of the Applied Physics Division of the Korean Physical Society, a Director of the Korean Vacuum Society, a Director of the Korean Synchrotron Radiation Users' Association and a Member of the Korean Academy of Science and Technology. Also, he is a Regular Member of the American Physical Society, an Adjunct Professor at Dalian University of Technology (China) and an Honorary Member, Institute of Metal Physics, National Academy of Sciences of Ukraine.

In: Frontiers in Optical Technology: Materials and Devices ISBN: 1-60021-084-8
Editors: P.K. Choudhury and O.N. Singh, pp. 73-95

Chapter 3

INTEGRATED OPTICS ON SILICON

Ulrich Hilleringmann
Center of Optoelectronics and Photonics, Sensor Technology Department
University of Paderborn, Warburger Str. 100, D-33098 Paderborn, Germany

Abstract

This review article describes various kinds of processes to integrate optical waveguides on silicon substrates, suitable for the visible to the near infrared spectral range. Different types of materials for the light guiding film and set ups for the waveguide structures are compared according to the propagation loss and the integration density. The link of the strip loaded rib waveguides to photo detectors, integrated in the same substrate, is performed by butt- or leaky wave coupling. Two kinds of processing for the monolithic integration of waveguides, photo detectors and CMOS circuits are presented, and an application as a pressure sensor is given. The last part of the article discusses optical waveguides in silicon suitable for data transmission in the 1.55 μm spectral range. Doped silicon waveguides and silicon-on insulator waveguides are presented as light guiding films. The capability of the silicon technology is demonstrated by a capacity controlled all-silicon modulator for 10 GHz.

1. Introduction

Although silicon is a semiconducting material with an indirect band gap, there is a huge interest in silicon technology for integrated optics [1]. In contrast to the compound semiconductors, it is a very cheap high quality material with excellent mechanical properties. Today the simulation of silicon devices and the processing of silicon wafers is well understood. Lithography, layer deposition and dry etching can be controlled with nanometer scale perfection. This has lead to strong research in integrated optics on silicon. First interests were located on the visible spectral range only, but today the infrared radiation at 1.55 μm used for data transmission in glass fibers is of strong interest, too.

The only property missing for applications in integrated optics is the light emission from silicon. The attempts to use porous silicon [2] or radiation caused by energetic electrons in *p-n* junctions (Bremsstrahlung) failed because of the very low efficiencies (far below 0.1%) [3]. Attempts to integrate silicon light emitting diodes (LEDs) with sufficient efficiencies failed up to now [4]. Nevertheless, only silicon is suitable for the integration of optical waveguides, photo detectors and high density CMOS circuits on one chip [5]. As already demonstrated, integrated optics on silicon is capable for high performance Mach-Zehnder interferometers and Michelson interferometers [6], which result in many other sensor applications like gas or chemical sensors [7]. Most common are silicon oxinitride layers as light guiding films, but during the last years, materials with higher refractive index (RI) like silicon nitride or aluminium oxide are of growing interest.

On the other hand, integrated optics using silicon itself as a light guiding film is applicable for wavelengths above 1.2 µm [8]. Silicon waveguides can give high quality passive devices in optical networks. Either early *n*-doped silicon waveguides or today's SOI-waveguides with high RI contrast result in excellent device characteristics.

2. Integrated Optics on Silicon in the Visible Range

Waveguides typically consist of a core with a higher RI n_1 than the surrounding cladding layer n_2. This structure is transferred to the planar technology for silicon wafer processing using different types of inorganic films, which are standard materials in MOS technology.

Silicon technology offers different transparent materials with a wide range of RIs which can be used for light guiding films and waveguide structures. The most common material for the cladding layers is silicon dioxide (SiO_2). It is normally used as a dielectric layer in MOS transistors. The RI is about 1.46, and the absorbance rate is extremely low in the visible spectral range. SiO_2 can be deposited by thermal oxidation of silicon or by chemical vapour deposition (CVD) using low pressure (LPCVD) [9,10] or plasma enhanced CVD (PECVD) [11,12]. Another high quality material is silicon nitride (Si_3N_4), which is often used as a diffusion mask for local oxidation of silicon in MOS processing. Its RI is about 2.0. The deposition is usually performed by LPCVD or PECVD; both methods give high quality films.

Beside silicon oxide and nitride, transparent films of aluminium oxide [13] or nitride can be deposited by sputtering. These films may be suitable for waveguides, however, there are only very few publications about aluminium based waveguides. The disadvantage of these aluminium based layers is the missing availability of a dry etching process for materials.

Most waveguides presented in literature use silicon oxinitride (SiON), a mixture of SiO_2 and Si_3N_4. The RI of this material can easily be adjusted in the range from 1.46 to 2.0 during the layer deposition in CVD processes by the control of the deposition gas mixture. Figure 1 illustrates the dependence of the RI of the film on the gas composition during the deposition process.

Optical waveguides on silicon can use different types of setup. Figure 2 depicts the most common structures of waveguides suitable for monolithic integration. The rib waveguide [fig. 2(a)] consists of a low refracting layer which is covered by a layer with a RI as the light guiding film. The top layer is partly etched to form a rib. The electromagnetic wave can propagate in the area of the rib only; outside the rib no mode can exist due to the reduced film thickness. Because of the roughness of the etched surface, the propagation loss of this kind of waveguide is large [15]. Figure 2(b) type of waveguides yield less propagation loss because the light guiding film is not etched itself. Only the cover layer is structured to ribs. This kind of waveguide can easily be integrated on silicon, but propagation loss is not optimized due to the roughness of the vertical rib side walls caused by dry etching. The strip loaded film waveguide [fig. 2(c)] exhibits very low propagation loss. The surface cladding layer is deposited on top of the light guiding film, and the etching process to form the strip does not affect the quality of the light guiding layer. So, the propagation loss is very low. The integration process is as simple as in case of the rib loaded waveguide.

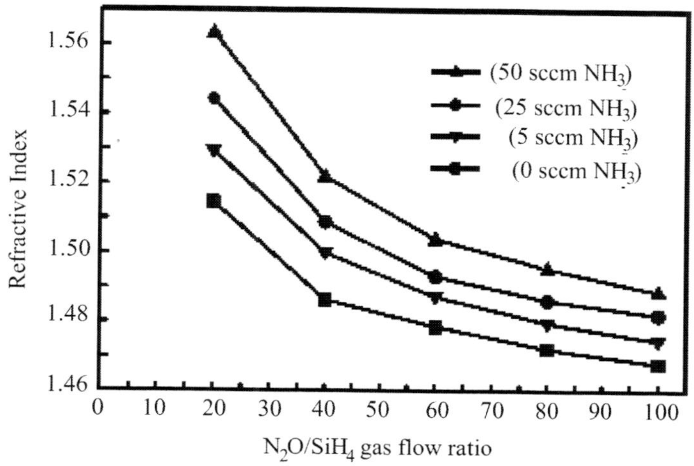

Figure 1. Refractive index of the deposited silicon oxinitride film depending on the ammonia flow for different ratio of N_2O/SiH_4 [14].

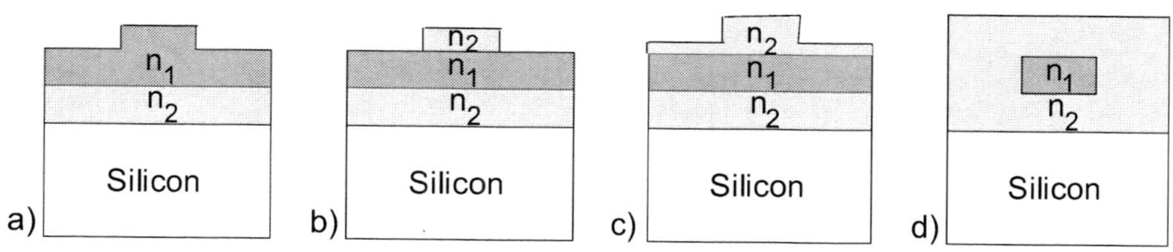

Figure 2. (a) Rib waveguide, (b) rib loaded film waveguide, (c) strip loaded film waveguide, and (d) core waveguide.

Figure 2(d) shows a core waveguide. Its shape can be symmetric or rectangular. The propagation loss will be determined by the quality of etching process applied to form the core layer. Neither wet etching nor dry etching will result in an absolutely smooth side wall of the core, and even nanometer scale roughness will cause large propagation loss. As a result, the propagation loss of an integrated waveguide will be high in all cases when the core layer is treated by an etching process. So, only the type of waveguide according to fig. 2(c) is suitable for high quality applications.

Figure 3 represents the structure of a typical monomode SiON waveguide for integrated optics on silicon in the visible spectral range. It consists of 2 μm of thermal oxide ($n_2 = 1.46$), 0.5 μm of SiON ($n_1 = 1.52$), and 0.5 μm of deposited silicon oxide. The top layer is etched by reactive ion etching to form a rib of 0.4 μm for light confinement in the SiON film below. The width of the rib is about 3 μm for monomode applications. The buffer oxide layer of 2 μm is usually grown by thermal oxidation in wet atmosphere at a temperature of about 1025 °C to 1100 °C. The SiON film deposition uses LPCVD or PECVD techniques. In the case of

LPCVD, a gas mixture of SiH_2Cl_2, NH_3 and N_2O is thermally decomposed at about 920 °C. In the case of PECVD, the deposition starts at 350 °C applying a gas mixture of SiH_4, NH_3 and N_2O [16]. In both cases, the rate of NH_3 to N_2O controls the RI in the deposited layer. Typical deposition rates vary between 8 nm/min and 50 nm/min depending on the equipment. Directly after the SiON deposition, the capping layer of SiO_2 grows on top by simply changing the deposition atmosphere of the CVD process. The NH_3 flow stops and the N_2O flow is increased. As the last step, optical lithography defines the rib size which is transferred into the top oxide cladding layer.

In the case of infrared light, the oxide layer thickness below the SiON film must be increased to about 8 to 10 μm to prevent leaky wave coupling to the silicon substrate. The SiON core layer is typically about 2 μm thick, and the surface cladding layer thickness is about 7 μm. The kind of waveguide used is of the ridge type, with an SiON rib height of 1.3 μm, as illustrated in fig. 3(b).

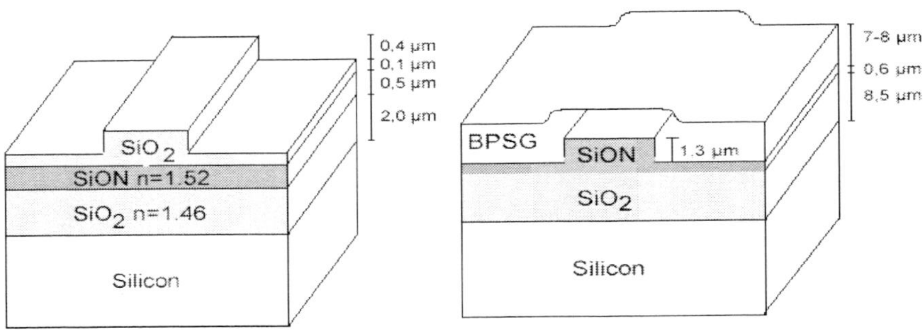

Figure 3. Structure of a typical SiON waveguide for (a) visible light, and (b) infrared light at 1.55 μm, as used in integrated optics on silicon.

3. Characteristics of Waveguides in the Visible Range

The waveguides are typically tested at 633 nm using a He-Ne laser. The light is coupled into the waveguide by a lens using a (x, y, z, θ) stage for proper alignment. The propagation loss of integrated waveguides on silicon depends on the thermal treatment of the oxynitride layer during and after deposition. It is about 1.1 dB/cm [17] to 1.8 dB/cm [18] at 633 nm for PECVD oxynitride layers, but even less than 0.25 dB/cm are claimed [19]. This is negligible with regard to the dimensions of a chip.

The integration of optical devices needs curvatures to form beam splitters, interferometers and couplers. Small radii result in radiation losses at the bending of the waveguide. But, due to the small size of a chip, it is necessary to prevent large radii for curvatures. So, test structures with different sizes of curvatures have been integrated and tested. Figure 4 shows the scattered light photograph of different waveguide curvatures using SiON (left) and Si_3N_4 (right) light guiding layers with a RI of 1.52 respective 2.0. Concerning the radiation loss in the curvatures, a minimum radius of about 200 μm is necessary in the $SiON/SiO_2$ system. In the case of nitride waveguides on silicon, even a radius of 20 μm is applicable, making thereby possible to increase the integration density by increase of the RI. The radiation loss in curvatures can be calculated as described in [20]. Figure 4 indicates that it is impossible to make small optical chips even with a relative high difference in the RI like $SiON/SiO_2$ waveguides with a $\Delta n/n$ of about 4%. Only a large index step enables highly integrated optical chips, e.g. the system Si_3N_4/SiO_2 with $\Delta n/n = 36\%$.

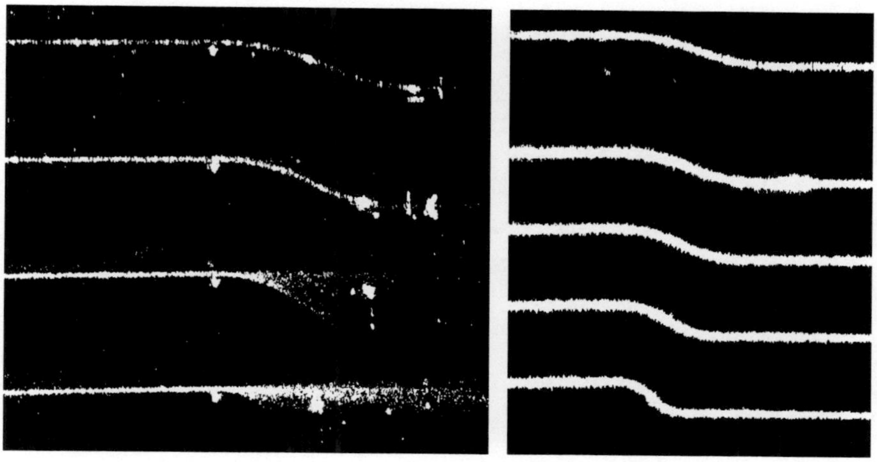

Figure 4. Different waveguide curvatures with radii of 500 μm, 200 μm, 100 μm, 50 μm, and 20 μm (left only), left for SiON ($n = 1.52$), right for Si_3N_4 ($n = 2.0$) ($\lambda = 633$ nm).

As an alternative, the curvatures may be replaced by mirrors consisting of a transition from the SiON waveguide to ambient air. In this case, a hole with rectangular or triangular geometry is etched into the light guiding film structure perpendicular to the wafer surface. As depicted by the scattered light photograph of the Mach-Zehnder interferometer in fig. 5, the light deflection at the mirrors and the beam splitting by a triangular shaped mirror operate well. Nevertheless, each mirror results in a loss of about 1 dB.

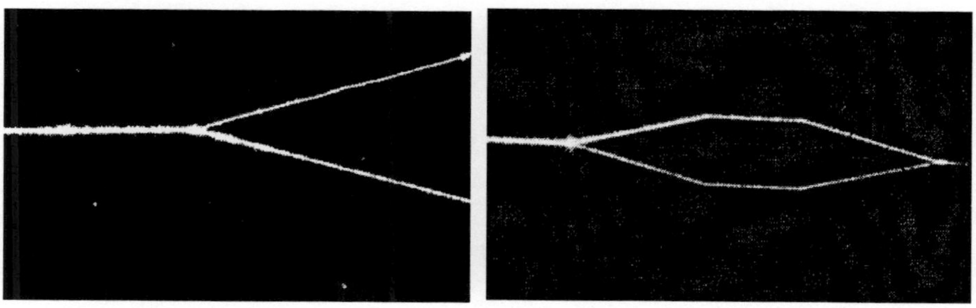

Figure 5. Mirror as a beam splitter, and Mach-Zehnder interferometer, consisting of two triangle shaped mirrors as beam splitters; four rectangular mirrors for light deflection.

Figure 6 depicts a beam splitter consisting of Y-splitters. The intensity is divided nearly uniformly into four waveguides. The propagation loss of the splitter strongly depends on the sharpness of the tip between the ribs etched in the center of the Y. Only high resolution lithography like e-beam guarantees low-loss beam splitting.

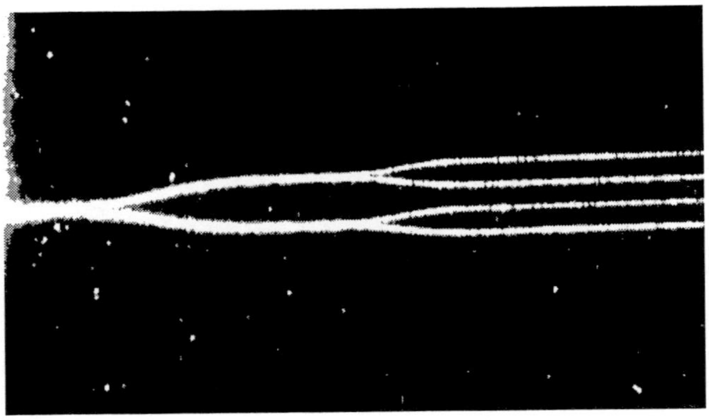

Figure 6. One to four beam splitting by three identical Y splitters with curvatures of about 1000 μm.

4. Integrated Optical Devices

Different kinds of switches have been integrated in SiON waveguides by applying micromechanical elements [21] or thermo-optical actuators [22]. A 1-to-8 switch using an arrayed waveguide grating wavelength multiplexer with thermal control is presented in [23]. It achieves high optical isolation of about 29 dB at $\lambda = 1.55$ μm between the active channel and the off state channels. Switching can be performed in 1 ms, whereas the power to control all switches simultaneously is about 6 W. In [18], an arrayed waveguide grating demultiplexer has been integrated in SiON technique in a very compact size of 5 mm × 2 mm. SiON add-drop filters for wavelength-division multiplexing have been demonstrated in [24]. The devices show high performance over a large spectral range. Today silicon devices for wavelength division multiplexing are based on resonant couplers with SiON waveguides [25]. The devices have a total loss of 6 dB from fiber-to-fiber, and the crosstalk for a single wavelength is less than −55 dB.

5. Optoelectronic Systems on Silicon

Integrated optics on silicon offers the advantage of monolithic optoelectronic chips with waveguide structures for sensor applications, and CMOS circuits for signal processing on the same substrate [26]. One condition for these applications is an interface between the optical and the electronic part of the system, e.g. a photodiode for the conversion of the optical information into an electrical signal. The integration of photodiodes is a standard in silicon MOS processing, as each *p-n* junction generates charge carriers in case of incident light. The main problem rises out of the thick oxide layer below the light guiding SiON film, which separates the guided photons in the SiON film from the semiconducting silicon substrate. Due to this oxide, the propagating electromagnetic wave cannot generate charge carriers in an integrated diode. There are different ways to get the light guiding film in contact to within silicon for signal transfer:

- applying photodiodes integrated in a mesa structure
- forming a planar surface between the diode surface and the thermal oxide surface

- tapering the oxide layer down to the silicon surface
- deflecting the light out of the waveguide to the silicon surface

5.1. Butt Coupling

Butt coupling uses a silicon mesa structure for the integration of the photodetector. The mesa rises out of the oxide layer, so the SiON waveguide film deposition takes place beside and on top of the mesa. So, the propagating light hits the side wall of the diode mesa, and can easily penetrate into the active silicon area, as depicted in fig. 7. Efficiencies of 84% have been demonstrated [27].

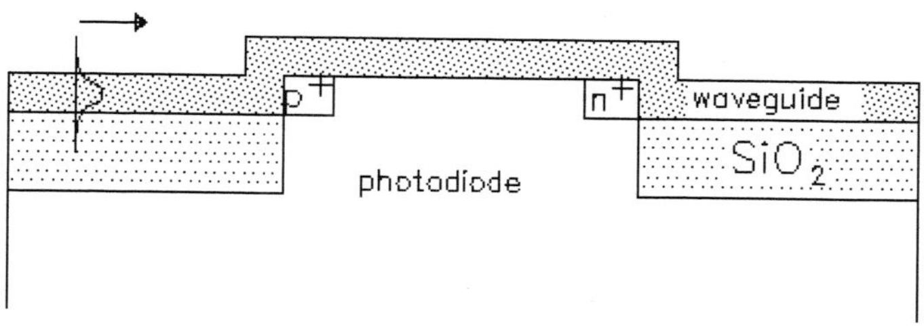

Figure 7. Butt coupling of an integrated waveguide to a photo diode on silicon.

The main difficulty in this kind of coupling is the processing of MOS transistors on the same chip. Due to the necessity of mesa structure for photo diode, all MOS transistors are integrated in mesas. This causes a complete change in MOS processing at least, which is not possible due to its complexity.

5.2. Leaky Wave Coupling

To avoid the mesa structures rising out of the oxide layer, another structure has been developed applying leaky wave coupling of waveguides and integrated photo diodes (fig. 8). In this case, the silicon oxide layer below the light guiding SiON film and the silicon surface are on the same level. As such, no silicon mesa rises out of the surface. MOS processing is affected only weak. The main difficulty of this process is to get a smooth transition from the oxide layer to the active silicon area. This is impossible without a chemical mechanical polishing (CMP) step. CMP is a complex and expensive process step, which easily causes defects reducing the yield. Without CMP, small steps occur at the transition. They result in reflection and scattering of the electromagnetic wave, reducing the effective intensity in the waveguide.

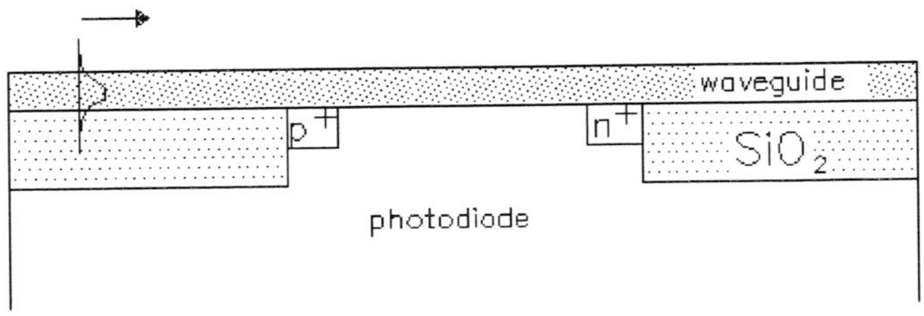

Figure 8. Leaky wave coupling of waveguide and photo diode applying a smooth transition from the oxide layer to the active silicon surface.

5.3. Taper Coupling

Another coupling technique uses an extended taper in the oxide layer below the light guiding film (fig. 9). Due to the small RI difference between the cladding layers and the SiON, the taper must be very smooth to prevent optical leakage. The processing of an extended taper is possible, but it is not easy to do. Even in the well developed silicon technology, it is expensive to etch a suitable structure for this kind of coupling. Additionally, the leaky wave coupling starts at the point of reduced oxide thickness. So, charge carriers are generated outside the photo detector, which results in low frequency operation due to charge carrier diffusion.

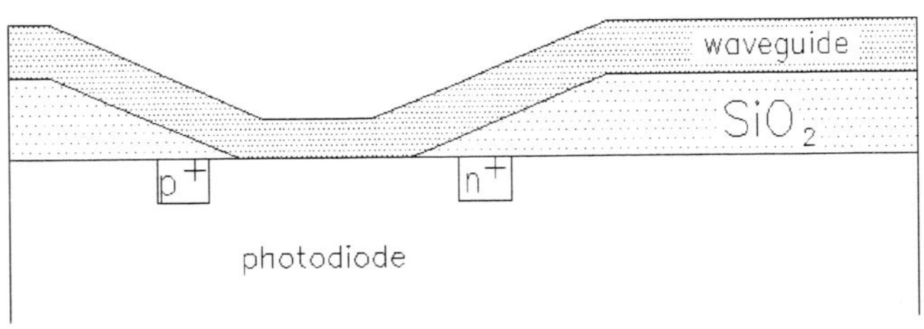

Figure 9. Coupling of waveguide and photo diode by an extended taper.

5.4. Coupling by Integrated Mirrors

The most promising technique for coupling of waveguides and photodiodes is a special kind of butt coupling using an integrated mirror to deflect the light out of the waveguide into the photo diode (fig. 10). Again a taper is necessary to get a sloped wall with a slope of about 45°. The propagating electromagnetic wave hits the mirror surface, consisting of the transition SiON to ambient air, and is reflected by total reflection. This technique is applicable on nearly any MOS circuit, because there is no special need for a planar surface at the transition from the silicon oxide to the active silicon area. The advantage of the mirror coupling technique results in the modular processing of the MOS structures and the waveguide layers. It can easily be implemented in any MOS technology. Figure 11 shows the scattered light photograph of a waveguide coupled to a photodiode by applying the process of mirror coupling.

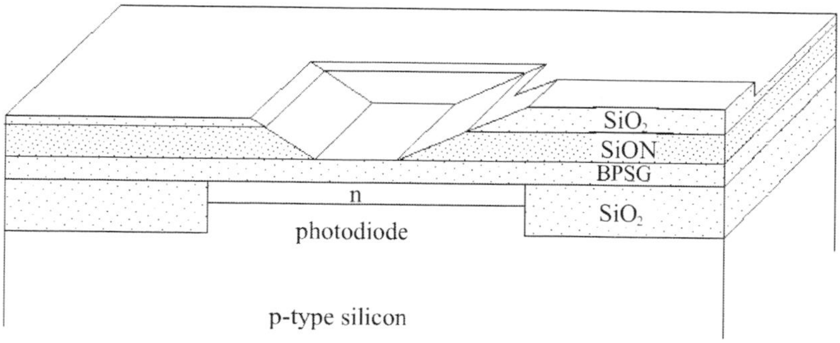

Figure 10. Coupling of a waveguide to an integrated mirror to deflect the light into the photo diode.

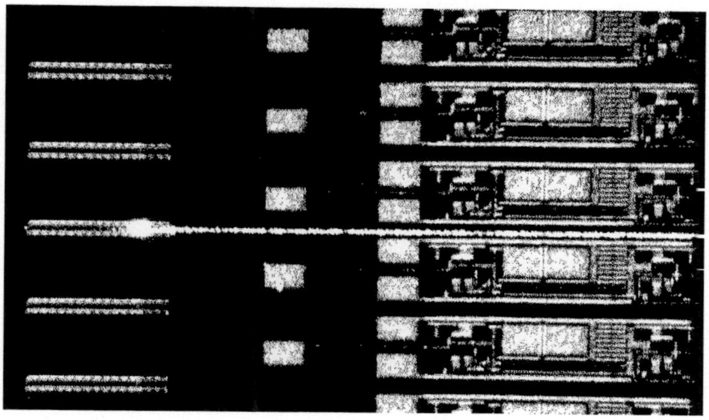

Figure 11. Scattered light photograph of a waveguide coupled to a photodiode applying mirror coupling.

6. Monolithic Integration of Waveguides and MOS Circuits

Today's MOS processes apply local oxidation techniques (fig. 12) or shallow trench isolation to separate the active transistor areas by silicon oxide [28]. These processes typically have a soft or step less transition from the oxide area to the active silicon surface, but there is no possibility to extend the active silicon area to form mesa structures for butt coupling.

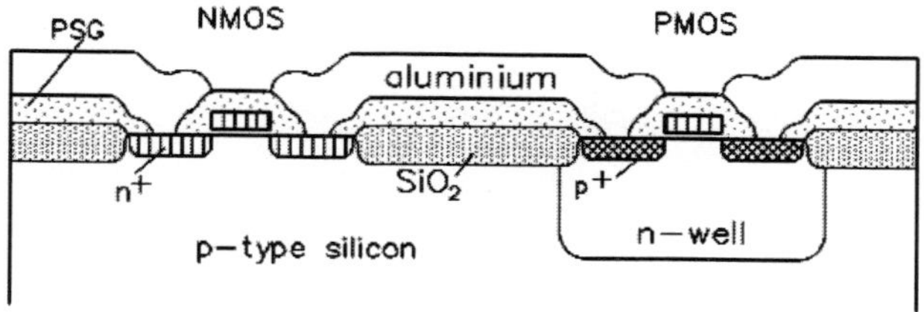

Figure 12. Cross-section of a normal CMOS structure in LOCOS technique.

The Side Wall Mask Isolated LOCal Oxidation of Silicon (SWAMI-LOCOS) technique [29] offers a simple way to get mesa structures in an oxidized silicon wafer. As depicted in fig. 13, after deposition and etching of the standard LOCOS mask, a deep silicon etch removes the crystalline material outside the active device areas. The depths of this etch defines the kind of coupling – in the case of 55% of the oxide thickness grown afterwards, it results in leaky wave coupling. In the case of about 75% of the thermal oxide thickness needed for the separation of the light guiding SiON film from the silicon substrate, butt coupling of the waveguides to the diodes results.

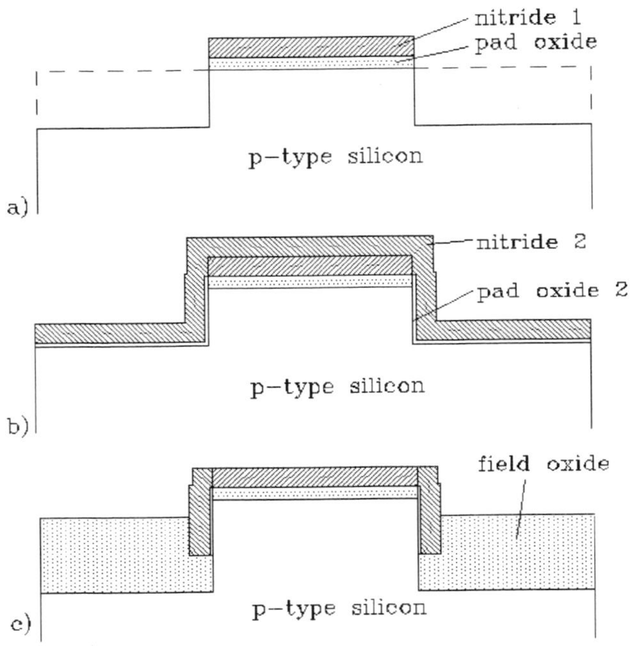

Figure 13. Formation of mesa structures using SWAMI-LOCOS technique.

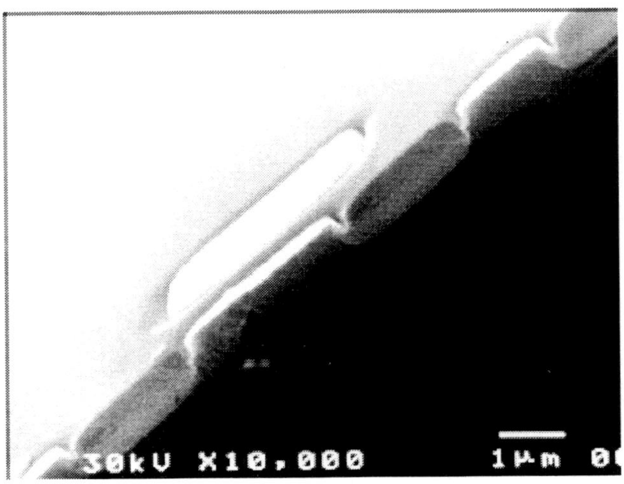

Figure 14. Surface planarity after local oxidation applying SWAMI technique.

Next, another nitride layer deposition masks the vertical walls of the etched mesas. This layer is etched back anisotropically by reactive ion etching. So, only the side walls and the top of the mesas are protected against the thermal oxidation step. During the thermal oxidation of the silicon, the oxide grows outside the mesa structure. After removal of the nitride mask, the surface is nearly plane, or a mesa rises out of the oxide layer. The planarity of the surface after 55% of the oxide thickness for silicon etching is depicted in fig. 14. The only disturbance of the flatness is caused by the side wall passivation nitride around the active silicon areas, which is removed during surface nitride etching.

In the following section, two integration principles are explained for monolithic integration of optical waveguides and CMOS circuits on one silicon chip applying SWAMI-LOCOS.

7. Layer-Adjusted Monolithic Integration Technique

The layer-adjusted monolithic integration technique uses the field oxide and the intermetal dielectric layer of the MOS process for the waveguide structure by the adjustment of the layer thicknesses and their RIs. If the field oxide in the MOS process is extended to 2 μm, and the SiON layer replaces the standard PSG or BPSG to cover the polysilicon, the typical strip loaded film waveguide structure is available in any MOS process. Instead of the simple LOCOS technique, the process must use SWAMI-LOCOS to form silicon mesas or a planar surface. So, as a starting point of MOS transistor integration, a field oxide of 2 μm covers the wafer surface outside the active transistor areas. Then MOS transistors can be integrated by gate oxidation, polysilicon gate deposition and ion implantations for the drains and sources of the devices. Afterwards, the SiON deposition for the film waveguide can start. It replaces standard PSG or BPSG. As the reflow characteristics of SiON are nearly zero, the aluminium metallization is difficult in the area of the polysilicon gates. The light guiding film covers the mesa structure of the integrated photo diode, and the waveguide hits the diode perpendicular from its side. The SiON film serves as the polysilicon/intermetal dielectric layer, and after contact opening the metallization process starts. Figure 15 illustrates a cross-section of this monolithic integration process.

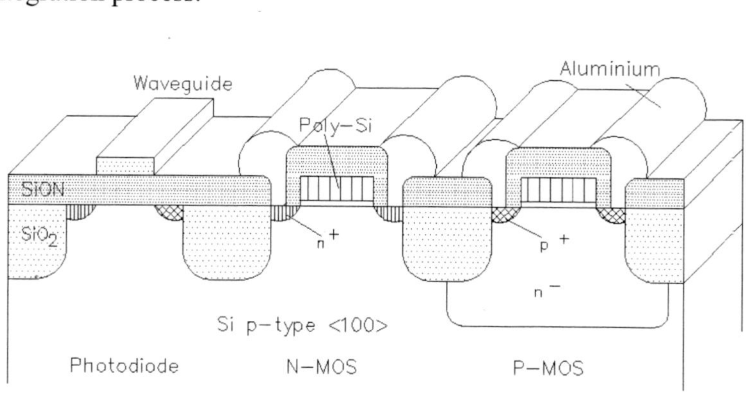

Figure 15. Cross-section of the combined monolithic integration process for waveguides and CMOS components on the same chip using either leaky wave coupling or butt coupling, depending on the step at the transition from the SiO_2 to the active silicon area.

The integration of the CMOS structures and waveguides using this kind of processing requires no additional layers, but the quality of the waveguides and of the transistors is rather low. Due to the reactive ion etching of the aluminium wires on top of the SiON film, the surface roughness increases, resulting in high propagation loss. This may be prevented by a high selective etching process for the metallization, e.g. by wet etching. Additionally, it is nearly impossible to get a smooth transition from the field oxide to the active diode area.

On the other hand, the parameters of the electronic devices depict large leakage currents in the *p-n* junctions. Even in the off-state, MOS transistors leakage is high, because of the boron segregation below the field oxide. This causes channels for a current flow below the oxide. In conclusion, this kind of monolithic integration results in low quality optical and electrical devices, although the process flow does not need any additional layers. So, the process flow must be developed completely new to get sufficient yield in electrical and optical devices.

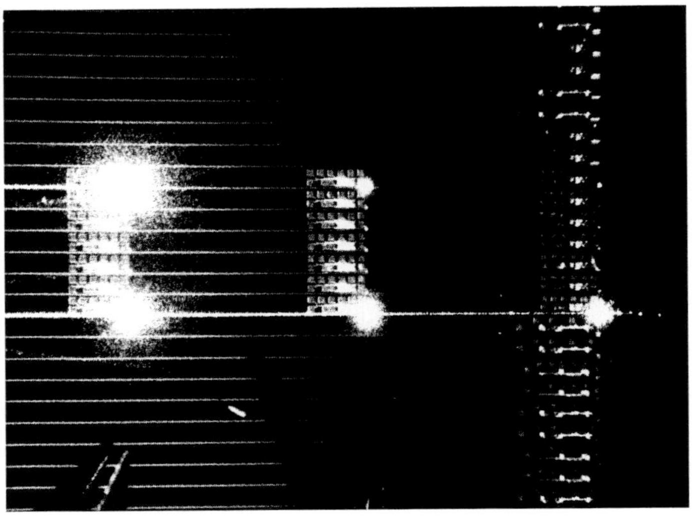

Figure 16. Waveguides coupled by butt coupling to three identical diodes with a length of: top 25 μm, bottom 3 μm.

Figure 16 shows a scattered light photograph of waveguides coupled to three identical photo diodes aligned in a chain. The diodes were integrated using the layer adjusted monolithic integration technique with butt coupling, as explained before. Top diodes consist of 25 μm active silicon below the waveguide, the bottom diodes have only 3 μm active diode area. Optical power input is 5 mW at 633 nm. The intensive spot at each transition from the field oxide to the active diode area characterizes a nonideal transition of the waveguide to the diode. Due to a non-perpendicular diode side wall caused by the SWAMI-LOCOS processing, a large amount of the light is scattered out of the waveguide, and only a part of the intensity is guided.

Figure 17 depicts the photo currents at the three diodes coupled to one waveguide for different diode lengths. The data are normalized to the current of the first diode due to the unknown intensity coupled into the waveguide. In short diodes, only a part of the intensity couples into the diodes, and results in charge carrier generation. Thus, the intensity at the second diode is much higher, and results in a relatively high photocurrent. With increasing diode length, the absorption raises, that means decreasing photocurrents at the second and third diodes.

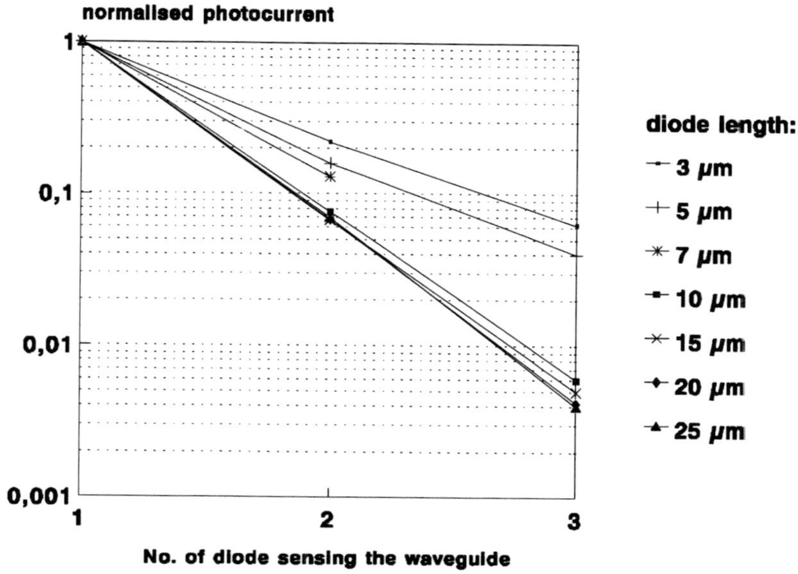

Figure 17. Photocurrent of three diodes in a chain in dependence on the active diode length.

Figure 18 indicates the coupling efficiency for leaky wave coupling. The photocurrents measured at the diodes increase at about 1000 μm diode length; below this value there are no clear tendencies detectable. The reason is the quality of the transition of the waveguide to the diode surface. Even a step of 40 nm causes a partial butt coupling. So, the photocurrents of the short diodes do not show the leaky wave photocurrent, but mainly the butt coupling caused photocurrents.

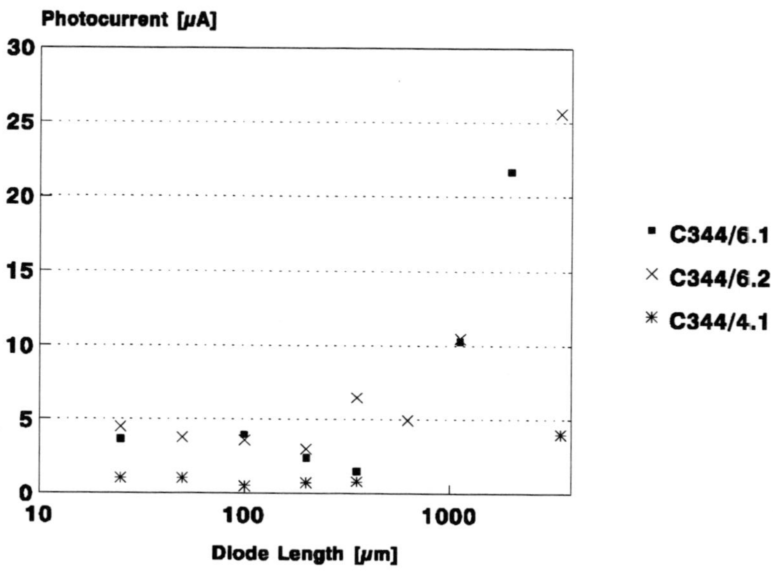

Figure 18. Photocurrent of diodes of different lengths coupled to a waveguide by leaky wave coupling.

8. Modular Processing

Modular processing of waveguides and MOS transistors starts with the integration of the electronic components. The only modification of the standard MOS process is the replacement of the standard LOCOS by SWAMI-LOCOS to get a plane wafer surface after standard field oxide growth. After transistor integration, the wafers may be tested electrically, and only if all parameters are right, the waveguide deposition can be added. Waveguides need 2 μm of isolation oxide below the light guiding film. So, an additional oxide deposition at low temperature must be performed, e.g. by PECVD, to increase the isolation layer to 2 μm. Then the SiON film and the rib layer are deposited by PECVD. As the last step, the rib is defined by lithography and etched into the top layer by reactive ion etching.

The modular process uses PECVD layers for waveguides due to the very low deposition temperature. LPCVD is not applicable, because a temperature above about 450 °C will destroy the electronic structures by the formation of an aluminium silicon alloy in the junction areas. Nevertheless, after layer deposition, the MOS structures are operating like standard transistors. There is no leakage in the *p-n* junctions, the field areas of the circuits or in the off-state of the transistors. The waveguide structures are of high quality too, because there is no reactive ion etching process on top of the SiON film. Only the coupling efficiency is low because of a rather thick oxide film of more than 1.3 μm between the light guiding SiON film and the diode surface. Leaky wave coupling is very low, so the electrical diode output signal is small.

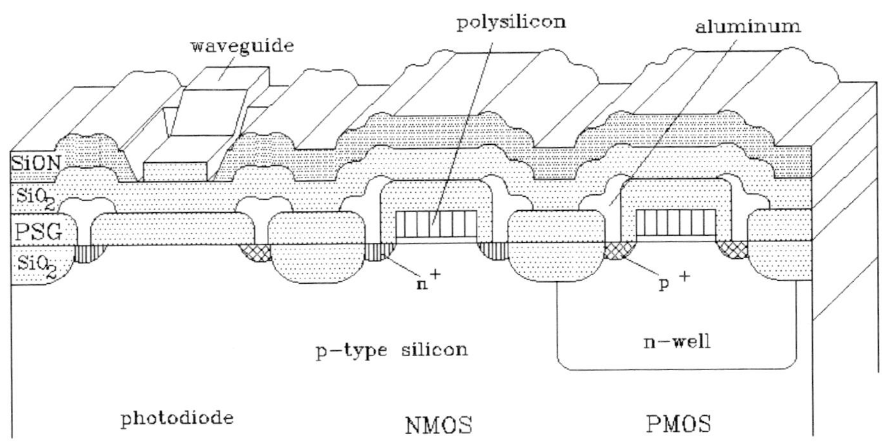

Figure 19. Modular integration of CMOS components and waveguides using PECVD waveguides after finishing the CMOS integration process.

To increase the coupling efficiency, it is necessary to etch a mirror into the light guiding film to get butt coupling by a mirror (fig. 19). The mirror etch uses a mixture of CHF_3 and oxygen during reactive ion etching; the amount of oxygen determines the slope of the mirror in relation to the wafer surface. At about 18% oxygen content in the plasma, during the dry etching, an angle of 45° will result (fig. 20). Figure 21 shows a photograph of mirror etched into a waveguide.

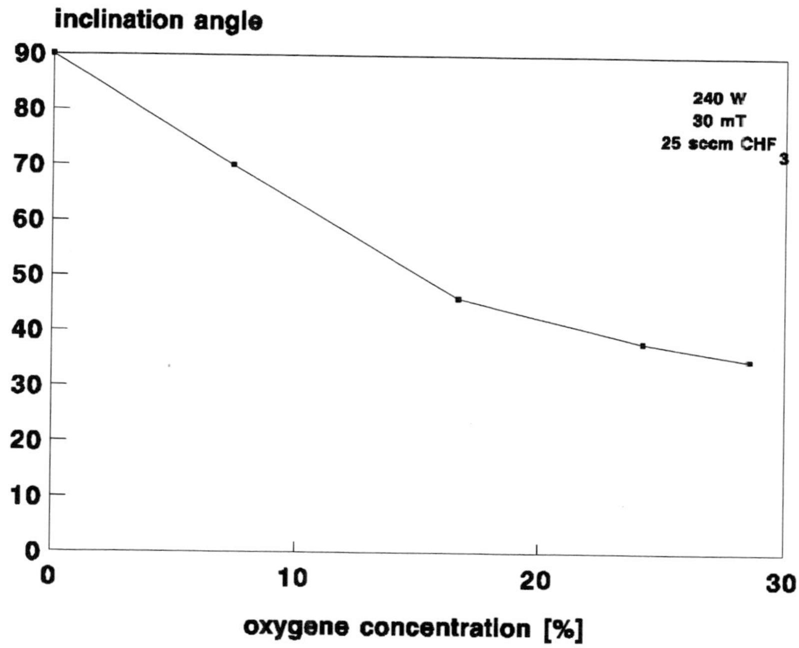

Figure 20. Slope of the mirror opening in dependence on the oxygen content of the plasma during reactive ion etching.

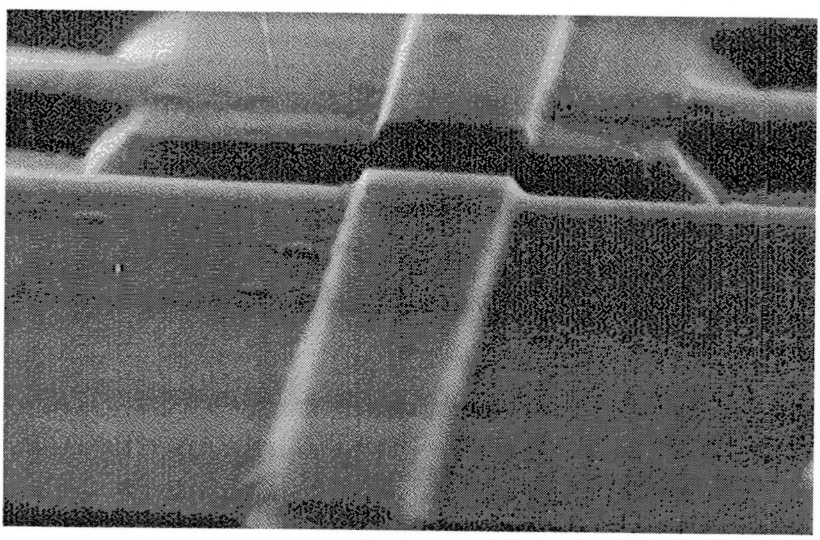

Figure 21. Mirror etched into a waveguide.

The coupling efficiency depends on the etching depth strongly (fig. 22). If only the rib in the surface cladding is etched, nearly no photocurrent is generated in the photo diode. If the light guiding SiON film is etched completely, almost all light is coupled into the diode.

The modular integration technique has been used for microsystem integration consisting of waveguides, photo detectors and amplifiers on one silicon chip [30]. It is applicable to frequencies up to 20 MHz, limited by the features of the CMOS compatible photo detector.

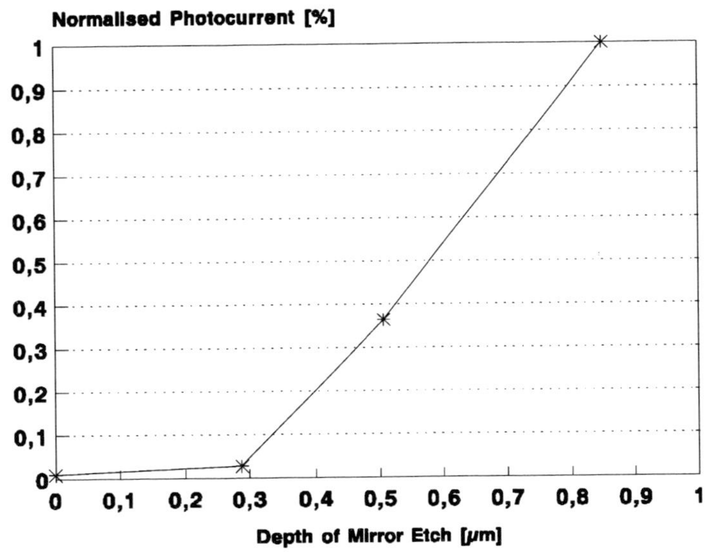

Figure 22. Photocurrent measured at a diode in dependence on the etch depth of the coupling mirror.

9. Micromechanic Add -Ons for Sensor Applications

Integrated optics give the opportunity to detect small elongations or bendings of membranes by interferometry [31]. It enables the fabrication of a microsystem consisting of a pressure sensor with optical readout and microelectronic circuitries, as presented in fig. 23. The main component is a Mach-Zehnder interferometer in SiON technique. Its measurement branch is located on top of a membrane, whereas the reference branch is placed beside the sensitive area. The membrane bending causes a change in the effective RI of the light guiding film, due to the mechanical stress. So, the phase of the propagating light in the sensitive branch is correlated to the external pressure applied to the membrane. The result is a small difference in the optical paths of the measurement and the reference branch of the interferometer, which causes an intensity change at the interferometer output.

Applying the modular technology for integrated optics on silicon (as explained above), it is possible to transform the optical information into an electrical signal by a photodetector and a MOS-amplifier, all integrated on the same chip. So, the magnitude of the external pressure is directly related to the interferometer output signal, respective to the photocurrent at the photodetector. The integration of such a microsystem can use the modular processing of waveguides and microelectronics with slight modifications for the mechanical components only. In this case, the micromechanic structures like cantilevers and membranes may be integrated after the test of the microelectronic circuits, prior to waveguide deposition. The add-ons require one resist mask and two reactive ion etching steps [32].

The photoresist defines an array of holes with diameters of about 1 µm on top of the field oxide in the membrane area (fig. 24). These openings are etched down to the silicon surface anisotropically, across the whole field oxide of the MOS structures. In the next step, the silicon substrate, which is buried below the dielectric films, is etched isotropically through the oxide holes using an SF_6 plasma. This is highly selective to oxide and photoresist. Due to the undercut of the oxide layer, small cavities arise below each opening. Finally, they grow up to one large cavity closed by a porous oxide membrane [33].

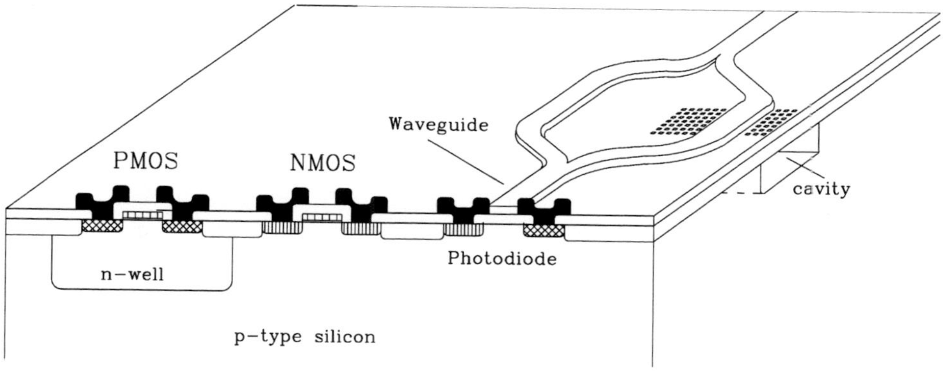

Figure 23. Cross-section of an integrated microsystem for pressure sensing, consisting of a membrane, a Mach-Zehnder interferometer, digital and analogue CMOS circuits, built on one silicon chip.

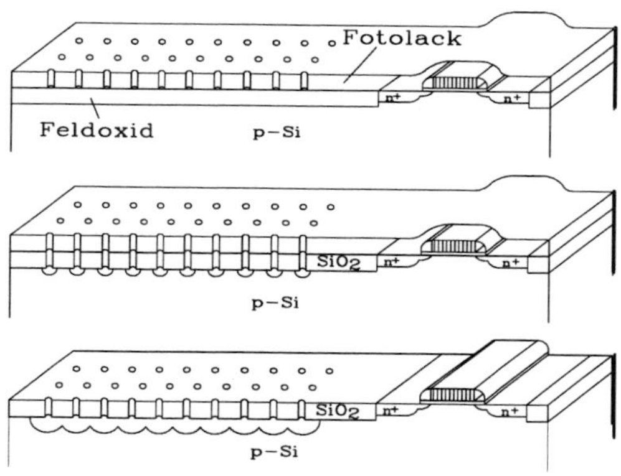

Figure 24. Flow chart for the integration of a membrane.

To get a closed cavity for pressure reference, a nearly conformal PECVD oxide film deposition of about 1.0 μm thickness is performed. In addition, this oxide serves as an additional layer to increase the optical insulating layer for the SiON waveguides. On top of this oxide, the light guiding film and the capping oxide is deposited by PECVD. Figure 25 shows the result of these depositions. Due to the low conformity of the deposition processes, the openings were not closed completely. Their cross-sections were reduced, but the membrane is still not sealed. A more conformal deposition process with ozone would help to solve this problem.

The process concludes by rib etching to form the waveguides. It results in an optoelectronic/micromechanic chip, which is capable of pressure sensing. To increase the sensitivity, it is possible to integrate some membranes in one interferometer. Figure 26 depicts a scattered light photograph of a microsystem consisting of four membranes in a Mach-Zehnder interferometer. The interferometer output hits a photo detector; its signal is amplified by an operational amplifier integrated on the same chip. The whole chip size is about 14 mm × 0.5 mm. The only external element is the laser diode.

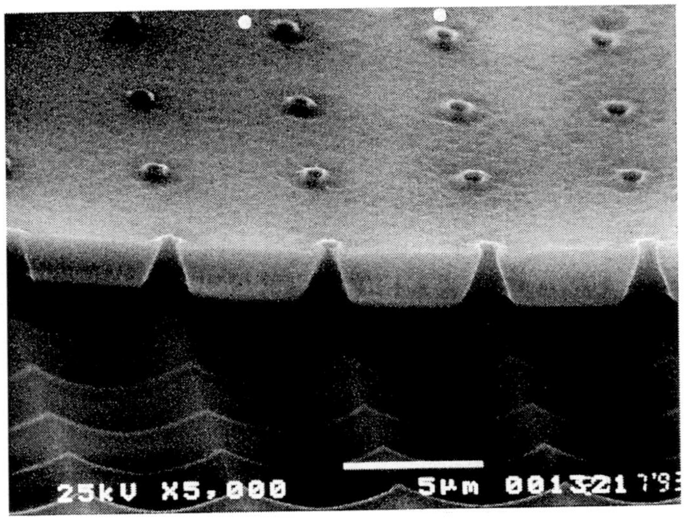

Figure 25. SEM cross-section of the membrane area after isotropic silicon etch and layers depositions.

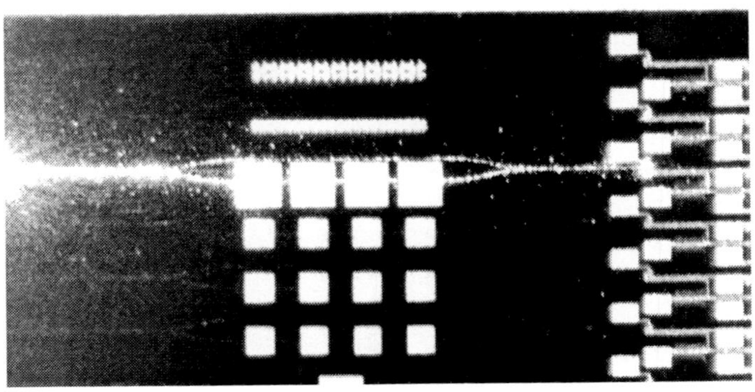

Figure 26. Top view of the monolithically integrated optoelectronic microsystem with micromechanic pressure sensor, and mirror coupling of waveguide and photodetector.

In this monolithic integration process, all steps were done from the front side wafer surface. Only materials typically used in MOS-technology, were applied for preparing the structures. The tested microsystems for pressure sensing consist of one or four membranes of 120 μm × 120 μm or 200 μm × 200 μm size, crossed by the sensitive branch of a Mach-Zehnder interferometer of 2500 μm length. A photobipolar transistor detects the output intensity of the interferometer; this signal is amplified by a transimpedance amplifier.

In fig. 26, He-Ne laser light, used for the parameter extraction, is coupled into the waveguide structure. The scattered light photograph shows a Mach-Zehnder interferometer crossing four membranes of 200 μm × 200 μm each. To rate the sensitivity of the optomechanical pressure sensor, the photocurrent was measured directly at the photodetector output, without any amplification. Membranes were stimulated by a pulsed stream of compressed air.

Figure 27 shows the dynamic output signal of the photodetector at 20 Hz, applying a pressure of 400 mbar to a system consisting of four membranes of 200 μm × 200 μm each. The amplitude of about 5.5 mV results in a signal rate of 13 μV/mbar. This signal may be amplified by the transimpedance amplifier integrated on the same chip. There is a small delay time between the valve voltage (CH2) and the output voltage swing (CH1), caused by the response time of the valve and the compressed air spread. The output signal is modulated by a frequency of about 180 Hz. This may be a resonance frequency of the whole microsystem, because the etch openings in the membranes under test were not completely closed. According to the theory of plates and shells [34], the membrane resonance frequency has to be much higher. Nevertheless, the resonance depends on the membrane size.

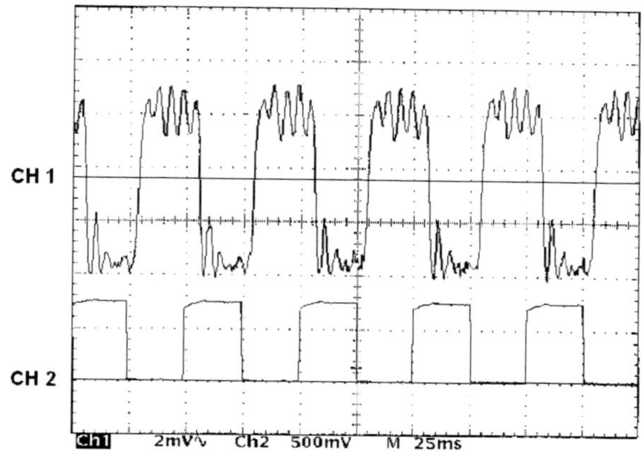

Figure 27. Top: photo detector output signal at dynamic air flow modulation, bottom: valve control voltage.

10. Infrared Light Waveguides in Silicon

In the visible range, it is necessary to use the above mentioned waveguides consisting of dielectric layers because of the strong absorption of silicon. But, at $\lambda = 1.3$ to $1.55\ \mu m$, silicon itself is transparent, and therefore, no absorption of the electromagnetic field occurs. The RI of silicon at 1.55 μm is 3.5 for undoped silicon, and only a very little higher value for *n*-type silicon. So, silicon can be used as the waveguide or the cladding layer itself. The typical structures of silicon waveguides is formed by phosphorous diffusion or implantation into a silicon wafer, as depicted in fig. 28(a).

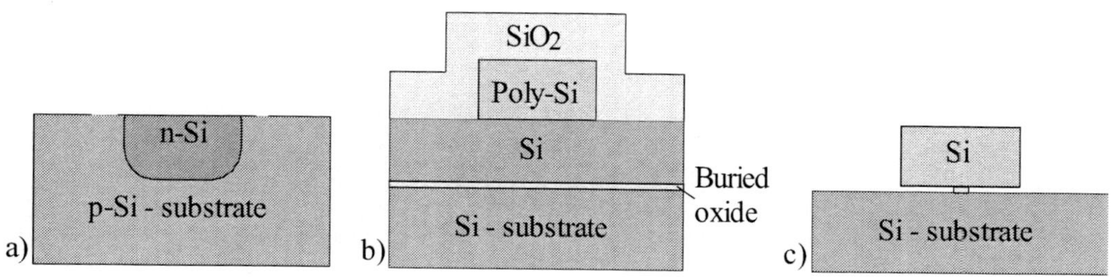

Figure 28. Cross-sections of silicon based waveguides: (a) doped surface core waveguide, (b) rib waveguide, and (c) column supported waveguides.

During the last years, interest in silicon on insulator (SOI) waveguides has been growing. An SOI waveguide consists of a 1.5 μm thick crystalline silicon layer on top of a buried oxide film. Beam confinement in horizontal direction is performed by a polysilicon rib deposited on top. In vertical direction, the buried oxide and a surface oxide serve as claddings. Another kind of waveguide consists of a rectangular silicon core, which is located on a few supporting oxide columns only. So, the ambient serves as the cladding. Figure 28 presents their cross-sections. These kinds of waveguides give the opportunity to integrate modulators even in silicon [35]. Although there is no linear electrooptical effect, and the Kerr effect is very small [36] too. A modulation up to about 1 MHz is possible by the thermo-optic effect [37]. Modulation frequencies up to the GHz range have been predicted by the use of the free carrier plasma dispersion effect in forward biased *p-i-n* diode devices [38]. In these devices, the operating speed is limited by the long carrier lifetime in silicon.

Another approach is a modulation of the RI in the silicon waveguide by a MOS capacitor integrated in a Mach-Zehnder interferometer. A voltage applied to the electrodes of the capacitor causes a charge carrier accumulation below the oxide at the waveguide surface. It causes a RI modulation leading to a phase shift, which results in an intensity modulation at the interferometer output. A cross-section of the phase shifter is shown in fig. 29. It has been demonstrated in [39] that a length of the device of about 8 mm will cause a phase shift of about 180° at an applied voltage of 10 V and a gate oxide thickness of 6 nm.

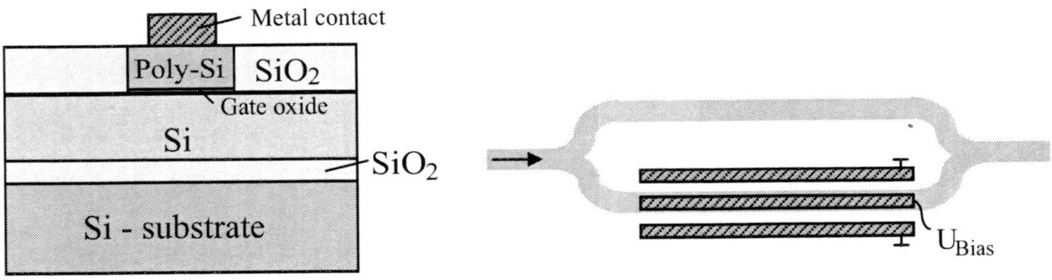

Figure 29. Silicon phase shifter and the complete Mach-Zehnder interferometer modulator.

The maximum operation speed in silicon modulators, using the principle like the device shown above, is 10 GHz reported up to now [40]. Optimising the design, even higher speed has been predicted by the authors. Although this is much lower as switching of today's $LiNbO_3$ devices, it is a promising modulation technique for data transmission.

Conclusion

It has been demonstrated that the indirect band gap semiconductor silicon is a valuable material for integrated optics in the visible and near infrared spectral range. In the visible range, SiON and Si_3N_4 waveguides can be integrated on silicon in a CMOS compatible technique using a special kind of local oxidation process. Different kinds of sensors demonstrate the capabilities of monolithic system integration with optical waveguides, photo detectors and CMOS circuits on one chip.

At a wavelength of 1.55 μm, silicon can be used as a light guiding medium itself. Although there is no electro-optic effect in silicon, an intensity modulator, based on a Mach-

Zehnder interferometer with capacitive charge carrier accumulation, demonstrates 10 GHz operating frequency. This opens up a wide area of further applications for integrated optics on silicon.

References

[1.] Soref, R.A. *Proc. IEEE* 1993, 81, 1687–1706.

[2.] Richter, A.; Steiner, P.; Kozlowski, F.; Lang, W. *IEEE Trans. Electron. Dev. Lett.* 1991, 12, 691–692.

[3.] Tsuchiya, T.; Nakajima, S. *IEEE J. Solid-State Circuits* 1985, 20, 325–332.

[4.] Ng, W.L.; Lourenco, M.A.; Gwilliam, R.M.; Ledain, S.; Shao, G.; Homewood, K.P. *Nature* 2001, 410, 192–194.

[5.] Hilleringmann, U.; Goser, K. *IEEE Trans. Electron. Dev.* 1995, 42, 841–846.

[6.] Yaffe, H.H.; Henry, C.H.; Kazarinov, R.F.; Milbrodt, M.A. *J. Light. Tech.* 1994, 12, 64–67.

[7.] Veldhuis, G.J.; van der Veen, L.E.W.; Lambeck, P.V. *J. Light. Tech.* 1999, 17, 857–864.

[8.] Rickman, A.G.; Ree, G.T.; Namavar, F.J. *Light. Tech.* 1994, 12, 1771–1776.

[9.] Modreanu, M.; Cosmin, P. *Int. Semiconductor Conf. Proc. CAS'95*, 1995, 323–326.

[10.] Peters, D.; Fischer, K.; Müller, J. *Sensors and Actuators A* 1991, 26, 425–431.

[11.] Ay, F.; Aydmli, A.; Roeloffzen, C.; Driessen, A. *IEEE Lasers and Electro-Optics Society 2000 Annual Meeting, LEOS* 2000, 2, 760–761.

[12.] Kapser, K.; Wagner, C.; Deimel, P.P. *IEEE J. Phot. Tech. Lett.* 1991, 3, 1096–1098.

[13.] Smit, M.K.; Acket, G.A.; Van der Laan, C.J. *Thin Solid Films* 1986, 138, 171–181.

[14.] Bona, G.-L.; Germann, R.; Offrein, B.J. *IBM J. Res. & Dev.* 2003, 47, 239–249.

[15.] Zhang, A.; Chan, K.T. *Proc. of the 6th Chinese Symposium on Optoelectronics* 2003, 124–127.

[16.] Hoffmann, M.; Voges, E. *Tech. Dig. 10th Int. Conf. on Integrated Opt. and Opt. Fiber Commun. (IOOC '95)* 1995, 5, 35–36.

[17.] Agnihotri, O.P.; Tyagi, R.; Kato, I. *Jpn. J. Appl. Phys.* 1997, 36, 6711–6713.

[18.] Schauwecker, B.; Przychrembel, G.; Kuhlow, B.; Radehaus, C. *IEEE J. Phot. Tech. Lett.* 2000, 12, 1645–1646.

[19.] Bezzaoui, H.; Baus, A.; Voges, E. Micro System Technologies; Springer: Berlin, 1990; 283–288.

[20.] Pregla, R.; Ahlers, E. *Electron. Lett.* 1994, 30, 1478–1479.

[21.] Hoffmann, M.; Kopka, P.; Groß, T.; Voges, E. *Electron. Lett.* 1998, 34, 207–208.

[22.] House, A.; Whiteman, R.; Kling, L.; Day, S.; Knights, A.; Hogan, D.; Hopper, F.; Asghari, M. *Proc. Opt. Fiber Commun. Conf.* 2003, 2, 449–450.

[23.] Flück, E.; Horst, F.; Offrein, B.J.; Germann, R.; Salemink, H.W.M.; Bona, G.-L. *IEEE J. Phot. Tech. Lett.* 1999, 11, 1399–1401.

[24.] Rotolo, S.; Tanzi, A.; Brunazzi, S.; DiMola, D.; Cibinetto, L.; Lenzi, M.; Bona, G.-L.; Offrein, B.J.; Horst, F.; Germann, R.; Salemink, H.W.M.; Baechtold, P.H. *J. Light. Tech.* 2000, 18, 569–578.

[25.] Offrein, B.J.; Germann, R.; Horst, F.; Salemink, H.W.M.; Beyeler, R.; Bona, G.L. *IEEE J. Select. Topics in Quantum Electron.* 1999, 5, 1400–1406.
[26.] Zimmermann, H. Integrated Silicon Optoelectronics; Springer: Berlin, 2000.
[27.] Wunderlich, S.; Schmidt, J.P.; Müller, J. *Appl. Opt.* 1992, 31, 4186–4189.
[28.] Sze, S. M. *VLSI-Technology;* McGraw-Hill: NY, 1988.
[29.] Kahng, D.; Shankoff, T.A.; Sheng, T.T.; Haszko, S.E. *J. Electrochem. Soc.* 1980, 127, 2468–2471.
[30.] Braß, E.; Hilleringmann, U.; Goser, K. *IEEE J. Solid-State Circuits* 1994, 29, 1006–1010.
[31.] Fischer, K.; Hoffmann, R.; Müller, J. Micro System Technologies '91 (Eds. Krahn, R.; Reichl, H.) VDE: Berlin; 1991, 472–481.
[32.] Adams, S.; Hilleringmann, U.; Goser, K. Micro System Technologies '92 (Eds. Krahn, R.; Reichl, H.) VDE: Berlin, 1992, 217–226.
[33.] Adams, S.; Hilleringmann, U.; Goser, K. ESSDERC '92, Microelectronic Engineering 1992, 19, 191–194.
[34.] Timoshenko, S.; Woinowsky-Krieger, S. Theory of Plates and Shells; McGraw-Hill: NY, 1959.
[35.] Reed, G.T.; Png, C.E. *J. Mat. Today,* January 2005, 40–50.
[36.] Soref, R.A.; Bennett, B.R. *IEEE J. Quantum Electron.* 1987, 23, 123–129.
[37.] Cocorullo, C.; Iodice, M.; Rendina, I.; Sarro, P. *IEEE J. Phot. Tech. Lett.* 1995, 7, 363–365.
[38.] Tang, C.K.; Reed, R.T. *Electron. Lett.* 1995, 31, 451–452.
[39.] Liao, L., Liu, A.; Jones, R.; Rubin, D.; Samara-Rubio, D.; Cohen, O.; Salib, M.; Paniccia, M. *IEEE J. Quantum Electron.* 2005, 41, 250–257.
[40.] Liao, L.; Samara-Rubio, D.; Morse, M.; Liu, A.; Hodge, D.; Rubin, D.; Keil, U.D.; Franck, T. *Opt. Exp.* 2005, 13, 3129–3135.

Ulrich Hilleringmann was born in Germany in 1958. He studied physics from 1978 to 1984 at University of Dortmund, Germany. From 1984 to 1985 he was with the Fraunhofer Institute of Microelectronic Circuits and Systems, Duisburg, Germany. Afterwards Hilleringmann changed to the Department of Electrical Engineering at University of Dortmund, where he wrote a thesis on "Laser Recrystallization of Silicon" in 1988. In 1994 he received the "venia legendi" at the same department with the corresponding thesis

"Integrated Optics on Silicon". Since 1999 he has been a Professor at the Institute of Electrical Engineering, Sensor Technology Department, University of Paderborn.

His main interests are silicon technology, nanometer scale devices, integrated optics on silicon, organic transistors and microsystem technologies.

In: Frontiers in Optical Technology: Materials and Devices ISBN: 1-60021-084-8
Editors: P.K. Choudhury and O.N. Singh, pp. 97-124

Chapter 4

PHOTONICS APPLICATIONS OF NANO-SILICON

Lorenzo Pavesi
Dipartimento di Fisica, Università di Trento,
via Sommarive 14, 38050 Povo (Trento), Italy

Abstract

The chapter reviews the photonic applications of nanostructured silicon. As we change the dimensionality of silicon, very fascinating and new optical properties of the material appear. In particular, light emission starts to be a very efficient process in nanostructured silicon and light emitting diodes with efficiency in excess of 1% have been fabricated. Optical amplification has also been observed when silicon nanocrystals are embeded into a dielectric matrix. This makes the system a potential candidate for laser action. In addition to electronic property variations, nanostructured silicon can also be used as a nanodielectric material, where controlled changes in the refractive index (RI) can lead to trapping or slow down of photons. Some examples will be discussed where photonic Bloch oscillations, photonic Zener tunneling and high quality microcavities are demonstrated. All these various phenomenons pave the way to a new field of applications for the "old dog" of microelectronics.

1. Introduction

Low dimensional silicon is a fascinating material which still has many unfold properties. The main motivation to study silicon comes from its success and dominance in modern technology, especially in microelectronics. Indeed, nanometer sized transistors are now switching in computers at working frequency exceedingly a few GHz. Convergence between microelectronics and telecommunications is looked for in order to couple the computing power of microprocessor with the transmission power of optical fibers. In this effort, merging of various semiconductor technologies is attempted in order to improve, on one side, the optical properties of microelectronic materials and, on the other side, to reduce the cost of photonic devices. In this respect, silicon photonics is playing a key role. We have already reviewed in the past several aspects of this emerging technology [1]. In this chapter, we aim to give an overview of our recent work towards the exploitation of low dimensional silicon to tune, on one side, its electronic properties, and on the other side, its dielectric properties.

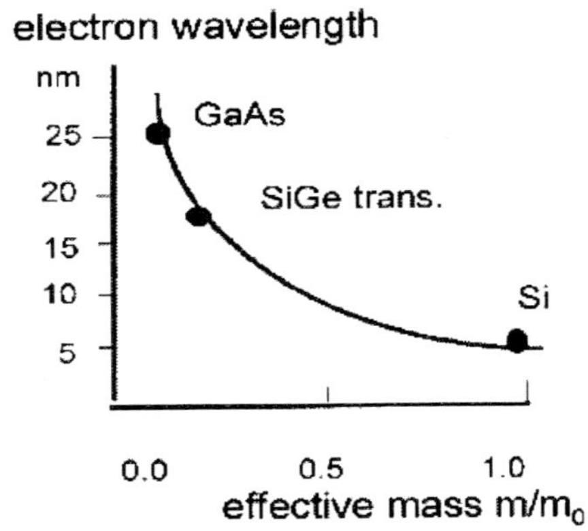

Figure 1. The De Broglie wavelength of an electron at 300 °K.

Let us define which dimensional scale we are concerned with. If one computes the De Broglie wavelength of electrons in silicon, one notices that it is about 5 nm (fig. 1). So, in this chapter, we are concerned with very tiny amount of silicon whose size is only a few nanometer. If the structure conserves a crystalline character, we will call it silicon nanocrystals; if not, we will call it silicon nanoclusters. In many instances this distinction is purely academic, so we will refer simply to Si-nc, meaning that we are not concerned by the crystalline or amorphous character of the particle.

This chapter is organized in three sections: the first is about the way to produce Si-nc and the modification of its electronic structure caused by the low-dimension, the second is about the improved optical properties of silicon which are observed when it is reduced to small particles, and the third is about the optical properties caused by the variation in its dielectric structures.

2. Low Dimensional Silicon

The best way to understand the effects of the reduced dimensionality on the energy spectrum of an electron is to resort to the particle-in-a-box problem so familiar in quantum mechanics. Within the effective mass approximation in the envelope wave function description, the transition energy of an electron-hole pair of effective reduced mass μ confined in a quantum dot of sizes L_x, L_y, L_z is

$$E = E_G + \frac{h^2}{8\pi^2\mu}\left[\left(\frac{n_x\pi}{L_x}\right)^2 + \left(\frac{n_y\pi}{L_y}\right)^2 + \left(\frac{n_z\pi}{L_z}\right)^2\right] \approx E_G + \frac{h^2}{8\pi^2\mu}\left(\frac{n\pi}{L}\right)^2 \quad (1)$$

where we have assumed that the confining energy barrier is of infinite strength, and that, E_G is the energy band-gap of the bulk semiconductor; n_x, n_y, n_z are quantum numbers to describe a given quantized energy level. Eq. (1) shows that the effective energy band-gap of the quantum dot increases as the quantum dot dimension decreases (fig. 2). When the dot is spherical with a diameter d, the transition energy scales with a $1/d^2$ law. More accurate calculations show that the scaling law does not follow an exact power 2 law, but instead, the exponent is lower [2]. A value of 1.7–1.8 seems to be more appropriate.

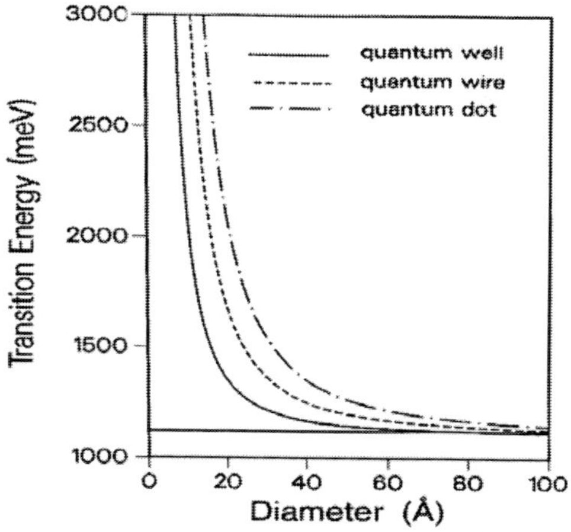

Figure 2. Optical gap in Si quantum wells, wires, and dots versus system diameter. The transition energy is calculated for the lowest electron and heavy hole energies for infinite confining potentials. From Ref. [3].

The assumption of having an infinite potential barrier is very crude since Si-nc are usually formed in a dielectric matrix. The matrix is usually a silica glass SiO_2, but more recently the use of Al_2O_3, Si_3N_4 or of oxynitride (SiON) is emerging as an alternative. Also, the effective mass approximation has strong limitation since the reconstruction of the lattice in the dots caused by the strain and the different elastic properties of the host should be considered. Finally, chemical bonds among the surface Si atoms of the Si-nc and the atoms of the host, strongly modify the energy spectrum of Si-nc. An example of a theoretical calculation is shown in fig. 3.

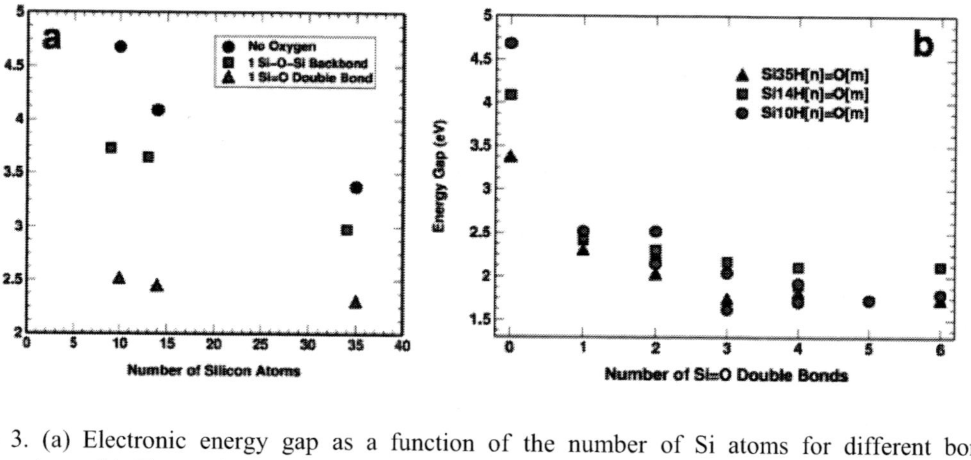

Figure 3. (a) Electronic energy gap as a function of the number of Si atoms for different bond configurations. (b) Electronic energy gap as a function of the number of Si=O double bonds at the cluster surface. Circles: $Si_{10}H_n{=}O_m$, squares: $Si_{14}H_n{=}O_m$, and triangles: $Si_{35}H_n{=}O_m$. After Ref. [4].

In addition to the variation of the energy spectrum, also the wavefunctions of the electron and of the hole are influenced by the quantum confinement in the dot. Overlap of the wavefunctions in real space and spreading of them in the k-space causes an increase in the optical transition probability (fig. 4).

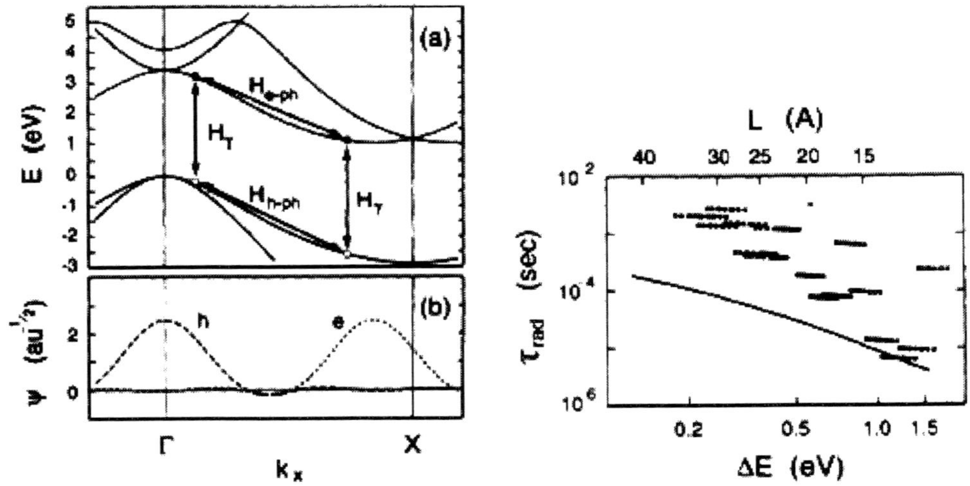

Figure 4. Spread of the electron and hole wavefunction in a quantum dot (left) and calculated radiative lifetime versus size for silicon quantum dots (right). The line corresponds to phonon-assisted recombination and the points to phonon-less recombination. After Ref. [5].

A quite indirect effect of the reduced dimensionality of Si-nc is the reduction of the effective RI n of the composite matrix where the Si-nc are formed. Indeed, in a simple scheme, one can consider that the RI is an average of the RI of Si-nc and of the host matrix. Thus, a composite layer formed by Si-nc with a large RI of 3.5 dispersed in a matrix such as SiO_2 with a RI of 1.45 has a smaller RI than bulk silicon. The exact value depends on the Si-content in the film and on the composition of the host dielectrics. A simple approximation is usually used, the Bruggeman approximation [6], where it is assumed that small particles of silicon are dispersed in an host of dielectric function ε_M:

$$f\frac{\varepsilon-\varepsilon_{eff}}{\varepsilon+2\varepsilon_{eff}}+(1-f)\frac{\varepsilon_M-\varepsilon_{eff}}{\varepsilon_M+2\varepsilon_{eff}}=0 \quad (2)$$

where f is the volumetric fraction of Si and ε is its dielectric function. With this formula, $n = (\varepsilon_{eff})^{1/2}$ can be calculated. The reduction in the RI has a positive influence when the Si-nc are used to enhance the emission efficiency of silicon. In fact, the extraction efficiency of light from a material depends on its RI: the smaller n is, the larger is the light gathered.

Various ways have been used in the past to form low dimensional silicon (fig. 5). We can divide them into three main approaches – (i) direct synthesis of silicon cluster, (ii) formation of a silicon rich layer and subsequent thermal induced phase separation, and (iii) electrochemical erosion of bulk silicon.

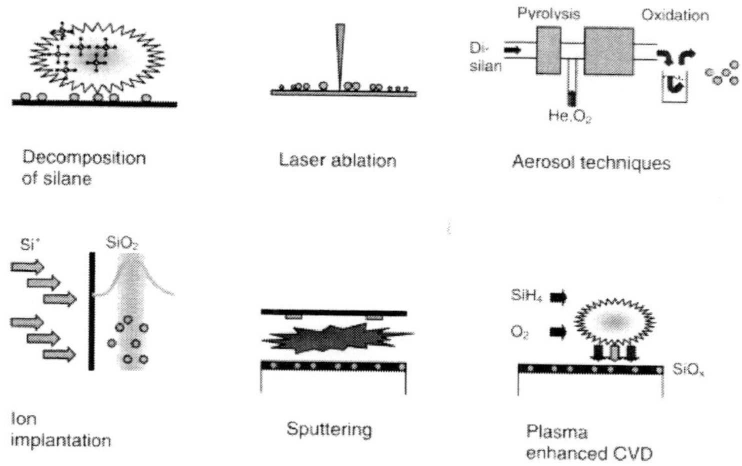

Figure 5. Silicon nanocrystal fabrication techniques. The three lower techniques produce Si-nc in a SiO_2 matrix. After Ref. [7].

2.1. Direct Synthesis of Silicon Clusters

Silicon nanoclusters can be directly synthesized by chemical reactions of suitable reactants. Methods of synthesis of Si colloids include pyrolysis of Si_2H_6 [8], formation by laser induced plasma in SiH_4 [9], combustion of SiH_4 [10], gas evaporation of Si [11] and controlled nucleation inside inverse micelles [12]. The end product is a colloidal suspension of Si-nc in a solvent which is usually ethanol.

Another way is to deposit Si-nc on a substrate after their formation in a molecular beam [13]. Within this approach, one can also mass filter the cluster in the cluster beam by time of flight methods which ends-up with a size selection of the clusters. However, during the deposition, it is not clear whether the cluster remains isolated or collapses with the formation of large mass of Si-nc

2.2. Silicon Clusters Produced by Phase Separation

The formation of Si-nc by phase separation of silicon rich dielectrics is the most widely used approach. Various techniques have been employed to form the sub-stoichiometric dielectric: ion implantation, plasma or low-pressure chemical vapour deposition (PE-CVD or LP-CVD), sputtering, and silicon evaporation. The differences among them are related to the degree of purity of the film (better for ion implanted samples), to the degree of defects incorporated, to the porosity of the deposited films (which is known as greater for the sputtered samples), to the control of the Si-nc content profile (better for the CVD films). After deposition, the films are thermally treated at high temperature to induce the phase separation between silicon and the dielectrics. At the used temperatures, silicon diffusivity is large enough to cause clustering, but it is small enough to avoid significant bulk reconstruction. This balance is clearly dependent on the annealing temperature. For example, if SiO_2 is used as the dielectric, typical temperatures for the formation of Si-nc are between 900–1200 °C. At lower temperatures, the Si-nc do not crystallize. Temperatures larger than 1000 °C are necessary to form small crystallites [14].

Figure 6 reports the evolution of the crystalline fraction as a function of the annealing temperature. While the Si-nc gets more crystalline, their size increases but their density stays almost constant. Only at the highest temperatures collapse among Si-nc occurs and the density decreases. The phase separation is not complete, which means that the density of Si-nc is lower than the number, which results by taking into account all the excess silicon deposited in the film. This, in turn, means that the host is not formed by pure SiO_2, but by either a SiO_x or by a composite oxide with a larger Si content near the Si-nc [15].

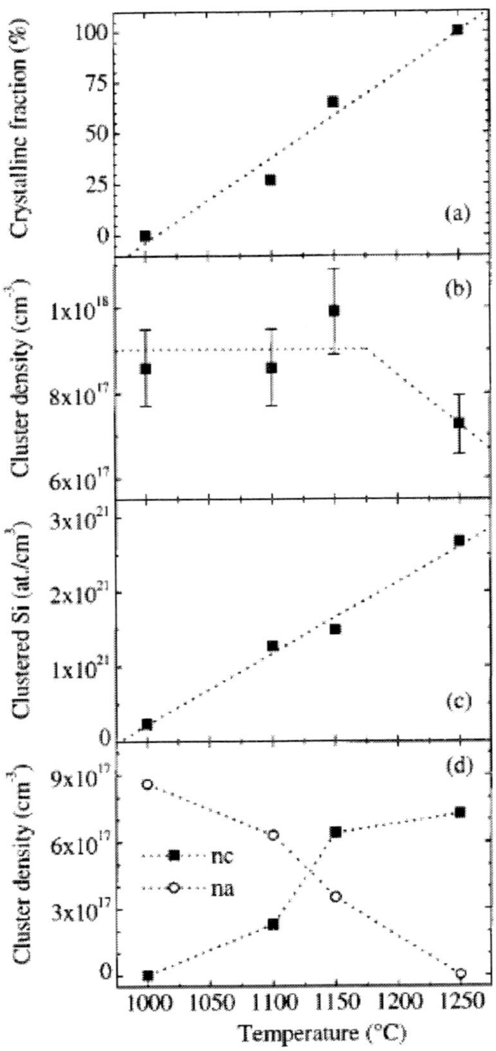

Figure 6. (a) Crystalline fraction (obtained from the comparison of energy filtered TEM and standard TEM), (b) density of Si-nc (obtained from EFTEM images), (c) concentration of clustered Si atoms and (d) density of amorphous (na) and crystalline (nc) Si-nc, as a function of the annealing temperature. The lines are drawn to guide the eye. From Ref. [14].

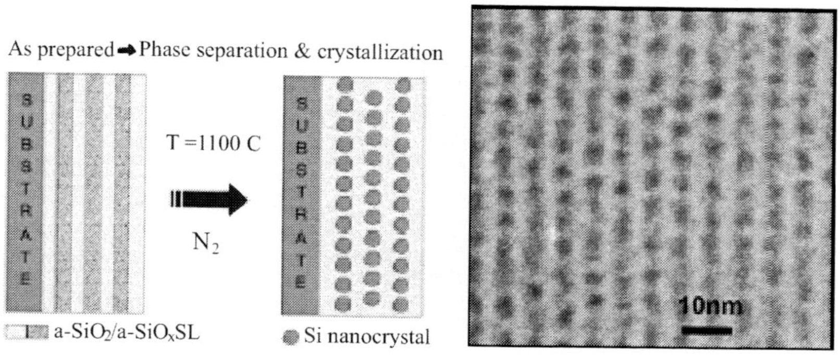

Figure 7. (Left) Fabrication of amorphous SiO/SiO_2 superlattice, and thermally induced phase separation and crystallization. (Right) Cross-sectional transmission electron microscope image of layer-arranged Si crystals (3 nm) closely separated by oxide. After Ref. [17].

The main problem related to these techniques of fabrication is the lack of control on the size dispersion of the Si-nc: large dispersion of sizes, larger than 25%, is usually obtained. Recently a method to avoid this dispersion has been proposed [16]. By depositing controlled thicknesses of Si rich amorphous silica layers separeted by pure silica layers, and by subsequent annealing treatments, Si-nc are formed whose size is fixed by the initial thickness of the Si-rich silica layers.

2.3. Electrochemical Etching of Silicon

In 1990 a paper boosted the research on low dimensional silicon [18]. This paper reported on intense room-temperature visible luminescence from porous silicon (PS) due to quantum confinement effects. Porous silicon is obtained by the electrochemical etching of silicon in an HF rich electrolyte [19]. Following the partial wafer dissolution, a porous structure is formed where the silicon skeleton is composed either by interconnected Si-nc or by thin silicon wires. What is astonishing in this processing is the fact that the etching process is self-regulated: once the porous layer is formed, no further etching of the porous layer occurs. The reason for this is the depletion of holes in the etched region of the samples. In fact, holes need to be exchanged with the electrolyte to achieve dissolution of Silicon. For this reason, the process is performed in the dark for *p*-type silicon, while it needs illumination for *n*-type silicon (fig. 8). Care has to be taken to limit the anodic current to values lower than the electropolishing current, above which electropolishing of the silicon occurs, and the final layer has a mirror-like aspect without any porous silicon on it.

The current as well as the doping of the wafer control the size of the pores, and in turn, the size of the low dimensional silicon in the silicon skeleton. The higher is the wafer resistivity, the smaller the Si-nc sizes. A rough measure of the Si-nc size is given by the porosity of the layers: porosity is defined as the void density in the film.

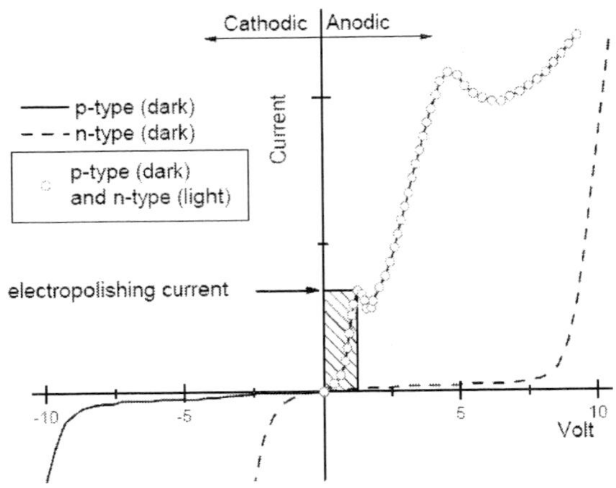

Figure 8. Typical *I-V* characteristics of an electrochemical cell for porous silicon fabrication. The shaded region corresponds to the useful regime where porous silicon can be achieved, assuming the *I-V* characteristic marked with hollow circles. In the anodic regime, the characteristics of a cell with *n*-type Si will lay in the region bounded by the characteristic in dark (dashed line) and in full light (hollow circles).

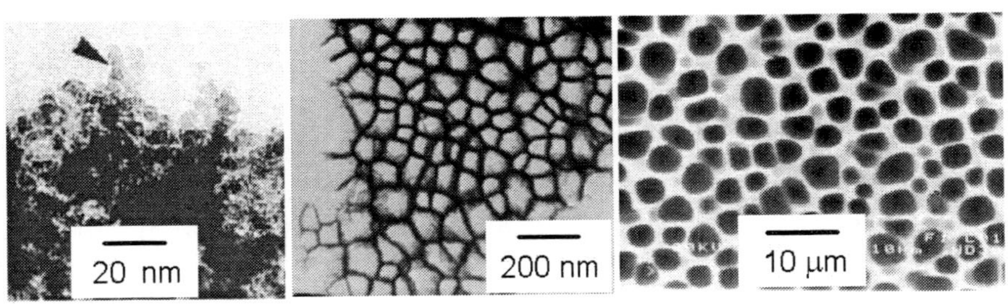

Figure 9. Examples of PS structures: microporous (left), mesoporous (center) and macroporous (right).

Table 1. Effect of anodization parameters on PS formation. As an approximate rule of thumb, the etch rate is of the order of 1 nm/s for each mA/cm^2 of the anodization current density

An increase of	Porosity	Etching rate results in a	Electropolishing threshold
HF concentration	decrease	decrease	increase
Current density	increase	increase	-
Anodization time	increase	almost constant	-
Temperature	-	-	increase
Wafer doping (*p*-type)	decrease	increase	increase
Wafer doping (*n*-type)	increase	increase	-

The porous structure of porous silicon has very appealing properties of being very sensitive to the environment. In fact, porous silicon is used to form the active material of gas and bio sensor [20]. This sensitivity to the ambient is also the main limitation of porous silicon because it causes time dependent properties (ageing effects). In addition, by suitable techniques, it is possible to form colloidal suspension of Si-nc by crumbling it into small particles. Si nanocrystal solutions have been obtained by sonication of *p*-Si in acetonitrile and toluene, in acetone and other solvents. For *in vivo* applications, however, it would be much more interesting to prepare Si-nc in water since most of their biological applications occur in aqueous environment. A simple sonication of naturally oxidized porous layer is shown to produce colloidal suspension in water [21].

3. Optical Properties of Low Dimensional Silicon

The first motivation to the study of silicon nanocrystals was the hope to get luminescent silicon. In fact, silicon has an indirect band-gap which causes a very long radiative lifetime (ms) for excited electron-hole pairs. Competing non-radiative recombinations prevail and cause most of the excited electron-hole pairs to recombine non radiatively. In addition, when the number of excited electron-holes increases, other non-radiative recombination processes start to play a role. These are Auger recombinations and free carrier absorption. In Auger recombination, an electron-hole pair recombines giving the excess energy to a third particle (electron or hole); free carrier absorption is a process for which the photon is absorbed by free carriers via an intra-band optical transition. Both are dependent on the density of excess free carriers, and dominate the recombinations for heavily excited or doped silicon [22].

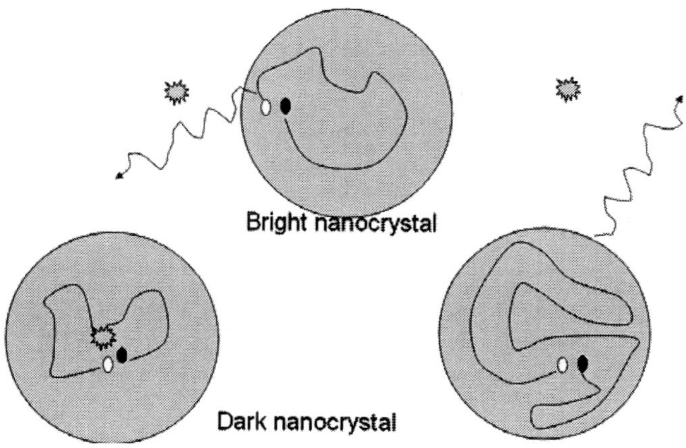

Figure 10. Schematic diagram of silicon nanocrystals in an amorphous matrix. Electron-hole pairs (dots) are locally excited: if the nanocrystal has a recombination centre (star), the electron and hole recombine non-radiatively. If the nanocrystal is free from recombination centres, the electron and hole recombine radiatively. The system shown in the figure has three electron-hole pairs excited, of which, one recombines non-radiatively and two radiatively, i. e. the internal quantum efficiency of this system is 2/3 ~ 67%.

The hope in using nanocrystals was to increase the radiative recombination rate by exploiting quantum confinement. However, another effect improved the emission efficiency of Si-nc. This is the spatial localization of excited electron-hole in a small region of the sample. If this region has a killer centre, the nanocrystal is dark. On the contrary, if it is free of killer centres, the nanocrystal is bright and the excited electron-hole recombine radiatively, even though with a long lifetime. In this case, the system has locally an internal quantum efficiency of 100%. This is shown in fig. 10.

3.1. Luminescence

Room temperature emission in Si-nc is routinely observed independently on the preparation method. The emission is usually characterized by a first band centred at about 500 nm whose position is independent on the processing parameters used to form the Si-nc and a second band in the wavelength range 600–900 nm whose exact spectral position depends strongly on the process parameters (fig. 11). The first band is defect related, and can be quenched by post-growth passivation with hydrogen. It is absent in sample of high quality. The second band is related to the presence of the Si-nc: when the Si-nc size decreases due to a low Si-content in the deposited film or to a low annealing temperature treatments, the emission band shifts to the blue. On the contrary, for high Si-content in the film or high annealing temperature, the emission band shows a red-shift.

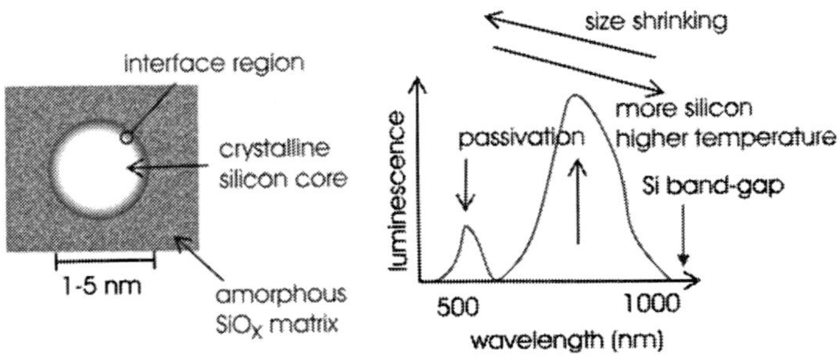

Figure 11. Schematic diagram of a Si-nc (left) and of the corresponding emission spectrum (right). The influence of the various processing parameters on the emission spectrum is shown by the arrows.

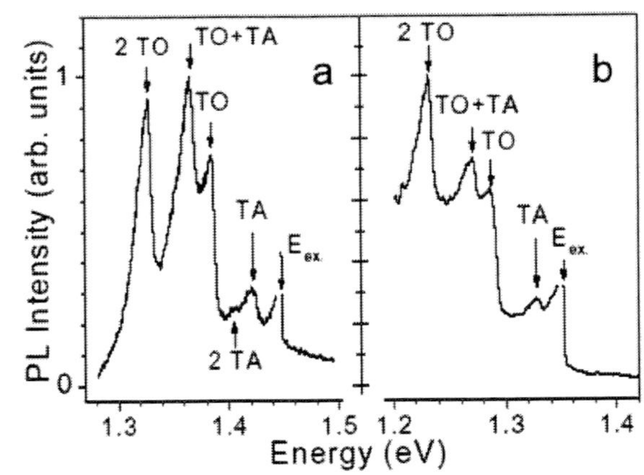

Figure 12. The resonant luminescence spectrum of naturally (a) and heavily oxidized (b) porous silicon. The arrows show the energy position of Si TA and TO momentum-conserving phonons with respect to the triplet exciton grounds state. From Ref. [23].

The exact origin of the emission is not clear. Certainly, quantum confinement effects are playing a crucial role. However, the role of the surface cannot be discarded. As we show in fig. 11, the structure of Si-nc is formed by three regions: the central region made of amorphous or crystalline silicon, the interface region made of substoichiometric and stressed silica and the embedding amorphous dielectric [15]. Resonant photoluminescence experiments have shown a structured emission spectrum which is interpreted in term of crystalline Silicon phonon assisted recombinations (fig. 12). These datas are interpreted in the frame of a pure quantum confinement process. Other experiments have shown that the emission is strongly dependent on the exposure to ambient oxygen which seems to support a key role played by silicon-oxygen bond in the emission processes (fig. 13). The most reasonable conclusion is that both mechanisms are co-present in Si-nc.

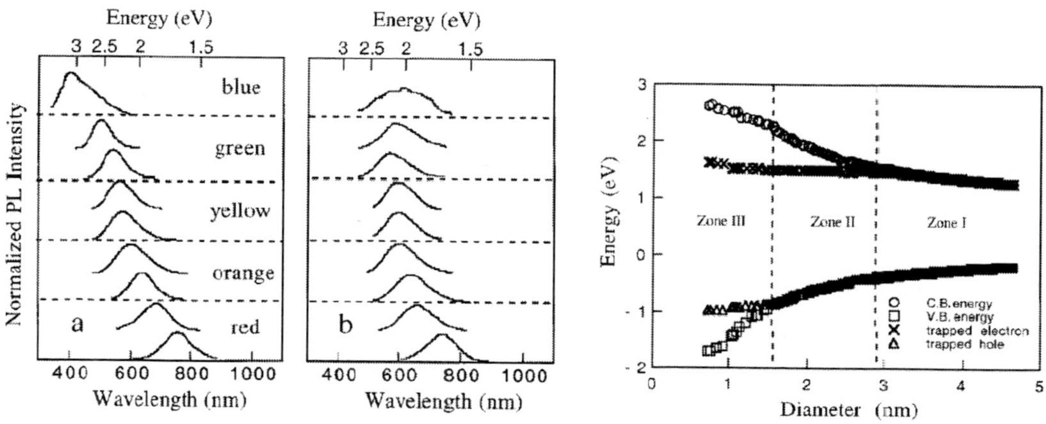

Figure 13. Measured photoluminescence spectra in various porous silicon samples before (left) and after (middle) exposure to oxygen air. The samples are labelled according to their emission wavelength. The average Si-nc size decreases from 4 nm (red) to < 2 nm (blue). Calculated conduction band and valence band energy levels (right) for H-passivated Si-nc and calculated energy levels associated with a trapped electron and a trapped hole at a Si=O bond at the Si-nc surface. As the size decreases, the band-gap increases by quantum confinement and the Si=O levels appear in the band-gap. Zone I, Zone II and Zone III mark, respectively, the regions where the Si-nc emission is predominantly band-to-band (Zone I), involves a trapped electron (Zone II) or involves a trapped electron and a trapped hole (Zone III). After Ref. [24].

3.2. Electroluminescence

From a device point of view, photoluminescence is interesting, but much more appealing is electroluminescence where light is generated by current injection into the Si-nc [25]. Here the problem is tough since carriers have to pass through a dielectric to excite the Si-nc. Indeed, in most of the reported devices, the electroluminescence is produced either by black-body radiation (the electrical power is converted into heat which raises the sample temperature which, in turn, radiates) or by impact excitation of electron-hole pairs in the Si-nc by energetic electrons which tunnel through the dielectric by a Fowler-Nordheim process (fig. 14).

Electron-hole pairs excited in this way recombine radiatively with an emission spectrum, which is very similar to that obtained by photo-luminescence. Fowler-Nordheim tunneling has very peculiar characteristics, such as a squared dependence of the current on the voltage. The problem with this kind of excitation mechanism is its inefficiency, and the damage it induces in the oxide is due to the energetic electron flow. LED based on such a unipolar injection mechanism were reported with low external efficiency of 0.1% and low turn on voltage of 5 V when the overall thickness of the active layer is lower than 25 nm [26].

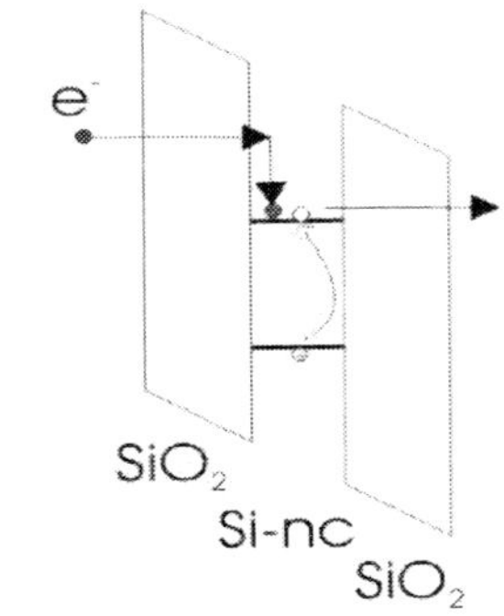

Figure 14. Schematic view of the process of generation of electron-hole pairs in Si-nc by impact excitation. Here an energetic electron tunnel trough the oxide under a strong electric field. When in the Si-nc, it losses energy by exciting an electron-hole pair. Then it tunnels again into another Si-nc driven by the high field strength. The impact-excited electron-hole pair in the Si-nc may then recombine radiatively.

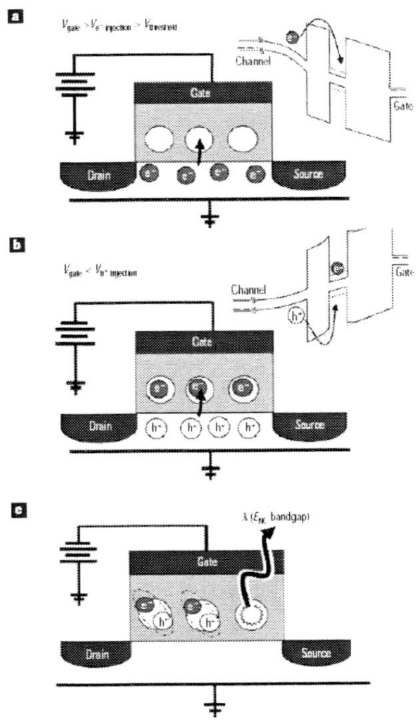

Figure 15. Schematic of the field-effect electroluminescence mechanism in a silicon nanocrystal floating-gate transistor structure. Inset band diagrams depict the relevant tunneling processes; (a)–(c). The array of Si-nc embedded in the gate oxide of the transistor can be sequentially charged with electrons (a) by Fowler-Nordheim tunneling, and holes (b) via Coulomb field enhanced Fowler–Nordheim tunneling to prepare excitons that radiatively recombine (c). From Ref. [28].

However, to get high electroluminescence efficiency, one should try to get bipolar injection. What most impedes this is the fact that the effective barrier for tunneling of electrons is much smaller than the one for holes. Such separate tunneling of electrons and holes is possible and well known in the literature (e. g., Ref. [27] and reference therein), and has been proved also in the recent paper [28]. Here by the use of a single layer of Si-nc formed in a very thin gate of FET transistor, bipolar injection has been achieved by changing the sign of the gate voltage (fig. 15). When the voltage is positive, electrons tunnel into the Si-nc; when the gate is negative, holes tunnel into the Si-nc. By periodically changing the gate voltage, the Si-nc are filled with electrons and holes, and electroluminescence occurs. After each switching of the gate voltage, electroluminescence goes to a maximum, and then decays as the other sign charge carriers are consumed.

Recently, a breaktrough of 1.6 % external quantum efficiency in Si-nc LED has been reported [29]. The idea was to decrease the tunneling barrier by replacing SiO_2 by Si_3N_4 and to block one charge carrier in the active region by using a suitable material on top of the diode; in this work amorphous SiC was used.

3.3. Absorption and RI

In order to exploit the optical properties of Si-nc in photonics, another parameter of paramount importance is the RI of the Si-nc rich oxide. A simple picture of it would be to assume an effective medium where the RI is given by the Bruggemann formula (eq. (2)).

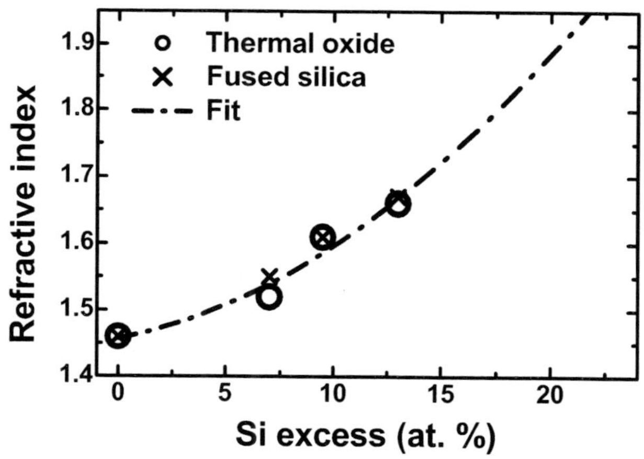

Figure 16. RI of the active layer as a function of the introduced Si excess. The values obtained by m-lines measurements at 632.8 nm have been interpolated by a polynomial fit. From Ref. [30].

However, this simple scheme is valid to get a rough estimate, but fails when accurate values are needed. In fact, the embedding matrix is not simply composed by SiO_2, and the clustering of excess Si into Si-nc influences the RI value. Figure 17 shows that the RI increases as the phase separation between silicon and silica proceeds. In addition, nitrogen is usually found in PECVD deposited Si-nc. This causes the formation of a silicon-oxynitride layer or of a three component matrix with Si, SiO_2 and Si_3N_4. The presence of nitrogen in the film increases the RI of the layer significantly (see Table 2).

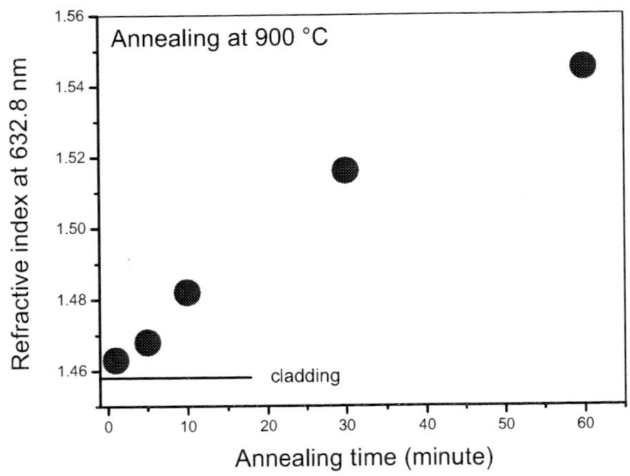

Figure 17. Samples obtained by sputtering with a Si excess = 6–7%, annealing temperature 900 °C. RI measured at 633 nm for various annealing times. Samples courtesy of Fabrice Gorbilleau (Caen).

Table 2. Atomic contents of silicon (Si), Oxygen (O) and Nitrogen (N) (from RBS measurements), measured RI at 1250 °C annealing temperature, estimated RIs from Bruggemann approximation (assuming Si-nc in SiO_2 medium or Si-nc in SiO_2-Si_3N_4 medium) [31].

Atomic Si content	**Atomic O content**	**Atomic N content**	**Measured RI for a 1250 °C annealing**	**RI from Bruggemann approximation**	
(Si at %)	(O at %)	(N at %)		Si-SiO_2	Si-SiO_2-Si_3N_4
46	45	9	2.00	1.91	2.00
42	48	10	1.82	1.75	1.80
39	49	12	1.66	1.64	1.66

3.4. Waveguides

The large RI of the Si-nc rich layer allows their use as core material in optical waveguides. Slab waveguides are easily realized as well as channel waveguides (fig. 18). More sophisticated designs have also been proposed, such as slot waveguides, which are formed by two nearby SOI ridge waveguides, and where the space between the silicon cores is filled with the active Si-nc rich layer [32].

One of the most important parameters in these waveguides is the propagation losses. Signal light, which travels trough the waveguide, can loose power due to various mechanisms, viz. direct absorption in the Si-nc, scattering due to roughnesses at the core boundaries (both between the core/cladding interface and at the stripe edges), radiation into the substrate, and scattering due to the composite nature of the core layer. This last loss mechanism is of worth some discussion. The presence of Si-nc, while increasing the average RI of the core layer, causes large local variations in the core RI. These variations occur on a length scale smaller than the wavelength of the optical mode. Thus, Rayleigh scattering occurs. In reality, as the RI mismatch between Si-nc and the dielectric is large, one has to use the complete Mie scattering theory to estimate the effect of this process on the optical losses.

Measurements on ridge waveguides formed by ion-implantation and optical lithography have shown that at 780 nm and for sufficiently wide ridges, most of the propagation losses are due to direct absorption in the Si-nc [30]. In fact, propagation losses of 11 dB/cm have been measured, which can be divided into 2 dB/cm due to Mie scattering, 2 dB/cm due to interface scattering and 9 dB/cm due to Si-nc absorption (α). Knowing that $\alpha = N_{\text{Si-nc}}\sigma$, an estimate of the absorption cross-section of Si-nc can be obtained: $\alpha_{abs} = 5\times10^{-19}$ cm^2 [30].

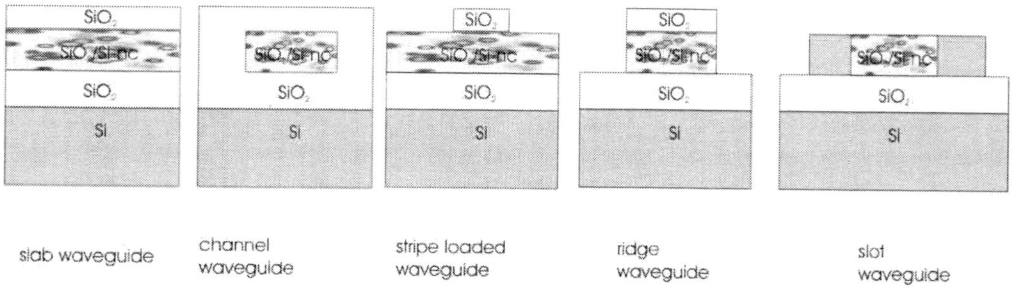

Figure 18. Various geometries of waveguides. The first is a planar waveguide where optical confinement is achieved only in the vertical direction. The others are two dimensional waveguides where optical confinement is achieved both in the vertical and in the horizontal directions.

3.5. Optical Gain

Active devices, such as lasers or amplifiers, need optical gain in the active material. So, the observation of optical gain is of paramount importance. Optical gain is related to population inversion. Indeed, in a simple scheme, one can write

$$g_M = \sigma_{em} N_2 - \sigma_{abs} N_1 \tag{3}$$

where g_M is the material gain, σ_{em} is the emission cross-section, N_2 and N_1 are, respectively, the excited and ground state densities of active centres.

To measure optical gain, one method is to use the luminescence of Si-nc as a probe beam, and look for the enhancement of it as it propagates in an optically pumped waveguide. A practical implementation of this idea is given by the variable stripe length (VSL) method.

Figure 19 shows on the left the idea of the VSL method: a pump beam is used to excite through the top of an active planar waveguide, the luminescence emitted by the waveguide is travelling through a region excited whereby stimulated emission is amplified. If one changes the extent of the pumped region, one will record an exponentially increasing luminescence signal coming out from the edge of the sample. This luminescence is due to the amplified spontaneous emission (ASE) when the optical gain is present. By modelling the system within a one dimensional amplifier scheme, the intensity of the ASE, I_{ASE}, as a function of the excited length l, is given by

$$I_{ASE}(l) = \frac{J_{sp}(\Omega)}{g_{\text{mod}}}\{\exp(\Gamma g_M - \alpha)l - 1\} \tag{4}$$

where J_{sp} is the spontaneous emission intensity, Ω the collection solid angle, g_{mod} the net modal gain which is related to the material gain g_M as $g_{mod} = \Gamma g_M - \alpha$, where Γ is the optical mode confinement factor (fraction of the optical power carried by the optical mode which is confined in the active layer) and α is the propagation loss in the waveguide [34]. When the active layer is not excited enough, no population inversion occurs, and the net modal gain is negative. In this case, I_{ASE} saturates as the pumping length increases. Right side of fig. 19 shows the evidence of the two regimes; red points refer to amplification while green points refer to absorption.

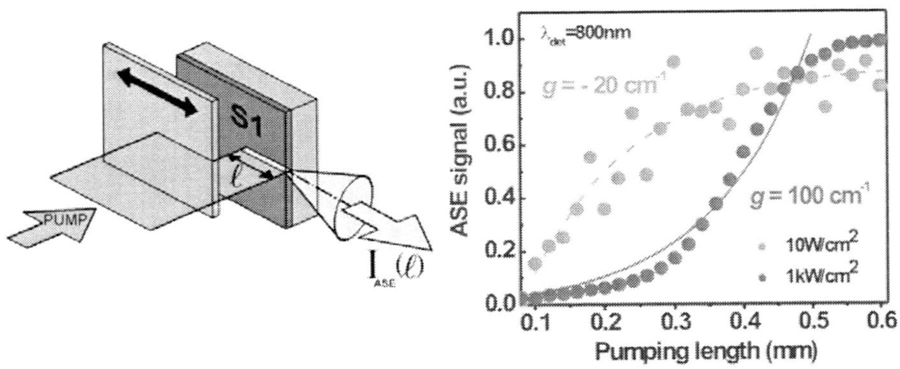

Figure 19. (Left) schematics of the VSL method. (Right) ASE versus the pumping length for two pumping powers at a wavelength of 800 nm and for a PECVD slab waveguide. Ref. [33].

Another experiment we used to measure gain is a signal amplification experiment [35]. We used a thin layer of Si-nc formed on a transparent quartz substrate. If one measures the transmitted intensity, when the system is optically pumped, (I_{ON}) subtracts the photoluminescence intensity (I_{PL}) and normalizes by the incident intensity (I_o). Finally, one can write

$$T = \frac{I_{ON} - I_{PL}}{I_0} = \exp\left[-\left\{\alpha_{QZ} d_{QZ} - g_{Si-nc}(J_P) d_{Si-nc}\right\}\right] \tag{5}$$

where α_{QZ} and d_{QZ} are, respectively, the absorption and thickness of the quartz substrate, d_{Si-nc} is the thickness of the Si-nc layer, and $g_{Si-nc}(J_P)$ is the pump power density; J_P being the dependent gain coefficient.

Figure 20 reports the results of such an experiment. As the pumping rate is increased, the transmission intensity increases too. For J_P = 1.4 kWcm^{-2}, the transmitted intensity is larger than 1. This means that overall amplification of the signal beam is achieved: the transmitted intensity is larger than the incident intensity, even accounting for the losses through the substrate [35].

A summary of the optical properties of the Si-nc system is shown in fig. 21 where absorption, gain and luminescence spectra are compared. It is worth noticing the large energy difference between absorption and emission spectra; when absorption is negligible, emission (either stimulated or spontaneous) is strong. Furthermore, the gain spectrum peaks at the high energy wings of the luminescence spectrum. These datas can be explained with a four-level model of gain where lattice relaxation of Si=O double bonds at the interface of the Si-nc provides the energy for the four level model. Shifts between luminescence and gain point to a different nature of the active centres in the two cases – either two populations of Si-nc are present in the system or interface radiative recombination is responsible for gain and band-to-band recombination in large S-nc is responsible for luminescence. The exact model is still under discussion in the literature.

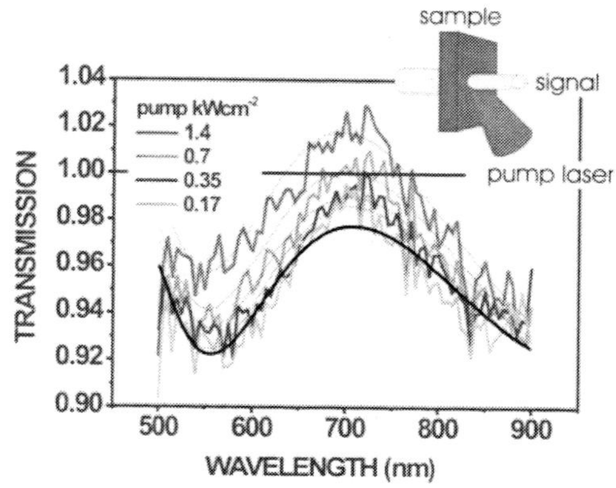

Figure 20. Transmitted intensity as a function of the pump power density. Dark line refers to the transmission of the sample without any pumping. The inset shows the schematic of the experiment. From [35].

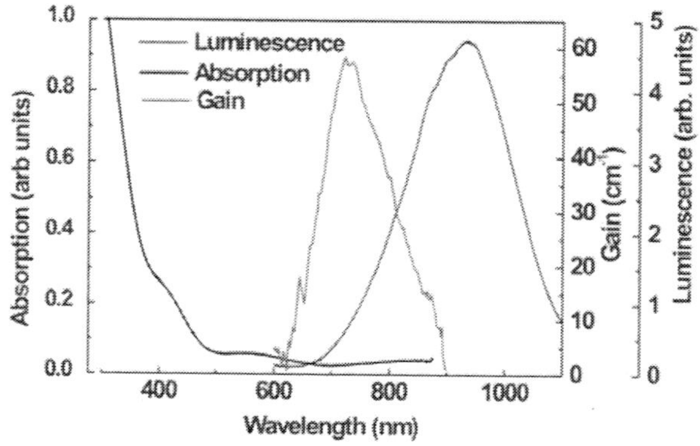

Figure 21. Summary of the optical properties of Si-nc. Black curve is the absorption spectrum, red curve is the gain spectrum, blue curve is the luminescence spectrum. Data from ref. [33].

Time resolved luminescence experiments in the waveguide geometry help to shine some further light on the stimulated emission process in Si-nc. Figure 22 reports the results of such an experiment. When the VSL geometry is used, and one measures I_{ASE} versus time, one can observe that a fast lifetime (in ns) is appearing under high pumping power and at long excitation length. This fast component is superimposed to a slower one (in μs) which is the only one remaining when either the pump power or the excitation length is decreased. This behaviour is typical for samples which show gain. Samples which do not show gain have a different behaviour, in particular, when high pump power is used, the fast component is observed for both long and short excitation lengths. These observations are due to a critical balance between stimulated emissions and non-radiative recombinations. Let us consider the lifetime due to stimulated emission,

$$\tau_{se} = \frac{4}{3}\pi R_{NS}^{3}\frac{1}{\xi\sigma_{g}cn_{ph}}.$$

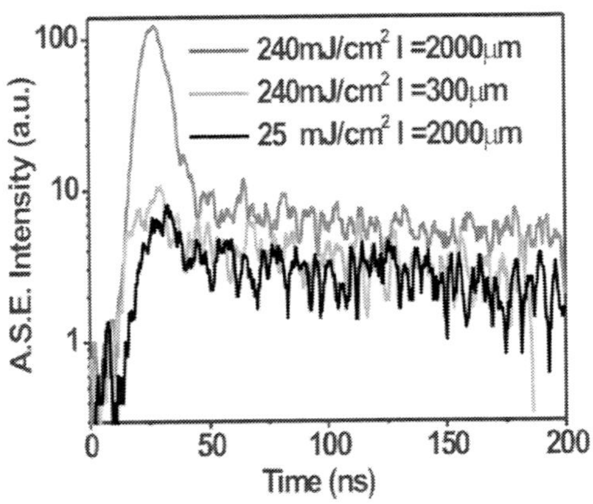

Figure 22. Time resolved edge emission from Si-nc waveguide in the VSL geometry. The ASE intensity as a function of time is plotted for different excitation conditions: black line – low power density and long excitation length, green line – high pump power density and short excitation length, red line – high pump power and long excitation length. From Ref. [36].

τ_{se} depends on the Si-nc properties, i.e. radius R_{NS}, packing density ξ, emission cross-section σ_g, and the photon flux density n_{ph} which is propagating in the waveguide. To change the photon flux in the waveguide, one has to increase the pumping rate so that to have more emitted photons, but one has to care also about the propagation losses of the waveguide which reduce the photon flux density, and thus, the stimulated emission probability ($1/\tau_{se}$). The competing and power dependent non-radiative recombination mechanisms are Auger recombination and confined carrier absorption. These processes are described by two lifetimes – $\tau_A = (2C_AN_3)^{-1}$ and $\tau_{CC} = (2C_{CC}N_3)^{-1}$, respectively [22]. Both lifetimes depend on the excited carrier density in the nanocrystals (N_3). Their probabilities can be increased by using high pumping power but not by the photon flux in the waveguide. Optical gain can be only observed when $\tau_{se}<\tau_A$, τ_{CC}. The fast component in fig. 22 is an evidence of stimulated emission in this sample: indeed, the fast lifetime appears when the pump power density and the excitation length are increased, and both these facts tend to increase the photon density in the waveguide by emission and by amplification. In the other samples, which do not show optical gain, we observed that the fast component was dependent only on the pump power, and not on the excitation length, that is only on the carrier density excited in the Si-nc. In this case, the fast component was due to Auger recombinations. This explains why in the literature sometimes gain has been observed and sometimes not. In fact, not only the properties of the Si-nc influences the stimulated emission lifetime, but also the waveguide parameters (such as losses and modal confinement) are affecting the value of τ_{se}.

4. Complex Dielectric Structures

When the dielectric function of a material is purposely varied in a periodic manner, the photon propagation properties in the material are drastically modified. Energy regions, where photon propagation along given directions is permitted or prohibited, appear and the system is described as a photonic crystal. However, not only periodic dielectric systems are of interest, but also more complex structure where the dielectric function is varied aperiodically or randomly. A simple system to produce these complex dielectric structures is porous silicon where one can vary the dielectric function in one dimension by controlling local variation of the porosity [19]. In this way, 1D photonic complex structures, can be achieved where new phenomenons on photon propagation can be observed.

In order to produce the complex dielectric structures, one has to change the current flow through the sample in a controlled manner while performing the electrochemical etching. A profile in depth of porosity is achieved, i.e. a profile in depth of the RI (fig. 23).

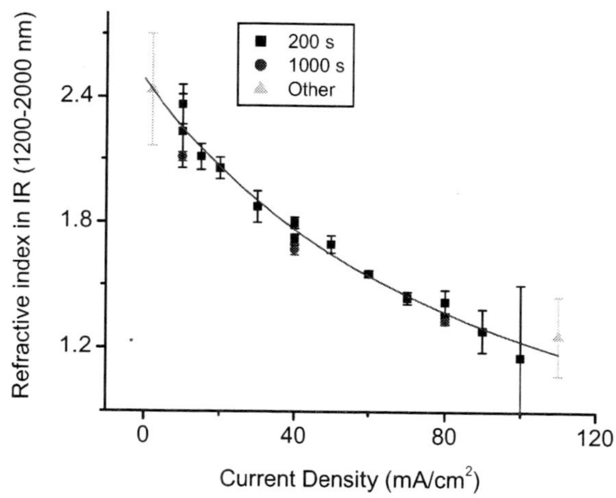

Figure 23. Measured RI versus current density used to form thin layer of PS. The various points refer to various PS thicknesses. Data are courtesy of C. J. Oton.

Multilayer dielectric filters are well known in optics [37]. An example is the dielectric Bragg reflector (DBR) mirror where, by stacking periodically $\lambda/4$ thick dielectric layers, a high reflectivity region is obtained around λ. The number of periods defines the reflectivity while the RI mismatch between each layers defines the width of the high reflectivity region. DBR can also be described as 1D periodic photonic crystals where the high reflectivity region is interpreted as the photonic band-gap. If one forms a defect in periodic crystals, or in classical optics terms, if one forms a microcavity by using two DBRs separated by a λ-thick layer, a high transmission mode appears in the stop-band, or a propagating photon state associated with the defect resonance appears in the photonic-band-gap. This is observed in the transmission spectra (fig. 24) as a high transmission peak in the 1D photonic crystal band-gap [38]. One can also notice, in the high resolution spectrum of fig. 24(b), the high quality of the structure since we got a Q-factor of 3380, which is extremely high for a microcavity based on a porous material where Rayleigh light scattering causes optical losses of 100 cm^{-1} at 1.55 μm. In the same figure, we compare the results with calculations made by using a transfer matrix method [39]. The extremely good agreement between calculations and measurements is important to validate the dielectric stack parameters (RI and thickness of each layer) used in the calculations. In this way, we can use the transfer materix method also to calculate the optical field inside the structures.

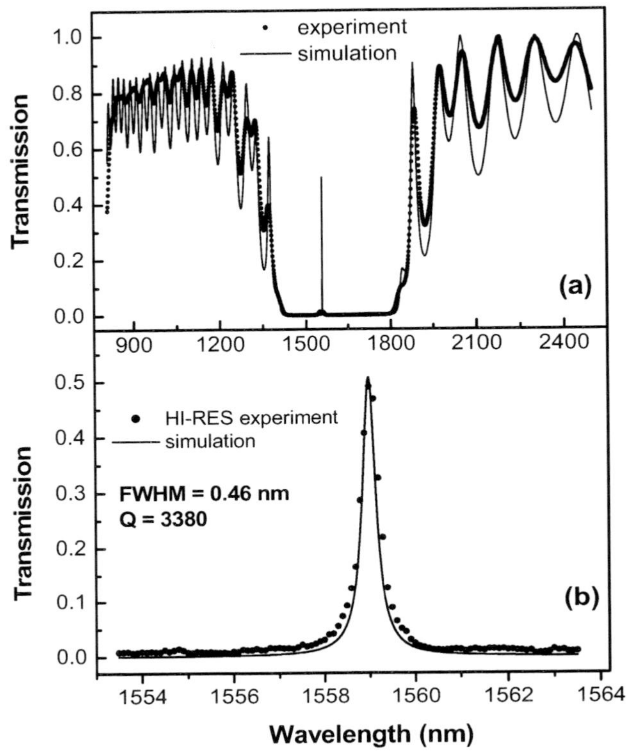

Figure 24. Transmission spectra of a single λ-thick microcavity: solid dots are the experimental data and the lines are the simulations. (a) Measurements with 1 mm spot in a wide wavelength range, (b) measurements with 35 μm spot with a high resolution (HI-RES) tuneable laser source. From Ref. [38].

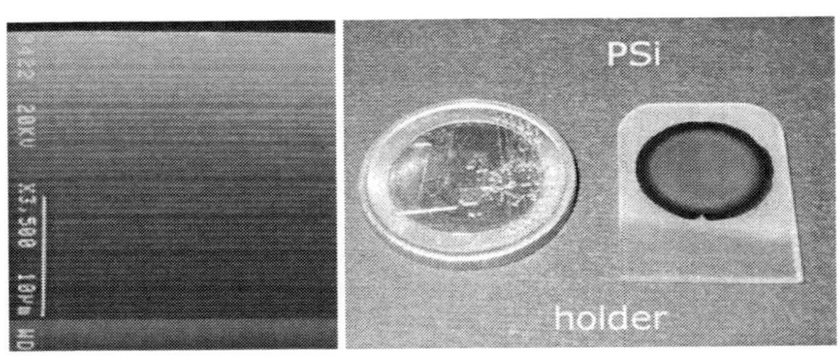

Figure 25. (Left) cross-section of a 7 coupled microcavity structure made in porous silicon. After the etching of the structure a high current pulse allows to detach the layer from the substrate so that a free standing film is obtained. The film is then glued onto a plastic holder to allow handling of the sample. (Right) optical photograph of a free standing microcavity. From Ref. [38].

The control of the ecthing process is so high that one can easily form multiple cavity layers, or in other terms, multiple defects in the photonic crystals. A SEM image of a cross-section of coupled microcavity structure is shown in fig. 25. Optically, if the reflectivity of the central mirrors is low, the single resonances couple, and a splitting of the resonance peak in a number of peaks occur.

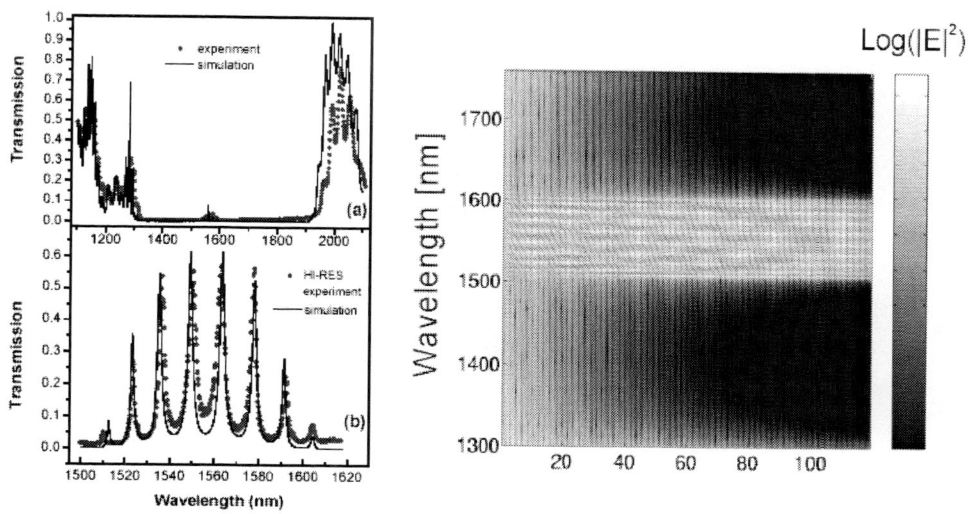

Figure 26. (Left) transmission spectra of a 10 coupled microcavity structure with $\lambda/2$-thick cavities sandwiched between 4.5-period external DBRs (dots) and numerical calculations (lines). (a) Wide range measurements with 1 mm spot size, (b) HI-RES measurement with 35 μm diameter spot; from Ref. [38]. (Right) results of transfer matrix calculations of the scattering state electric field associated with the coupled microcavity structure.

This is observed in fig. 26 where the transmission of a 10 coupled cavity sample is reported. In the wide stop band region, a transmission band forms [fig. 26(a)]. A high resolution spectrum of the sample shows that this band has high transmission, and that, it is split into a number of nearby peaks – one for each cavity. Also, for this sample, the agreement between calculations and experimental data is remarkably good. We used the same parameters to get the optical mode intensity map inside the sample. This is shown in fig. 26 right for a wavelength region inside the photonic band-gap region [40]. It is observed that a band of photonic coupled states forms inside the photonic band-gap. Photons can propagate through the stop band due to the presence of coupled resonances. It is very interesting to compare these systems: single microcavity and multiple coupled microcavities, with their electronic counterparts: single isolated quantum well and superlattices. In both cases, the presence of a layer within barrier materials (DBR in the photonic case, high band-gap material in the electronic case) causes the appearance of a state in an otherwise forbidden energy region. In both cases, if one stacks more cavities/wells, coupling of the single states occurs, and a miniband is formed where particles can freely move. For this reason, we call the coupled microcavity system as *optical superlattice*. If one uses this analogy further, it is tempting to try to reproduce optically electrical phenomenons. The key ingredient will be to develop the optical analogue of an electric field, i. e. an external means to tilt the energy band of the optical superlattice. To this aim, we notice that the resonance wavelength position is given by $\lambda = nd = \delta$. So, if one change n or d or both through the structure, one is able to change locally the resonance wavelength. If the optical path δ is increased/decreased linearly, then $\lambda + \Delta\lambda = \delta + \Delta\delta$, which ends with the tilting of the optical superlattice miniband [41]. We call $\Delta\delta$ as the optical path gradient. This simple reason is confirmed by the calculation of the optical superlattice with a gradient. This is shown in fig. 27, where an optical gradient of 14% induces a significant tilting of the photonic structures.

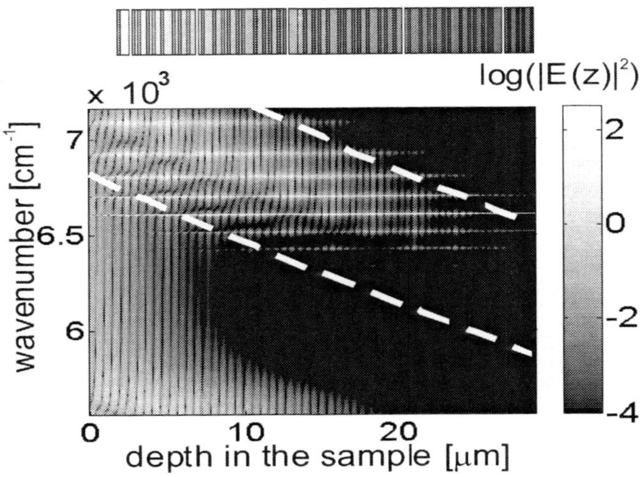

Figure 27. Transfer matrix calculations of an optical superlattice with a 14% optical path gradient applied along the sample. The lines show the tilting in the photonic gap structures. It is worth noting that the effect of the applied optical path gradient causes the formation of a Wannier Stark ladder of photonic states inside the photonic band-gap. From Ref. [41].

In the first paper on electronic superlattice, it was suggested that the demonstration of the formation of a miniband of electronic state was the observation of electronic Bloch oscillations [42]. These are due to the fact that, when an electron is accelerated by a static electric field, it increases energy up to the band edge where Bragg reflections take place. So, in the presence of a static electric field, the electron starts to oscillate back and forth. A static bias produces an oscillating current. It is very difficult to observe Bloch oscillations in a bulk material due to the high-energy bandwidth of the conduction bands. The situation is more favourable in a superlattice where the folding of the Brilluoin zone causes the formation of minibands few hundreds of meV wide. This phenomenon (predicted in the mid seventies) has been observed only at the beginning of the nineties due to the small coherent time of carriers in semiconductors [43]. Photons have the advantage of longer coherent times. The possibility of tilting the photonic miniband in an optical superlattice makes possible the observation of Bloch oscillations. Photons are Bragg reflected by the tilted miniband edges, which is observed as the appearance of a train of pulses when a short pulse is injected into the photonic structures. Figure 28 reports the first measurements of this oscillating light produced by a single incoming pulse [41].

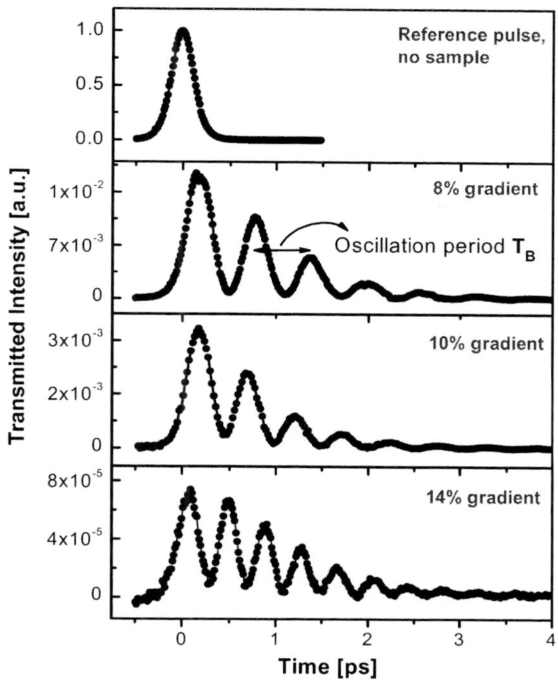

Figure 28. Time resolved transmitted intensity of a short photon pulse through a biased optical superlattice. The different panels refer to different samples with an increasing value of the optical path gradient. Note the drop in the transmitted intensity as the gradient increases, which is an evidence of the increased localization of the photons inside the tilted superlattice. The photons get Bragg reflected inside a thinner region as the gradient increases, and hence, the tilting of the miniband increases. This is also observed in a decrease of the Bloch oscillation periods. From Ref. [41].

The anlogy between electrons and photons can be tested even further if one considers now the effect of a bias (optical path drift) on a two miniband optical superlattices (fig. 29, which reports calculations). For zero bias, the photonic structure is characterized by two minibands of propagating states. A transmission experiment will measure two transmission bands in the photonic band-gap. If one starts to bias the optical superlattice (increase the optical path gradient), a ladder of localized photonic states appears (fig. 29(b)). A transmission experiment will measure a drastic drop in the transmission with the appearance of a series of equally spaced weak peaks. This spectrum reflects the lack of propagating states in the structure. If the bias is increased even further, the first Wannier Stark ladder state of one miniband resonantly coulpes with the last Wannier Stark state of the other miniband. This coupling causes the formation of an extended state through the whole structure, and photons can tunnel through the sample. Indeed, transmission experiments show that a strong peak appears. This fact is a manifestation of the resonant photonic Zener tunneling. The electronic counterpart of this phenomenon is the Zener tuneling of the electrons in a biased superlattice or, more lousy, the Zener tunneling of electrons from the valence band to the conduction band in a reverse biased diode in breakdown.

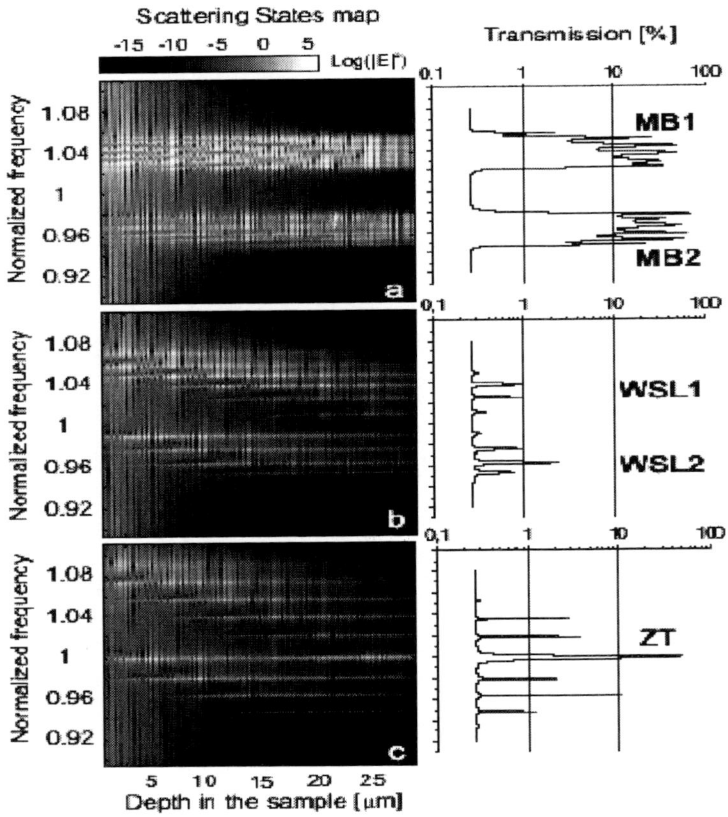

Figure 29. Transfer matrix calculations of the intensity distribution of the light inside the sample. The intensity is plotted as a colour scale versus the normalized frequency ω_0 and depth inside the sample, with ω_0 =192.3 THz. (a) Flat miniband situation, $\Delta\delta = 0\%$, (b) optical Wannier Stark ladder, $\Delta\delta = 6.7\%$, (c) resonant Zener tunneling (ZT), $\Delta\delta = 10.3\%$. The calculated transmission spectra are shown in the right panels (a slight offset of 0.27% is used to allow for the use of a logarithmic scale). The blue shift of the spectral features with increasing gradient is due to the reduced optical thickness of the layers. From Ref. [40].

The experimental evidence of Zener tunneling is shown in fig. 30. The various pannels refer to different applied bias (optical path gradient). For no bias, the transmission spectrum shows two transmission bands; when bias is applied, the transmission drops due to the formation of localized Wannier Stark ladder states. For a critical bias value, a resonant coupling (Zener tunneling) between Wannier Stark states occurs and a strong Zener tunneling peak appears. As the bias is increased further, the coupling is removed and the transmission drops again. If we plot the transmission intensity at the resonat wavelength as a function of the optical path gradient value, a plot similar to an *I-V* characteristics of an inversed biased diode in breakdown is observed (fig. 31).

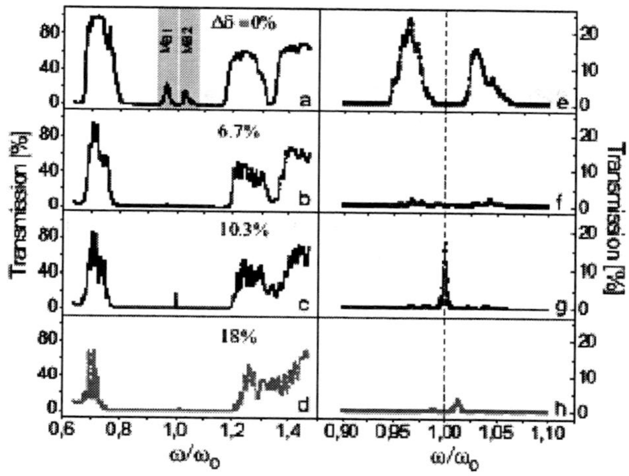

Figure 30. Transmission spectra of optical superlattices: (a) flat band situation $\Delta\delta = 0\%$, (b) $\Delta\delta = 6.7\%$, (c) $\Delta\delta = 10.3\%$, (d) $\Delta\delta = 18\%$. (e)–(h) a zoom of the wide range spectra around the miniband region. Resonant Zener tunneling is observed in (c) and (g) as an enhanced transmission peak in the centre of the minigap. From Ref. [40].

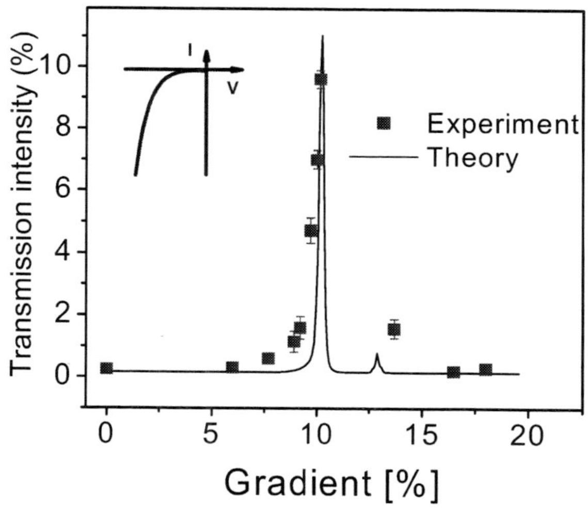

Figure 31. Transmission intensity at the resonant wavelength as a function of the optical path gradient for an optical superlattice. At a critical gradient value, a strong increase is observed. Points are experimental data, line is a numerical calculation. The inset shows qualitatively the *I-V* charcteristic of a reverse biased diode. From Ref. [40].

5. Conclusion

In this review, some of the many applications of Si-nc in photonics are described. Both quantum size effects, the new chemestry which occurs at the Si-nc sruface, the tunability of the dielectric functions by changing the composition of the systems, all allow to generate new phenomenons which can be eventually used to add new functionalities to silicon.

Acknowledgements

I acknowledge the help of many co-workers both from my resarch group and from international collaborations. They can be recognized in the cited literature. The investigations discussed in this review have been financially supported by many programs, viz. from the EC the research program SINERGIA, SEMINANO, PHOLOGIC, from MIUR through FIRB (RBNE01P4JF and BNE012N3X) and COFIN (2004023725) projects and by PAT through PROFILL project.

References

[1] Pavesi, L.; Lockwood, D. (eds.) Silicon photonics; Topics in applied physics; Springer-Verlag: Berlin, 2004, Vol. 94.

[2] Ossicini, S.; Pavesi, L.; Priolo, F. Light emitting silicon for microphotonics; Springer Tracts in Modern Physics; Springer-Verlag: Berlin, 2003, Vol. 194.

[3] Lockwood, D.J.; Aers, G.C.; Allard, L.B.; Bryskiewicz, B.; Charbonneau, S.; Houghton, D.C.; McCaffrey, J.P.; Wang, A. *Can. J. Phys.* 1992, 70, 1184.

[4] Luppi, M.; Ossicini, S. *J. Appl. Phys.* 2003, 94, 2130.

[5] Hybertsen, M.S. *Phys. Rev. Lett.* 1994, 72, 1514.

[6] Theiβ, W. *Surf. Sci. Rpts.* 1997, 29, 91.

[7] Bisi, O.; Campisano, S.U.; Pavesi, L.; Priolo, F. Silicon based microphotonics: from *Basics to applications;* IOS Press: Amsterdam, 1999, pp. 47–86.

[8] Littau, K.A.; Szajowski, P.J.; Muller, A.J.; Kortan, A.R.; Brus, L.E. *J. Phys. Chem.* 1993, 97, 1224.

[9] Fojtik A.; Giersig, M.; .Henglein, A. *Ber. Buns. Phys. Chem.* 1993, 97, 1493.

[10] Fojtik, A.; Henglein, A. *Chem. Phys. Lett.* 1994, 221, 363.

[11] Iwasaki, S.; Ida, T.; Kimura, K.; *Jpn. J. Appl. Phys.* 1996, 35, L551.

[12] Wilcoxon, J.P.; Samara, G.A.; Provencio, P.N.; *Phys. Rev.* B 1996, 60, 2704.

[13] Leduox, G.; Guillois, O.; Porterat, D.; Reynaud, C.; Huisken, F.; Kohn, B.; Paillard, V. *Phys. Rev.* B 2000, 62, 15942.

[14] Iacona, F.; Bongiorno, C.; Spinella, C.; Boninelli, S.; Priolo, F. *J. Appl. Phys.* 2004, 95, 37231.

[15] Daldosso, N.; Luppi, M.; Ossicini, S.; Degoli, E.; Magri, R.; Dalba, G.; Fornasini, P.; Grisenti, R.; Rocca, F.; Pavesi, L.; Boninelli, S.; Priolo, F.; Bongiorno, C.; Iacona, F. *Phys. Rev. B* 2003, 68, 085327.

[16] Zacharias, M.; Heitmann, J.; Scholz, R.; Kahler, U.; Schmidt, M.; Bläsing, J. *Appl. Phys. Lett.* 2002, 80, 661.

[17] Zacharias, M.; Yi, L.X.; Heitmann, J.; Scholz, R.; Reiche, M.; Gösele, U. *Solid State Phenomena* 2003, 94, 95.

[18] Canham, L.T. *Appl. Phys. Lett.* 1990, 57, 1046.

[19] Bisi, O.; Ossicini, S.; Pavesi, L. *Surf. Sci. Rpts.* 2000, 264, 1–126.

[20] Pancheri, L.; Oton, C.J.; Gaburro, Z.; Soncini, G.; Pavesi, L. *Sensors and Actuators B* 2003, 98, 237–239.

[21] Froner, E.; Adamo, R.; Gaburro, Z.; Margesin, B.; Pavesi, L.; Rigo, A.; Scarpa, M.J. *Nanoparticle Res.* (2006).

[22] Pavesi, L. *Materials Today* 2005, 8, 18–25.

[23] Kovalev, D.; Heckler, H.; Polisski, G.; Koch, F. *Phys. Stat. Sol.* B 1999, 215, 871.

[24] Wolkin, M. *et al. Phys. Rev. Lett.* 1999, 82, 187.

[25] Gaburro, Z.; Pavesi, L. Light emitting diodes for Si integrated circuits *In* Handbook of Luminescence, Display Materials and Devices. Vol. 3, *Display devices* (Nalwa, H.S.; Rohwer, L.S. eds.) American Scientific Publishers: Stevenson Ranch, USA, 2003, 101.

[26] Iacona, F. *et al. Appl. Phys. A: Mater. Sci. Process.* 2002, A74, 1.

[27] Cai, J.; Sah, C.-T. *J. Appl. Phys.* 2001, 89, 2272.

[28] Walters, R.J.; Bourianoff, G.I.; Atwater, H.A. *Nature Materials* 2005, 4, 143.

[29] Kwam, S.C.; Park, N.-M.; Kim, T.-Y.; Kim, K.-H.; Sung, G.Y.; Shin, J.H. *Appl. Phys. Lett.* 2005, 86, 071909.

[30] Pellegrino, P.; Garrido, B.; Garcia, C.; Arbiol, J.; Morante, J.R.; Melchiorri, M.; Daldosso, N.; Pavesi, L.; Schedi, E.; Sarrabayrouse,G. *J. Appl. Phys.* 2005, 97, 074312.

[31] Prakash, G.V.; Cazzanelli, M.; Gaburro, Z.; Pavesi, L.; Iacona, F.; Franzò, G.; Priolo, F. *J. Mod. Opt.* 2002, 49, 719.

[32] Almeida, V.R.; Qianfan, X.; Barrios, C.A.; Lipson, M. *Opt. Lett.* 2004, 29, 1209.

[33] Negro, L.D.; Cazzanelli, M.; Daldosso, N.; Gaburro, Z.; Pavesi, L.; Priolo, F.; Pacifici, D.; Franzò, G.; Iacona, F. *Physica E* 2003, 16, 297.

[34] Negro, L.D.; Bettotti, P.; Cazzanelli, M.; Pavesi, L.; Pacifici, D. *Opt. Commun.* 2003, 229, 337.

[35] Negro, L.D.; Cazzanelli, M.; Danese, B.; Pavesi, L.; Iacona, F.; Franzò, G.; Priolo, F. *J. Appl. Phys.* 2004, 96, 5747.

[36] Negro, L.D.; Cazzanelli, M.; Pavesi, L.; Ossicini, S.; Pacifici, D.; Franzò, G.; Priolo, F.; Iacona, F. *Appl. Phys. Lett.* 2003, 82, 4636.

[37] Born, M.; Wolf, E. Principles of optics, Cambridge University Press: Cambridge, UK, 1980.

[38] Ghulinyan, M.; Oton, C.J.; Bonetti, G.; Gaburro, Z.; Pavesi, L.; *J. Appl. Phys.* 2003, 93, 9724.

[39] Oton, C.J.; Negro, L.D.; Bettotti, P.; Pancheri, L.; Gaburro, Z.; Pavesi, L. *In* Radiation-matter interaction in confined systems (Andreani, L.C.; Benedek, G.; Molinari, E. eds.) *Società Italiana di Fisica:* Bologna, 2002, 303.

[40] Ghulinyan, M.; Oton, C.J.; Gaburro, Z.; Pavesi, L.; Toninnelli, C.; Wiersma, D. *Phys. Rev. Lett.* 2005, 94, 127401.

[41] Sapienza, R.; Costantino, P.; Wiersma, D.; Ghulinyan, M.; Oton, C.J.; Pavesi, L. *Phys. Rev. Lett.* 2003, 91, 263902.

[42] Esaki, L.; Tsu, R. *IBM J. Res. Dev.* 1970, 61, 16.

[43] Feldmann, J. et al. *Phys. Rev. B* 1992, 46, 7252; Leo, K. et al. *Solid. State Comm.* 1992, 84, 943.

Lorenzo Pavesi is a Professor of Experimental Physics at the University of Trento (Italy). Born the 21st of November 1961, he received his Ph.D. in Physics in 1990 at the Ecole Polytechnique Federale of Lausanne (Switzerland). In 1990 he became Assistant Professor, an Associate Professor in 1999 and Full Professor in 2002 at the University of Trento. He teaches several classes both at the Science as well as at the Engineering Faculties of the University of Trento. He founded the research activity in semiconductor optoelectronics at the University of Trento and started several laboratories of optical spectroscopy, growth and advanced treatment of materials. During the last years, he concentrated on Si-based optoelectronics where he has been working on porous silicon and on silicon nanostructures. His most important achievements are the demonstration of the first all porous silicon optical microcavity, the fabrication of Si LED in a fully CMOS compatible environment, the first evidence of optical gain and stimulated emission in Si quantum dot nanocrystals, the observation of photonic Bloch oscillations and Zener tunneling. All of these results opened the way to a Si-based optoelectronics and to the future fabrication of a Si-based laser. In silicon photonics, he is one of the worldwide recognized experts. He organized several international conferences, workshops and schools, and is a frequent invited speaker. He manages several research projects, both national and European. He is an author or co-author of more than 250 papers, author of several reviews, editor of 8 books, author of 1 book and holds five patents.

In: Frontiers in Optical Technology: Materials and Devices ISBN: 1-60021-084-8
Editors: P.K. Choudhury et al, pp. 125-158

Chapter 5

COMBINING ENERGY EFFICIENCY WITH AESTHETIC APPEAL USING ADVANCED OPTICAL MATERIALS

Geoff Smith
Department of Applied Physics and Institute of Nanotechnology
University of Technology, Sydney (UTS)
PO Box 123, Broadway NSW 2007, Australia

Abstract

Buildings, internal spaces and urban environments, which are not only very energy efficient, but also pleasant to occupy and look at, are needed to accelerate the rate of reduction worldwide in CO_2 emissions, and to raise living standards. New materials, whose optical properties are tailored to this task, are making this goal possible. Lighting, visual, and thermal control properties of windows, skylights, paints and lighting systems are treated. These three functions must be considered together, not independently. The way a range of new polymeric and nano-structured materials enable the optimised spectral management and distribution of incoming solar or lamp radiation, and outgoing thermal radiation, is presented.

Most of the materials discussed are composites whose properties can be varied in a controlled way, if the physics is understood. The basic models linking structure, composition and optical response are presented, for example, from among four classes of optical materials: opaque, transparent, translucent, and the group – light piping, mixing and homogenisation. Issues for plasmon resonant nanoparticles in visibly transparent glazing include resonance tuning, and differences in short wavelength (Rayleigh) and plasmon resonance scattering. For energy efficient glare control and uniform daylighting, for integration of light piping and continuous illumination, and for short length light mixing and homogenisation, the use of dopants, which are close in refractive index (RI) to the host, are analysed. These are found to be ideal to use with many emerging LED applications and daylight. For LED's they enable very energy efficient white light lamps based on RGB arrays at low cost, and also, enable uniform illumination despite the small area of the source. Finally, some special spectrally selective paints are discussed; those that can be deep colours or even black and still either reflect or absorb much solar energy (whichever is desired), and those that enable efficient radiative cooling or water collection from the atmosphere.

1. Introduction and Background

1.1. Energy Savings and Building Ambience

Energy efficiency in interior spaces concerns supply of lighting needs and maximising thermal comfort, with minimum use of electrical power from the grid, and of fossil fuels. It is relevant to all classes of buildings, and also, to transport. The world is faced with two apparently conflicting demands right now, a rapid growth in demand for better living standards and lifestyles, and an urgent need to cut greenhouse gas emissions. If this "conflict" can be eliminated or softened, then the process of scaling back our negative impact on the environment will accelerate. If it cannot, all living standards are at risk in the long term. Such a changeover, in common with past technology driven shifts in human activity, will also generate wide ranging opportunities for economic growth in all regions of the world. There is much new science needed to optimise these technologies, and optics is playing a central role. Examples of two science based systems for better use of natural lighting are in fig. 1.

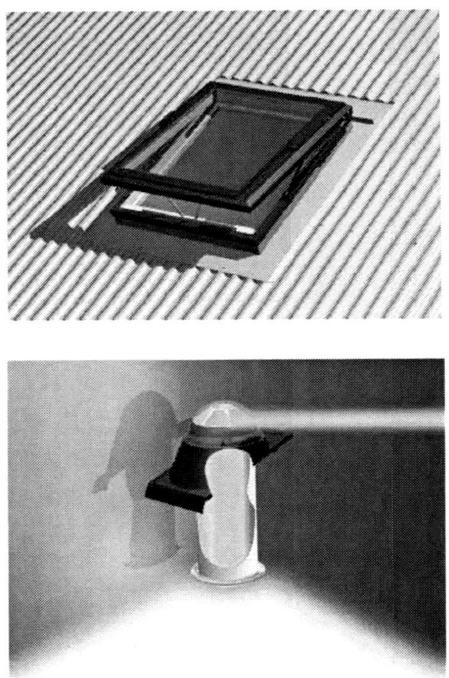

Figure 1. A spectrally selective roof window and an angular selective mirror light pipe – two ways of combining daylighting with thermal control. They exemplify building technology based on new optical materials. (Courtesy Skydome Skylight Systems Pty Ltd, www.skydome.com.au).

The human element, not just energy, is thus central to this study, and impacts on the science needed. Our perception of how a space feels, looks and how it performs, are strongly influenced by the optical and radiative properties of the materials at its surfaces. These surfaces may be opaque, transparent or translucent. Optical characteristics of the radiation sources contributing to light and energy fluxes in the space must also be taken into account when choosing and designing these surfaces. The sun, and various lamps are thus of interest in this study. These characteristics also depend on sky conditions and geographic location, the luminaires which house the lamps, and light pipes or other devices which are used to distribute the light or reduce solar heat gain.

Lighting is a core human need, but the complexities of lighting science and, in particular, the many difficulties in using our abundant daylight resource, especially in warm climates, has meant that lighting technology has evolved slowly, and there remains much scope for better lighting in terms of energy efficiency, greater use of daylight, visual comfort, decorative appeal and versatility. Quoting from Balcomb [1] "In building design strategies the use of natural light to replace artificial light - fills a unique niche. It stands alone as the most important design issue".

Specially engineered optical materials provide us with the ability to control thermal energy and light flows between the environment and interior spaces, in combination with the ability to control the ambience, appeal and enjoyment of these spaces. Issues to be considered include colour and colour rendering, glare, the exterior view from inside, the exterior appearance from outside, light distribution, lighting dynamics, functionality, occupant productivity and health, and design flexibility. Dealing with so many issues at once appears formidable. We will show in this article that, starting with a systematic basic approach to what is needed in spectral, directional, and scattering properties of materials, then looking at how such properties can be realised in novel composite or structured materials, makes an overall approach to these issues feasible and has led to much recent progress and new products. While the science is novel and sometimes complex, we are only interested in technology which is easy to realise in large volumes or areas, and hence, is cost effective.

Examples which have come up in my own and other solar materials research groups in recent years, may help to clarify the nature of these goals.

- A good performance with partially coloured, as opposed to the usual black, solar selective absorbers for better roof integration?
- A visibly black surface which is a reasonable solar reflector?
- Arbitrary coloured completely uniform sources of light, including white, from mixing output from multiple LEDs with almost no energy loss?
- A simple transmitting surface which blocks light from some directions of incidence while transmitting others, without distorting the view?
- Use of those components of solar radiation which are not wanted in a window or skylights transmission, to produce useful heat or electric power in the window itself?
- Reduced cooling loads or condense useful amounts of atmospheric water with special radiative coatings?
- Simple extraction of selected incident radiation components, then channelling and concentrating them for various remote uses.

1.2. Optical Properties and Materials

In this field, required optical properties must be evaluated according to a subjective human critique, as well as energy performance. Much related recent research by designers and lighting scientists appeared in the literature [2, 3], but detailed linking of measurable physical parameters to subjective response is still in its infancy. Standards and regulations must often be satisfied. They rely on measurable quantities, are based on experience, and are typically defined by task or activity requirements and safety, rather than overall appeal or energy issues. Energy fluxes, luminance (candela), illuminance (lux), and colour can each be measured using standard instruments. They can also be simulated in terms of material properties, radiation sources and the geometric properties of the spaces, using programs such as Radiance [4]. As an example, we used Radiance to simulate daylighting in the 2000 Sydney Olympic Stadium using the measured properties of four differently doped translucent panels distributed over the large roof area. The needs of spectators, the media and the athletes had all to be considered, in thermal and glare terms. Such subjective issues inside buildings include our response to the interplay between temperature, humidity and air flows; glare, the dynamics of light and temperature, lighting patterns, visual clarity, and modification of external views. Display lighting performance is primarily subjective and depends on brightness, homogeneity, colour, contrast when daylight or other strong sources are present, and field of view. Speciality lighting in hospitals, for safety on stairs and in aisles, for signalling, and for inside refrigerators and coolers, can also be made optimal in terms of energy use, appeal and functionality.

Paints also have an important role to play, and interesting recent developments can help in their energy performance [5, 6] while broadening decorative options. While the focus in this article is on passive environmental control systems, some of the material and optical properties we analyse can also be adapted for use in active energy saving systems such as solar water heaters, air heaters, radiative cooling, water distillation and improved solar cells. Indeed, it will be seen that it is possible to even combine some of these active functions with the passive control of lighting and heat flows, for a very high degree of multi-functionality in a single building element.

1.3. Polymer Composites and Nano-Structured Materials

Two rapidly developing and overlapping areas of materials and optical science are helping to achieve these goals – new polymer optical systems, and nano-materials. They can combine attractive cost with new approaches to saving energy. Flexible polymer light guides and polymer rods, both with unusual optical characteristics are one feature in this work. Nanoparticle doped polymers are another. Coatings based on vacuum and pyrolytic deposited thin films have well established capabilities in the windows area [7], while important developments are also occurring with switchable films, some of which also rely on nanostructures. In general, however, the established technologies have less chance of sizeable cost reductions. In the area of energy efficient windows, single and multi-layer thin film systems with low thermal emittance are now commonplace, and often mandatory in colder climates, where minimising heat loss is paramount. A major need is for simpler and cheaper glazing and daylighting systems, optimised for cooling dominated climate zones. An example of warm climate needs is in the rapid growth in uptake of air conditioning. It is causing emerging summer electricity demand problems in many parts of Australia and Asia, and is an established problem in California. Comfort is possible with much lower loads on these cooling units, often without them.

1.4. Optical Phenomena of Interest for Energy Efficient Materials

The optical and material physics, needed for analysis and development of both energy efficiency and aesthetic appeal, include scattering of nanoparticles and microparticles, surface plasmon resonance on conducting nanoparticles, nanostructured conductor surface optics, optical homogenisation in nanostructured materials, and light redirection and transport mechanisms intrinsic to materials. Mirrors and refractive optics are often used for redirection, but except possibly for micro-optic systems on surfaces, we have problems for building systems, such as a need for tracking and cleaning, which we wish to avoid. Micro-optics is currently too expensive for the scales we are after, but that may change. We will also touch on other materials which are difficult to realise cheaply at present, but further research could change that. Phenomena including optical anisotropy [8], and one-dimensional photonic crystal multi-layers [9] apply in these. Details will not be given on those thin films for energy efficiency, which have been well reported elsewhere, except to draw some comparisons. Switchable films also have an extensive review literature [7,10] and recent developments, such as gasochromics [11], which have reached the product stage, are relevant, but will not be reviewed.

We will conclude with a look ahead based on the question – What are some possible “ultimates” in building skin technology, especially with paint, polymers, windows and skylights? What degree of multi-functionality can we achieve through attention to both using and controlling solar and thermal radiation in each of four spectral zones UV, visible, near infra red and thermal infra red, for different times of day and year? This then leads to a final question – can we cost effectively supply all our energy needs from the walls and roofs of buildings, while keeping them attractive or even looking much the same as they always have?

1.5. Spectral and Directional Issues: the Sun and the Sky

The characteristics of incoming solar energy and daylight, and of outgoing thermal radiation, which we must account for in our materials design, are determined by the sun and the sky conditions. The solar spectrum at the surface of the earth is illustrated using spectral irradiance as a function of wavelength in fig. 2. For discussion of spectral management, this spectrum can be conveniently divided into three spectral regions UV (300–400 nm), visible (400–750 nm), and near infra red (NIR) (750–2500 nm), as shown in fig. 2. In fig. 2, to aid discussion, we have further subdivided the NIR, because little solar energy comes in beyond 1300 nm. Thus, for NIR blocking mechanisms which are narrow band, as with selected nanoparticles or some organic dyes, we should centre them around 1000 nm. If the skin is directly irradiated, as often occurs in a car, some argue [12] that human comfort depends also on wavelengths beyond 1300 nm because skin absorptance is much higher there. This is due to our body’s absorption bands due to water, but atmospheric water has the same absorption bands which are what makes the solar intensity weak in these spectral zones.

Material optical properties modulate the incoming radiation in these different spectral zones. The amount of solar heat transmitted is T_{sol}, reflected is R_{sol}, or absorbed is $A_{sol} = 1 - T_{sol} - R_{sol}$, and the amount of visible radiation transmitted is T_{vis}. T_{sol} and T_{vis} are defined in eqs. (1) and (2), respectively

$$T_{sol} = \int_0^{\infty} d\lambda S(\lambda)T(\lambda) \tag{1}$$

$$T_{vis} = \int_0^{\infty} d\lambda S(\lambda)T(\lambda)Y(\lambda) \tag{2}$$

with $T(\lambda)$ as the total or global spectral transmittance, $S(\lambda)$ as the total solar energy spectral irradiance incident *on the surface of interest* at wavelength λ, and $Y(\lambda)$ as the photopic sensitivity of the eye. Reflectance counterparts are found by exchanging R for T in eqs. (1) and (2). $S(\lambda)$ is most commonly taken as that for air mass 1.5 (fig. 2) (reference spectra and tables downloadable at http://rredc.nrel.gov/solar/spectra/am1.5 or as the ASTM G-173-03 standard). If T is for normal incidence, as is common when comparing different glazing products, then the implicit assumption is that the radiation is specular not diffuse. To simulate performance in actual installations, we must know T and S for all angles, or directions of incidence.

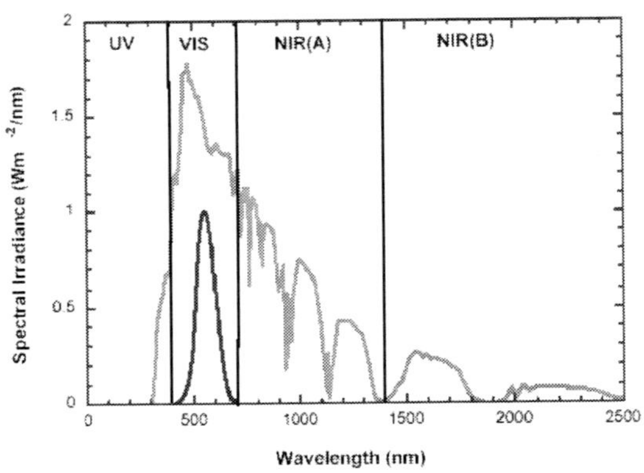

Figure 2. Air mass 1.5 solar spectral irradiance showing useful sub-divisions for spectral management purposes.

For light fittings, we simply change the function $S(\lambda)$ in eq. (2), to that produced by the lamp, or the lamp and its associated reflectors. The ways of representing the directional dependence of existing specular glazing products with algorithms for easy simulation has received a lot of interest in recent years, and is largely resolved [13,14]. Diffuse materials require much more work to reach this utility in simulation. They require a BRTF (Bi-Directional Reflectance-Transmittance Function) which connects incoming radiance for each direction with outgoing radiance in each direction. There has been some recent progress [15,16] but all too often made a Lambertian or "ideal diffuser" approximation, which is far from the behaviour of most commonly used diffusers in daylighting and lighting. The ideal situation would be a first principle scattering model for each generic system, rather than empirical models. Then changes in *BRTF* with say dopant concentration can be readily explored. In limited cases only, we can do this, since multiple scattering is itself usually too complex. One such case is the special diffusers discussed in Section 4. In that case, the ability to generate output light profiles from first principles for each particle has enabled much progress with the related technologies.

In discussing total heat management, we must include the thermal IR (or black body range) as it applies to radiating surfaces at moderate temperatures up to ~120 °C, as needed in buildings and some solar devices. The thermal emittance ε is the most widely used parameter for studying radiative losses as defined in eq. (3). However, for many situations, when any surface "sees" the sky to any significant extent, one can subdivide the thermal IR band further to better engineer radiative losses. This is because we want to either channel (to cool) or block (to retain heat) radiation loss to outer space which occurs in the 8–13 μm band. The parameter of main interest is then ε_A (atmospheric emittance) as given (un-normalized) in eq. (4).

$$\varepsilon = \int_0^{\infty} [1 - R(\lambda)] S_{bb}(\lambda) d\lambda \tag{3}$$

$$\varepsilon_A = \int_0^{\infty} [1 - R(\lambda)][1 - A_{atmosphere}(\lambda)] S_{bb}(\lambda) d\lambda \approx \int_{8\mu m}^{13\mu m} [1 - R(\lambda)] S_{bb}(\lambda) d\lambda \tag{4}$$

Here $A_{atmosphere}$ gives the atmospheric absorption and S_{bb} the black body intensity. For best cooling, when the sky is involved, it is thus actually best to have a high value of R at all wavelengths except in the 8–13 μm band, R should be very low [17], since the atmospheric absorption is high elsewhere. This is only for clear skies. The next best for cooling is to have the more common occurrence where $R(\lambda)$ is low across the whole 2.5–30 μm band, and high across the solar spectrum. This radiation band is the complete black body region at typical temperatures in windows, walls and roofs.

The thermal radiation band starting at 2.5 μm, has almost negligible overlap with the solar spectrum. This is where the original solar spectral selective concept originated for use on thermal collectors. For day-night radiative cooling, as just noted, the exact opposite qualitative spectral response to selective absorbers is required. Today we broaden the concept of spectral selectivity to include windows, and require one or more large qualitative switches in optical properties near some critical wavelength such as 750 nm where the human eye ceases to sense, or around 2.5 μm, where solar radiation vanishes and thermal radiation starts, or in the range 8–13 μm for sky cooling. An example is a switch from high to low transmittance in a window at 750 nm. Spectral and angle of incidence dependence of key spectral properties, which control energy flows, are transmittance $T(\lambda,\theta)$, reflectance $R(\lambda,\theta)$ and absorptance $A(\lambda,\theta)$; these must be known to track both selectivities. Often wavelength and angle of incidence impacts are not independent since the physical mechanisms, such as anisotropy in surface plasmon resonance, can cause both special angular effects and spectral shifts. Normal incidence is assumed hereon if we just define integral properties without reference to direction. For materials which are labelled "angular selective" spectral shifts of properties with incident angle or plane of incidence may also be quite large and polarisation sensitive.

2. Materials and Applications

It is useful to first catalogue the various materials and systems of interest according to four generic optical and visual groupings – (i) transparent, (ii) opaque, (iii) translucent, and (iv) light redirection, capture and transport. In later sections, selected promising examples of each will be discussed in detail. For basic systems including clear or tinted glass and polymer, with and without simple coatings, the underlying optical science is either quite mature or relatively simple, and we will not go into detail on these. It is, however, still useful to see these established systems placed into the context of the newer energy efficiency approaches, which are our focus. They all appear in Table 1 and examples of spectra and light distribution characteristics pertinent to a selection follow in figs. 3, 4, 5 and 6. Some of the opaque systems in Table 1a have a long scientific history, but interesting and useful new developments and possibilities are still emerging. It can be seen in Table 1 that energy efficient materials and advanced lighting materials are now a large and growing field. In column 1, the wavelengths, around which qualitative shifts occur, are included. New opportunities are expanding rapidly with developments in nano-materials, polymer optics, and new light sources such as LEDs.

Table 1. Examples of multi-purpose optical systems which combine energy efficiency with functionality and subjective appeal

(1a) Transparent – view or image preserving systems with low or negligible haze

Optical function	Generic systems	Specific examples
Block UV, admit daylight and provide clear view, high solar heat gain (400 nm)	Modified clear polymer, glass sheets, and glass/polymer laminates	Organic additives in polymer sheets and films, nanoparticles of ZnO or TiO_2, thin films SnO_2, ITO [18,19]
Reduce solar heat gain, admit daylight and provide clear view (750 nm)	NIR absorbing additives which do not scatter, 1D photonic crystal using bi- and tri-polymer unit cell	Iron doped glass, polymers doped with ITO, ATO or LaB_6 nanoparticles [20,21,22], PMMA/polyester multilayers [9,23]
Reduce solar heat gain, admit daylight, provide clear view, and improved thermal insulation (750 nm, 2500 nm)	Thin films and coatings which reflect both NIR and black body IR energy	Polymers doped with ITO, ATO or LaB_6 nanoparticles and exterior coat of a transparent conductor, TiO_2/Ag/ TiO_2/… thin film multi-layers [7], TiN thin films and multi-layers [24,25]
Block UV, admit daylight and provide clear view, high solar heat gain and improved thermal insulation (400 nm, 2500 nm)	Transparent coatings which reflect only thermal IR (low emittance coatings) on glass or polymer	Pyrolytic SnO_2 on glass [26], ITO on glass or polymers [18,19]
Block solar heat and direct sunlight from high incidence angles, admit diffuse daylight, provide clear view, redirect daylight	Angular selective thin films, angular selectivity by total internal reflection in a structured polymer, microstructured surfaces, holographic films	Oblique columnar cermet films (Ag/SiO_2, Cr/Cr_2O_3) [8,27,28], anisotropically microstructured polymer surfaces by lithography [29,30], laser cut polymer sheets [31], holographic sheets [32]

Table 1. Continued

(1b) Opaque systems – exterior surfaces in buildings and vehicles

Optical function	Generic systems	Specific examples
High solar absorptance and low thermal emittance (2500 nm)	Thin films on metal substrates, nanostructured metal surfaces	Ni doped anodic alumina on aluminium, Mo/WO_3 [33], Cr/Cr_2O_3 cermet films on metal [34], surface nanopatterned Ni or Mo [35]
Coloured and black layers which reflect the NIR solar heat component and have high thermal emittance (750 nm, 2500 nm)	Thin insulating films on metal flakes and dyed paint layers on metal	Thin films of SiO_2, or the bi-layer Fe_2O_3/SiO_2 on Al, visibly absorbing dye layers which are NIR transparent in binder on metal [5, 6]
Coloured layers which have high solar absorptance and low thermal emittance (2500 nm)	Nano-thin insulating layers on some metals, mixed pigments in thin layers on metal	Anodised or oxide coated stainless steel (blue or violet only) [36], green and black pigments in thin binder layer on Al [37,38]
High solar reflectance with (a) high thermal emittance (i.e. low black body IR reflectance-2500 nm) or (b) high emittance confined to the "sky window" 8–13 μm (8 μm)	(a) Thermal IR absorbing white pigments in dense paint (b) thin films or paints which are transparent except for high absorptance in the "sky window" (on metal)	(a) Dense TiO_2 pigmented layers [39], ZnO, ZnS pigmented polymer or binder layers [40,41], (b) silicon oxynitride films [42]

(1c) Translucent – high light transmittance, non- and partial view preserving

Optical functions	Generic systems	Specific examples
High forward transmittance, negligible backscattering with controlled light spreading	Large TRIMM (transparent RI matched microparticles) in polymer sheet, certain co-polymers	Cross linked PMMA spheres in PMMA sheet [43,44], co-polymers with close index [45]
Block solar heat and direct sunlight from high incidence angles, admit diffuse daylight	Microstructured polymers	Laser cut polymer sheets [46], microstructured polymer surfaces [29,30]

Table 1. Continued

(1d) Light capture, redirection, transport and emission

Optical functions	Generic systems	Specific examples
Light transport by total internal reflection over short and long distance with controlled emission rates from the guide	TRIMM microparticles in clear flexible polymer light guides or side lit flat polymer layers	Cross linked PMMA spheres in flexible acrylic cylindrical pipes and rectangular sheets at doping levels controlled according to transport distance [47,48]
Multi-source light mixing with homogenisation and low loss to side or back	TRIMM microparticles in clear polymer rods	Low concentration of PMMA microparticles in clear PMMA rods or clear rods plus doped sheets [49,47]
Fluorescence conversion and its emission trapping in part of a light guide	Dye doped polymer sheet optically bonded to clear polymer light guide	BASF Lumogen dye doped PMMA sheet optically glued to a PMMA ribbon light guide [50,51]
Near field coupling of (a) nanoparticle scattered radiation (b) plasmon resonant emission (c) fluorescent dye molecule emission into a close coupled waveguide	Oxide, metal or compound conductor nanoparticles deposited onto a thin transparent film or semiconductor layer to achieve (a) greater absorption in thinner underlying layer or (b) lateral transport to layer edges	Silica, Ag, LaB_6 nanoparticles onto an SiO_2 or Si layer. Film thickness controls waveguide modes and for plasmons, particle resonance wavelengths [52].

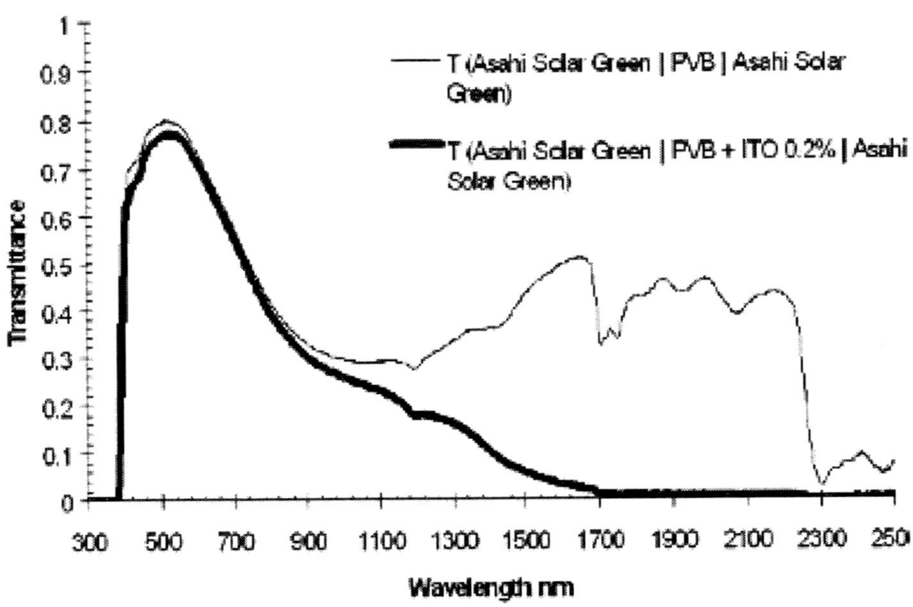

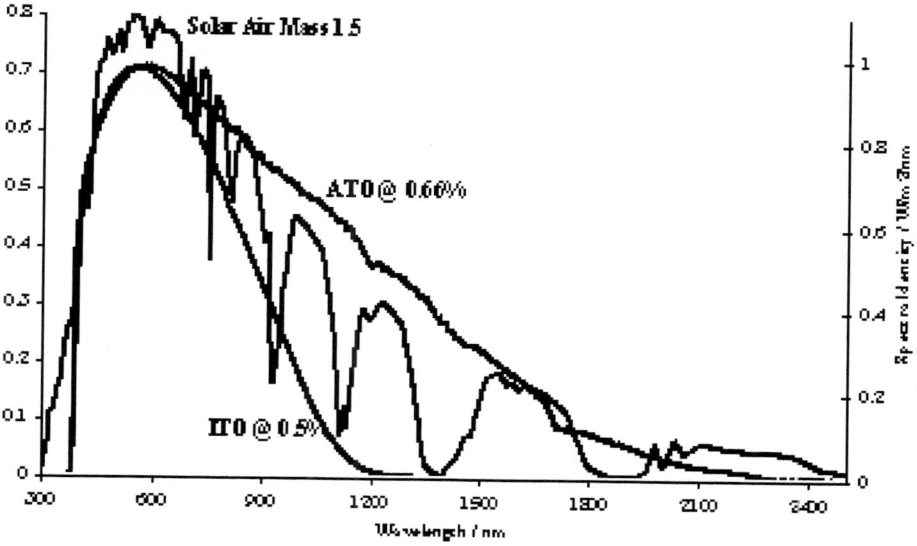

Figure 3. Spectral transmittance of thin polymer sheet between glass doped with nanoparticles of ITO and antimony tin oxide [20].

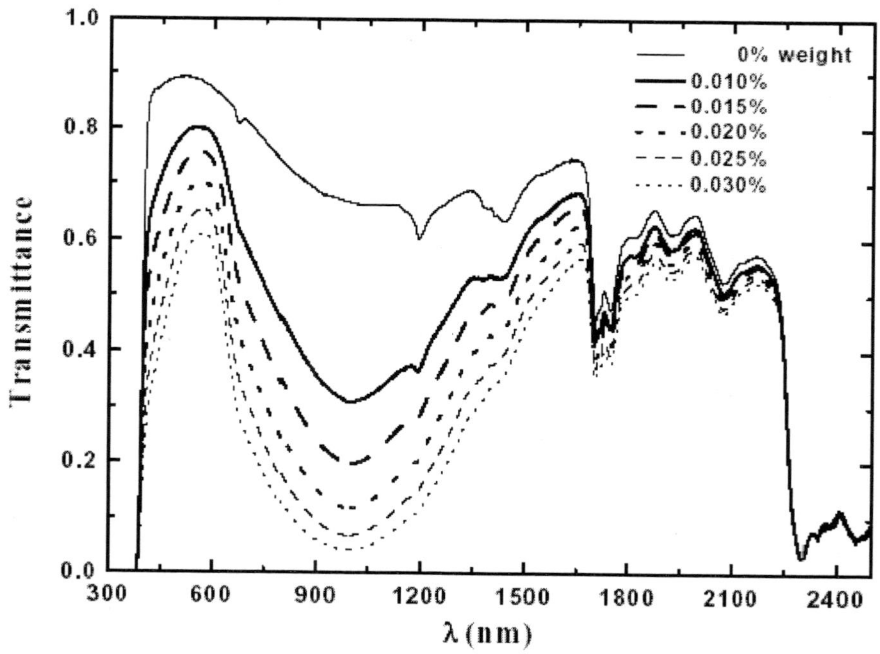

Figure 4. Spectral transmittance of laminated glazing doped with LaB_6 nanoparticles [21,22].

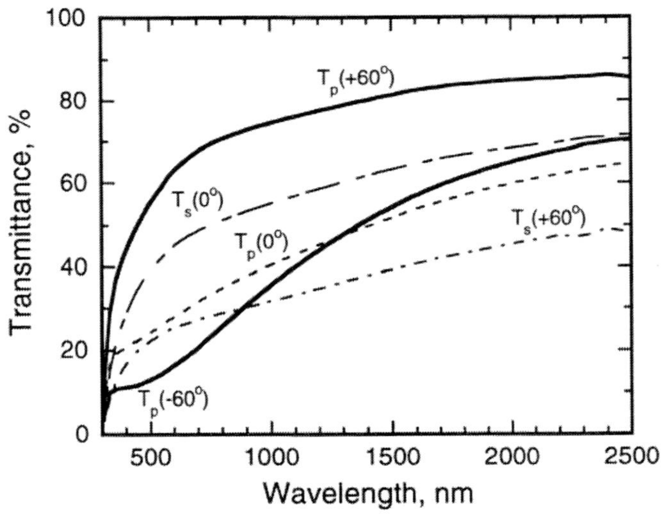

Figure 5. Angle of incidence and polarisation dependence of spectral transmittance for obliquely deposited Ag nano-columns mixed with oblique silicon oxide. Strong angular selectivity is evident.

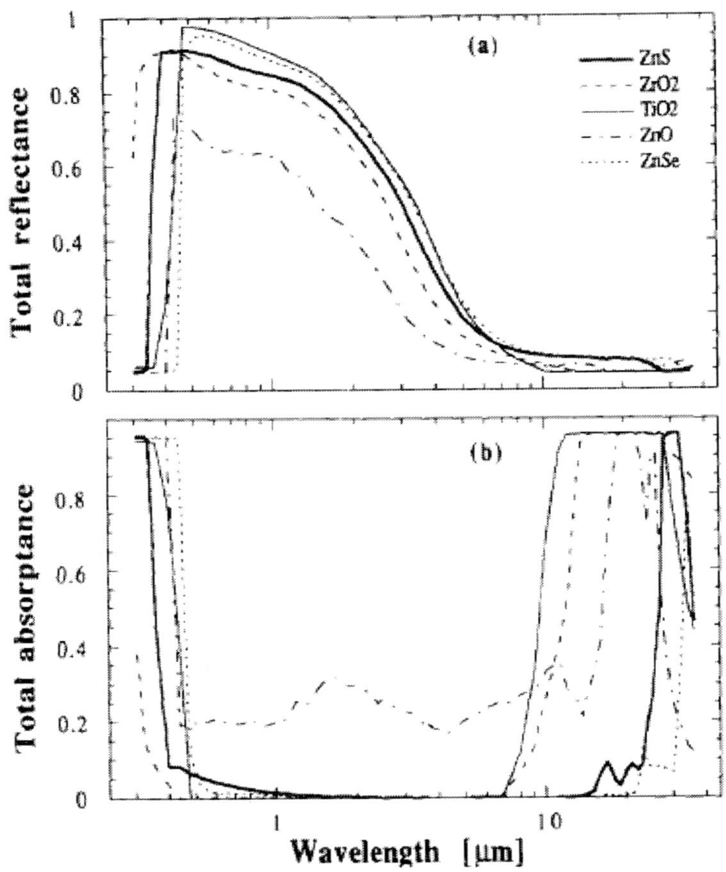

Figure 6. Spectral selectivity for radiative cooling purposes in polymer foils doped with inorganic various pigments [40,41]. Reproduced by permission of Elsevier.

3. Nanoparticle Plasmon Resonances for Spectral, Directional and Haze Control in Glazing and Skylights

Most glazing is clear and transparent, though it may be visibly tinted to reduce glare. Admission of daylight, provision of a clear view, and thermal insulation, provide the best combination for energy efficiency in cool climates. For temperate and warm climates, it is also desirable to block the NIR component of solar radiation to reduce cooling loads. Table 1b gives various established approaches to the latter goal using thin films. An alternative approach is now available using nanoparticle doped thin polymer layers which are typically either laminated between glass sheets or stuck onto the glazing surface. The particles in combination with their associated glass must satisfy three criteria, viz. (i) absorb strongly within the desired wavelength (NIR) range, (ii) produce negligible visible haze, and (iii) have an acceptable colour rendering impact on transmitted light.

The surface plasmon resonance (SPR), which can occur in conducting nanoparticles when they have a negative real part ε_c' of the dielectric constant of conductor, can fulfil all three criteria provided

1. the material used has a plasma frequency which is not large enough to cause visible SPR resonances,
2. the particle is small enough that Rayleigh scattering in the blue wavelength range is negligible,
3. the particle is small enough that plasmon resonant scattering centred in the NIR is strictly dipolar [22] (multipoles not present),
4. particle shape effect on resonant frequency gives desired resonant frequency [21], and
5. the medium in which the particles are embedded gives a suitable SPR location

Plasma frequency ω_p determines at what wavelength ε_c' becomes negative. ω_p is given by eq. (5) with n_e as the carrier density and m^* as the carrier effective mass.

$$\omega_p^2 = \frac{n_e e^2}{\varepsilon_o m^*} \tag{5}$$

Gold, silver, aluminium and titanium nitride have carrier densities which cause too strong attenuation in the visible, but lowering n_e below that in metals (such as gold and silver) shifts the plasma frequency down as desired. It is also desirable that inter-band absorption terms are negligible or weak in the range of interest, otherwise plasmon effects may be washed out or not present. This means we can effectively use a Drude model, though in practice, we always start with known dielectric constants. Residual interband effects, while not adding absorption, can cause polarisation, and hence, have an important influence. They are usually included in a modified Drude model (eq. (6)) via the constant ε_∞. This is because they have a weak frequency dependence in the range of interest, but do cause some screening of plasmon response, and hence, a lower SPR frequency

$$\varepsilon_c = \varepsilon_\infty - \left[\frac{\omega_p^2}{\omega(\omega + i\omega_\tau)} \right] \tag{6}$$

Here ω_τ is the relaxation frequency, and the nanoparticle sizes of interest (< 60 nm) must be modified from bulk values to account for charge collisions with the particle surface [21,22]. The effect of shape is given by eq. (7) assuming a spheroidal particle with depolarisation factor L for field along the axis of interest.

$$\varepsilon_c' = -\varepsilon_h \left(\frac{1}{L} - 1 \right) \tag{7}$$

Here ε_h is the dielectric constant of the host matrix. For spheres, the bracket in eq. (7) reduces to 2 since then $L = 1/3$. Substitution of the real part of eq. (6) into eq. (7) leads to the material (via n_e, ε_∞) and shape (via L) impacts on resonance energy in one expression. The carrier densities and location of inter-band terms in transparent conductors such as indium tin oxide (ITO) and antimony tin oxide (ATO), means they can be used for NIR blocking, but are inefficient because n_e is so low that the sphere resonance peak lies around 2–2.5 μm while we need most absorption (figs. 3, 4) from 0.75–1.3 μm. To date, a material identified to bridge the SPR frequencies of metals and the conducting oxide materials, and hence, give good nanoparticle results for solar control, is LaB_6. But LaB_6 also has limitations; shape must be correct and not too variable because of the combined broadening effects of the collisions with nanoparticle surfaces and shape impact on resonance location. LaB_6 spheres, for example, cause too much absorption in the visible red region, but slightly oblate ellipsoids, oriented as in typical extruded sheets, give a shift in the SPR peak to near 1 μm, whereas spheres resonate at 900 nm [21]. Resonance broadening can also arise from a wide spread in shapes as then there is a spread in SPR locations given by eq. (7). Thus, LaB_6 is very efficient in attenuating NIR solar energy, typically needing, for example, just 0.02–0.03% by weight in a 0.7 mm thick polymer foil for large NIR attenuation while maintaining good daylight gain. In contrast ITO requires 0.2–0.5% for similar levels of NIR solar attenuation, because it relies on resonance tails well away from the SPR peak to block solar NIR. The problem with LaB_6 is that visible attenuation does increase as concentration increases since the SPR spills over into the visible range (figs. 4, 7b). A detailed generalised analysis [22] indicates an ideal location for the SPR around 1.5 μm to achieve a combination of low particle concentration, no colour impact or visible attenuation, and excellent NIR blocking. If, however, visible tinting for glare reduction with some daylight is not a problem, as in much architectural glazing, then LaB_6 is fine.

Modelling of optical response can be done quite accurately in these systems provided particle shapes are approximately known and particles are well dispersed. Lack of dispersion also cannot be tolerated because clusters cause haze. The concentrations are so small that we can neglect inter-particle interactions in modelling absorption or scattering. For particles under 50 nm diameter we can also neglect scattering in the visible region, and it is also weak in the NIR. The absorption can be found from either calculating the combined dielectric constant ε^* from adding up the quasistatic induced dipole moment of all particles in a unit volume, then finding the complex effective RI $N = (n^* + ik^*) = \sqrt{(\varepsilon^*)}$. Equivalently, the absorption coefficient α is given in the dilute limit by

$$\alpha = f\frac{2\pi\sqrt{\varepsilon_h}}{\lambda}\mathrm{Im}\left[\frac{\varepsilon_c - \varepsilon_h}{\varepsilon_h - L(\varepsilon_c - \varepsilon_h)}\right] \tag{8}$$

where f is the particle volume fraction. This model fits data well, as shown in fig. 7(a). Discrepancies are probably mainly due to some scattering in the NIR.

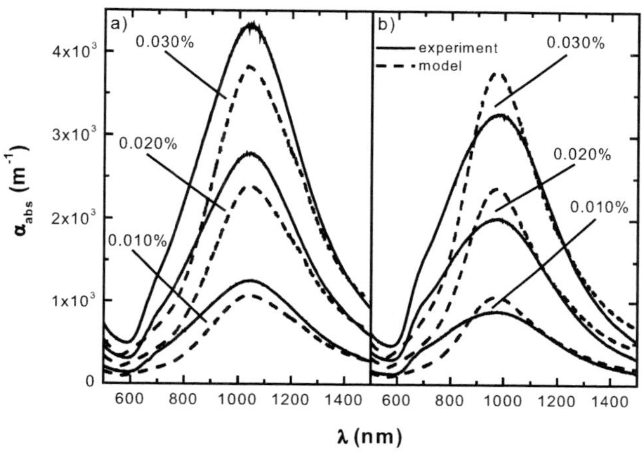

Figure 7. (a) Plasmon resonant absorption spectra in thin polymer sheets doped with LaB_6 nanoparticles [21].

Figure 7. (b) LaB_6 nanoparticle glass laminated samples with various doping levels against a bright blue sky.

It should be noted that, for the sizes of interest of up to around 80–100 nm, scattering is actually stronger in the NIR due to the SPR than it is due to normal polarisation effects at blue wavelengths, as shown in fig. 8. The latter gives classical Rayleigh and Rayleigh-Gans scattering. The fact there is a gap between the two scattering zones, as in fig. 8, is technologically important. There are also some important qualitative distinctions between the two types of scattering, which is not often realised. The Rayleigh-Gans term is only isotropic at the very small particle limit (~10 nm) where it is a pure dipole, but becomes mainly forward even by 80 nm diameters and almost all forward scattering by 150 nm. In contrast, the SPR scattering in the NIR remains quite isotropic up to 150 nm [22]. This is because dipole terms dominate in the latter until much larger particles. It is thus the onset of blue haze which limits the particle size for transparent glazing applications. In diffusing skylights, nanoparticle contribution to haze is less important, but one still needs to avoid colour related effects, which can be caused by Rayleigh or SPR scattering.

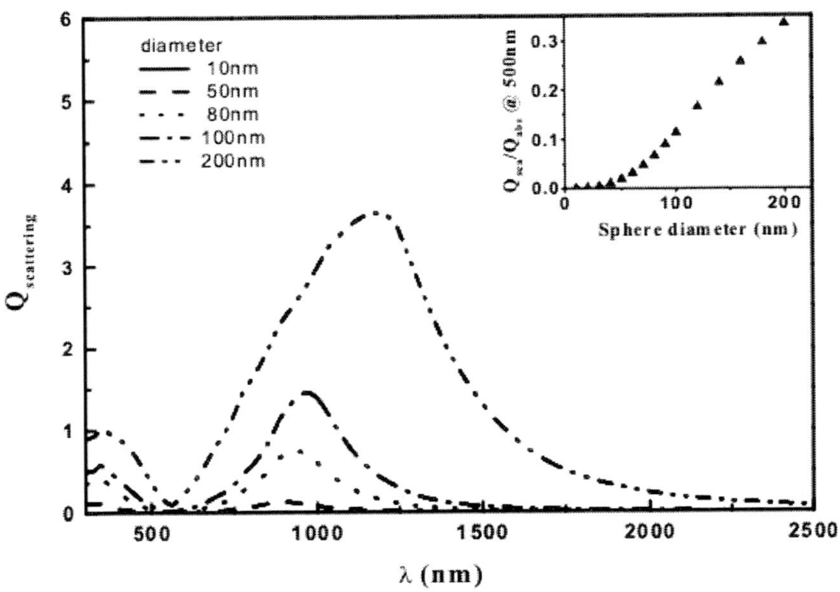

Figure 8. Scattering efficiency (and inset scattering to absorption ratio at 500 nm) as a function of wavelength for different diameter nanospheres in polymer showing scattering due to both NIR plasmon resonance and short wavelength classical Rayleigh scattering [22].

Another issue is that scattering changes affect the resonance width, and hence, spectral selectivity. This was recently highlighted by Schelm and Smith in the context of dielectric core-metal shell plasmon resonances [53], but it is also important in single material conducting particles. Once multipoles come into play, the extinction resonance broadens substantially. It is the dipole absorption resonance by itself that gives the narrow spectral response, and makes spectral control manageable.

Once nanoparticles start to interact, eq. (8) no longer applies. If scattering is still absent, various quasi-static approaches to model the effective dielectric constant ε^* are used, depending on the order if any, in the array, local clustering and percolation exist. The simplest and most widely used approximation assumes an array of dipoles in which each dipole interacts with a homogeneous average field due to all other dipoles. The 3-fold degeneracy of polarisation modes in an isolated sphere is not lifted in this approximation, which yields the Maxwell-Garnett (MG) model [54]. The MG model is inaccurate once any order sets in, or particles touch, as then multipoles also contribute to the dipole moment on each particle and the core assumption fails. Lattice [55] and local structure [56] models can then be useful. For a sphere once such interactions occur, the simple 3-fold degeneracy is lifted, and sets of symmetric and anti-symmetric polarisation modes should be observed for particle arrays. One of the most interest for solar work is the anti-symmetric mode in uniform particles which, relative to the original isolated dipole, moves to longer wavelength or lower energy. In practice, multiple modes are rarely observed as most samples do not have ideal lattice structures, so such effects are blurred out. However, some ordering is recently easily achievable using metal spheres coated with dielectric to form dense arrays of separated plasmon resonant spheres [57].

Mode splitting can also occur in a single inhomogeneous nanosphere [58] as in dielectric core-metal shell nanoparticles where plasmon resonance degeneracy is lifted. Then, however, it is the symmetric mode which is at the lowest energy, so it is one of the potential use in NIR blocking. The symmetric mode there and in other shell systems, such as nanotubes, has like charges at opposite sides of the thin conducting layer. This leads to the lowest energy for the whole sphere. The associated local field distribution is also quite unusual near resonance with the maximum field in the sphere shell at a position normal to the applied field [59] in the symmetric mode, and thus, near field profiles can be quite different to those on a uniform sphere at resonance. This has potential uses for molecular sensing.

In nanoparticle systems of high density, it is now possible to create arrays of conducting nanoparticles which fill most of the composite but remain separated. Thus, it then becomes possible to explore effective medium response and models for closely spaced but separate particles. One way is to coat the particles with an oxide of uniform thickness, and then pack the particles densely [57]. This creates an ordered array of metal particles, and results diverge from MG model predictions at fill factors of around 0.4–0.5, as expected. Another approach suitable for gold is to self assemble the nano-spheres in solution using organic di-thiol linker molecules to both join them up and keep them apart [60]. Spacing depends on linker length, and shorter linker molecules can provide high densities of nanoparticles. The nanoparticles must be small (< 8 nm). In these composites randomness is retained, and it appears that simple dipole response for the gold is retained as a result. In both examples, there are actually three mediums present since voids are also retained in the structures as they assemble. In the linker case with the self assembly occurring in solution before film deposition, the voids tend to phase separate from the linker-nanoparticle composite. A simple effective medium approach for the random three-component system works well only if the model is created in two steps which are consistent with the observed morphology: first a mix of gold in linker (to give the effective medium labelled AuL), then a mixture of the AuL medium with the voids. The AuL phase seems to be sufficiently small not to cause scattering.

The results on this system confirm a prediction of the random unit cell concept [61], namely, provided the randomness is retained, quite high densities of nanoparticles will not display multipolar response.

4. Energy Efficient Scattering Materials: Applications in Attractive Lighting Design Skylights, Display, Light Mixing and Homogenization

4.1. Sheet and Thin Layer Formats

Translucent polymer sheets and other sheets used in light fittings, skylights for glare reduction and more uniform illuminance, and some display work, are either traditionally pigmented with inorganic additives such as $CaCO_3$, TiO_2 and $BaSO_4$ or have special surface treatments. The RI of particles relative to the polymer host, their specific micron scale sizes and the amount added, combine to produce significant spreading of transmitted light. This spreading may be tailored as desired to reduce glare when the transmitting surface is viewed directly, and to provide more uniform lighting when used in skylights or luminaire covers. Traditional Mie scattering theory, modified to include multiple scattering effects, is able to semi-quantitatively describe their spectral transmittance and reflectance, and light spreading [62]. Thus, the emphasis on the use of these materials is on visual comfort, glare control and better use of the available light. Only the latter one has a clear impact on the overall energy efficiency in lighting. Most skylights are diffused to give more uniform illumination. But these traditional diffusers have an intrinsic negative impact on the energy efficiency. Pigment additives at concentrations high enough to achieve a reasonable spread of transmitted light cause a significant degree of backscattering, as is observable from their whiteness or brightness, when viewed in reflection. Consequent reduction of hemispherical transmittance and loss of lumens from lamps or the sun is commonly in the range 20–50%. It should also be noted that backscattering could be used to control light and solar energy throughput in some switchable window types (such as thermo-tropics) which switch from clear to diffuse on heating [63]. Strong backscatter usually requires small (sub-micron) size particles. Light diffusion with surface prism features on a sheet is another approach, but this often causes glare and less overall light spreading.

An easily implemented solution to this problem of eliminating backscatter while achieving controlled light spreading has recently become available in the form of a particular physical class of additive, which we now discuss in detail. This virtual elimination of backscatter comes with the associated optical properties, which provide not only energy efficiency but a whole new range of geometric configurations for lighting systems, new ways of achieving energy efficient "white" light lamps, new approaches to aesthetically pleasing light patterning, and new display and signage options. These will be covered in the next section, after we address the basic optical properties.

The key is in the combination of two features: (i) a large enough particle size for which Rayleigh-Gans scattering is weak and geometric optics is a good approximation, and (ii) a particle material whose RI is very close to that of the host material. The host is PMMA or related acrylics in our examples and in many applications. We have given to this class of additives, the generic acronym TRIMM [43, 47] which stands for transparent RI matched micro-particle. The particles used are normally spherical but the basic principles apply to other shapes. The key parameters are thus R (radius) and m (RI ratio) with $m = n_{particle}/n_{host} = 1 + \mu$, for $n_{particle}$ and n_{host} as the respective individual RIs. Thus, in TRIMM, m is close to 1, and μ is close to zero. In the two systems we deal with explicitly the same cross-linked

PMMA particles in two different matrices, one solid PMMA and the other a flexible clear acrylic. The respective values of μ are 0.0114 and 0.0182 at 590 nm. At normal incidence, such a material has

$$R \approx \frac{\mu^2}{4} < 10^{-4}.$$

At the median angle (45°) a ray strikes the spheres, and when $\mu = 0.0114$, polarisation averaged reflectance is 6×10^{-5}. Thus, many particles have to be struck before significant backscattering occurs. The reflectance that does occur, which is still weak, is predominantly for the more oblique rays falling onto the sphere surface, so they go forward and are of use. For these large spheres (diameter > 10 μm), the light striking a sphere near its edges is diffracted, but this is a tiny fraction of the total incident radiation. Thus, the wavelength does not enter explicitly into the light spreading equations in this geometric limit. All wavelength dependence of any consequence comes about through the wavelength dependence of μ. The ray deviation angle δ, and impact parameter h are shown in fig. 9. The deviation angle for a sphere of radius R is given by

$$\delta = 2\left[\sin^{-1}(h/R) - \sin^{-1}\left(h/R(1+\mu)\right)\right] \quad (4)$$

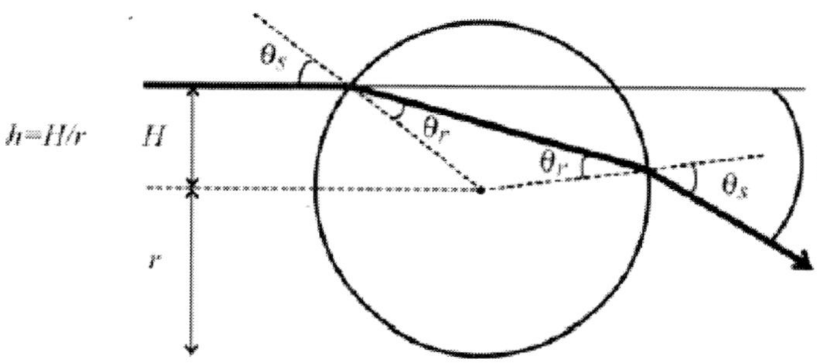

Figure 9. Ray deviation as it passes through a large sphere.

The spread of deviation angles for one sphere and the mean deviation angle are easily calculated [64]. If the probability of a deviation angle between δ and $(\delta+d\delta)$ is $f(\delta)d\delta$, the probability density function $f(\delta)$ can be found analytically first in terms of h, and hence, in terms of δ from eq. (4) [64]. For $\mu = 0.0114$, f is plotted versus δ in fig. 10. Also shown in this figure is the distribution of deviation angles found by Monte Carlo simulation. The mean value of δ, which is quite small at 1.8°, is also plotted on the curve. Note the long but weak high angle tail.

For the flexible matrix with $\mu = 0.0182$, the mean deviation is 2.1°, and for small μ, the mean deviation angle is approximately linear in μ. Such a small but significant linear deviation on average per sphere intersection makes light spreading control relatively easy, since average spread is then controlled through the average number of spheres with which a ray intersects. This is, in turn, easily controlled through particle concentration or sheet thickness. It is easily shown that the average spacing p between particles of radius r ($2r \sim 17$ μm in our samples) is given in terms of volume fraction V_f of spheres in the matrix by

$$p = \frac{4r}{3V_f} \tag{7}$$

where p is the main geometric parameter needed in simulation of multiple scattering. Thus, we have close to an ideal diffuser from an energy savings viewpoint because we can engineer the degree of light diffusion while losing hardly any light.

The other important aspect is colour shift of light on transmission due to TRIMM scattering. Since this just depends on how μ varies with λ, it is much less wavelength sensitive than Rayleigh scattering. For most sheets that we have studied, the colour rendering index of white light is not changed until quite large thicknesses. Only if doping and thickness are high, it becomes significant. All of these issues can be seen in the spectral transmittance plots of fig. 11, where total transmittance and the specular and diffuse components are shown together for four different thicknesses, namely, 1 mm, 2 mm, 3 mm, and 4 mm, all at the same composition.

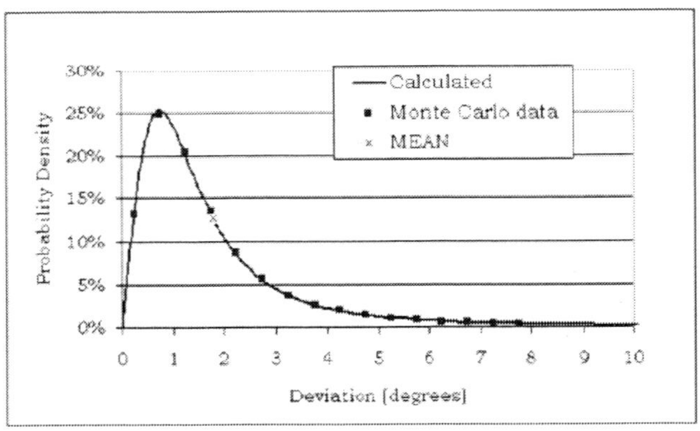

Figure 10. Probability density distribution function for angular deviations of all rays striking a TRIMM sphere in polymer, with $\mu = 0.0114$. [C. Deller, PhD dissertation, UTS, 2005].

While the specular and diffuse admix changes as thickness, and hence average number of particle-ray intersections increases, the total hemispherical transmittance at visible wavelengths is almost unchanged. In addition, there is little fall off in total T with wavelength λ, so the transmitted light usually appears unchanged in colour. However, the relative shift in diffuse and specular admix with wavelength indicates that μ is varying with wavelength enough to spread blue more than red, but the shift in μ never gets large enough to cause any significant backscattering. Once the diffuse component dominates, there is apparently a slight fall off in total T at short wavelengths (fig. 11). This is not due to backscattering but rather due to the measurement set up, since light, as shown in fig. 12, can be transported internally in the sheet by total internal reflection away from the integrating sphere aperture. In a large area sheet, total transmittance does not fall off as shown in fig. 11, at blue wavelengths. The variation in degree of diffuseness with λ correlates with direct measurements [64] of the dispersion of n_{host} and n_{TRIMM}, and their ratio, as the wavelength decreases. In practice, once rays have undergone a moderate number of interceptions with spheres, the diffuse component dominates. Considering all of these issues, it is clear that the trio; colour engineering, light spread control and low loss, is achievable in combination simply with TRIMM particle concentration control.

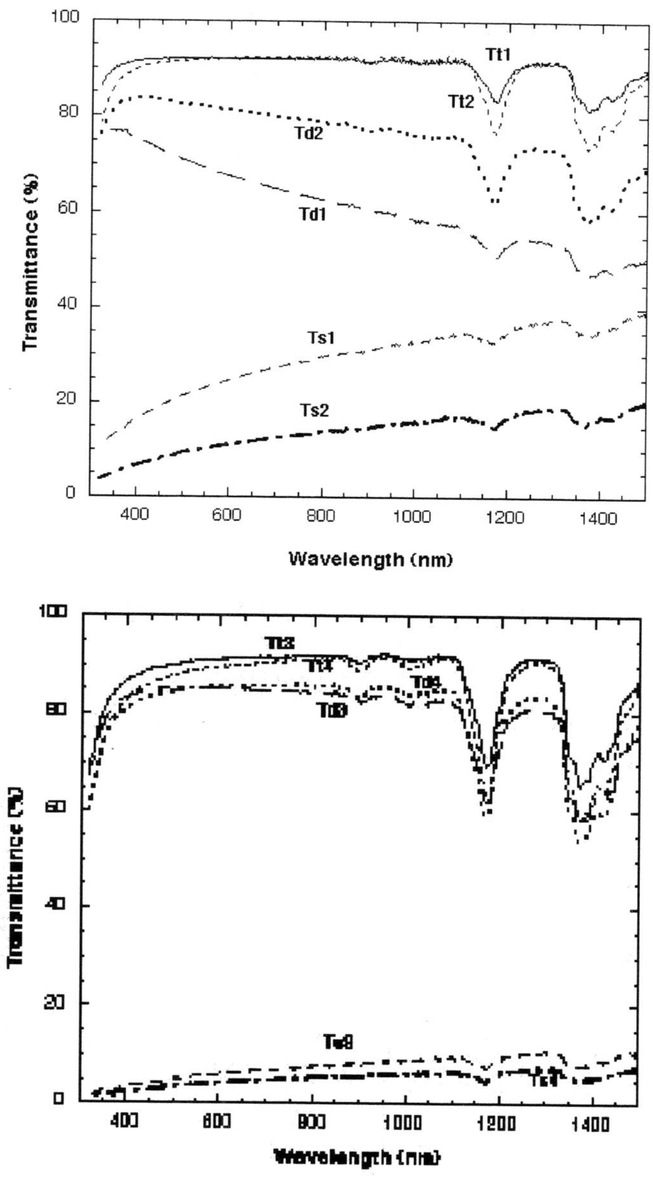

Figure 11. Total hemispherical transmittance, and specular and diffuse components of transmittance, for 1, 2, 3 and 4 mm thick PMMA sheet moderately doped with TRIMM spheres [43]. d = diffuse, s = specular.

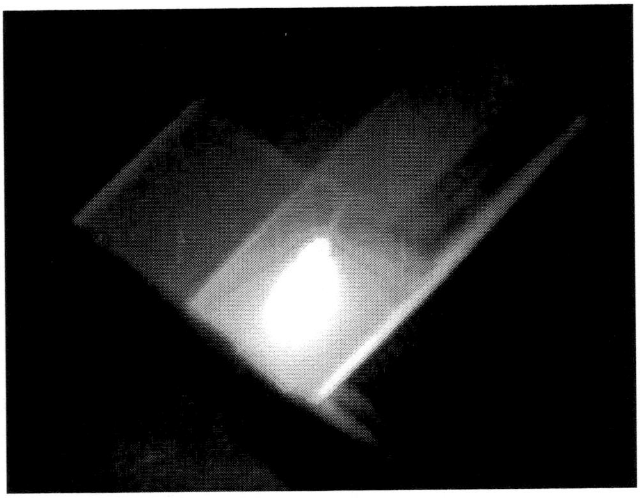

Figure 12. Photograph of exit light dispersion from a small illumination spot on a TRIMM doped sheet due to internal light guiding transport resulting from deviations that occur before emission. Note the light which makes it all the way to the edges [43].

4.2. Integration of Light Transport and Controlled Side Emission Using Doped Polymer Fiber Optic Cable

The ability to diffuse light not only by backscatter, but also to diffuse it only a small amount at each interaction, means it is possible to have a diffuser in which some internal light can be transported to long distances, so the fiber is continually emitting at a rate that is consistent with the maintenance of transport. This can be manifested in a fiber optic light guide, which can then be used for continuous illumination or display along its length with light input at one or both ends. The concentration, or more specifically the spacing parameter p, must be adjusted for the length. In contrast, doping a light guide with conventional Mie scattering diffuser pigments leads usually to two outcomes in addition to much backscattering loss. Either light is lost from the guide sides very quickly, so only short lengths are feasible, or if concentration is kept low enough to achieve moderate transport lengths, then side output at any point is quite weak.

Interest in fiber optic based lighting has increased significantly with the growth in bright LED availability and performance in recent years, especially for moderate diameter light guides. If small diameter guides are used, fiber bundles are required which are quite expensive, especially if made of glass fibers, or extruded polymers. Single core high transmittance fibers, however, are relatively cost effective, and can be made with good flexibility in their diameters. These are cast with high optical quality in single lengths up to 200 or more metres long. The transmitting jacket material is low index. Smaller diameters are usually extruded, but with these, it is less easy to couple in a lot of lumens into a single fiber.

Figure 13. Set up for measuring side-light output as a function of length along a TRIMM doped flexible polymer optical fiber with a single core. Continuous bright emission along the fiber length is obvious.

To extract light from the side of a fiber usually meant treating its surface, for instance, roughening or painting or screen printing one side. This has to be done intimately with the guide surface, so any existing jacket or low index coating has to be either removed where light is needed, or deliberately damaged in some way. This is messy and costly. If the guide itself can integrate transport and emission, lighting engineering possibilities are simpler, of lower cost, and more diverse. Thus, TRIMM doped single core light guides have vast potential in all areas of lighting and display. They also have another important capability for end lighting as opposed to side lighting, when lightly doped, which normal undoped single fibers do not. Good quality clear fibers produce caustics or bands of light at their exit end, which are sometimes called "hot spots", the details of which depend on the source beams characteristics [65]. For sources designed to efficiently couple into an optical fiber using focussing or parallel beams, the output caustics tend to be worse and quite visually uncomfortable. The way TRIMM doping can remove this problem without loss of energy is discussed in detail in the next section on end lighting, along with several important applications.

Here we focus on side lighting from single core polymer fibers, and the technology that flowed from this work we called "supersidelight". Though a recent development, it is available from Poly Optics Pty Ltd in Australia (www.fiberopticlight.com) who supported this research, and it is starting to be used by various specialist lighting businesses. Measurements of total output as a function of length have been carried out with an integrating sphere as shown in fig. 13, where the continuous output can also be seen. Fall-off plots based on spot photometer measurements agree with those from the broader distance sampling integrating sphere. The fall off in brightness is approximately exponential, and the exponent's dependence on concentration is complex [64]. The first TRIMM doped long fiber optic cables were made by us in 2000, and first reported openly in 2002 after earlier patents [66]. A French chemical group also reported on a similar system in 2002 [67]. Our models show their use of single event Mie scattering to describe the output is not appropriate to most of these systems. Typically, many scattering events occur before escape. In essence, since $\mu = 0.0182$ at 590 nm for TRIMM in the flexible matrix, the only material parameter to adjust is the particle density in the optical fiber. The axial particle number a in fiber length L (i.e. a = the total number of particles in length L) is the important parameter in terms of both end and side light applications. The related particle spacing parameter is $p = L/a$ with p defined in eq. (7) determining either the degree of homogenisation for end light, or the emission rate for side lighting. The smaller p means the more interactions in a given guide length, and hence, larger average deviation. As more rays approach the critical angle, more will exit. The initial light angular distribution profile just inside the fiber is dictated by the source input luminance profile, and the numerical aperture of the fiber. This profile gradually evolves with distance along the fiber. After a critical length L_c dependent on p and guide diameter $2R$, a steady state distribution is reached, and from then on, the rate of emission is a fixed fraction of the remaining light. This has been simulated in a special Monte Carlo ray tracing program using measured profiles on each source in a photogoniometer as input, known p values, and eq. (4). Poisson statistics are used to determine whether a particle is struck in terms of p and length l travelled from the last interaction. The ray path is determined by three parameters – r (the radial distance from the axis of the last interaction and the direction defined by (θ, ϕ)), θ (the angle to the axis) and ϕ (the rotational angle about the axis). If a particle is struck, a new ray direction is calculated using spherical geometry [68]. If a wall is struck, the angle of incidence K must first be defined to determine if the ray exits, or if it is totally internally reflected. In the latter case, the new ray direction is simply found, and free path distance l continues to accrue. For a circular cross-section guide of radius R, in terms of (r, θ, ϕ), K is given by

$$K = \cos^{-1}\left[\sin\theta\sqrt{1-\frac{r\sin\phi}{R}}\right] \tag{8}$$

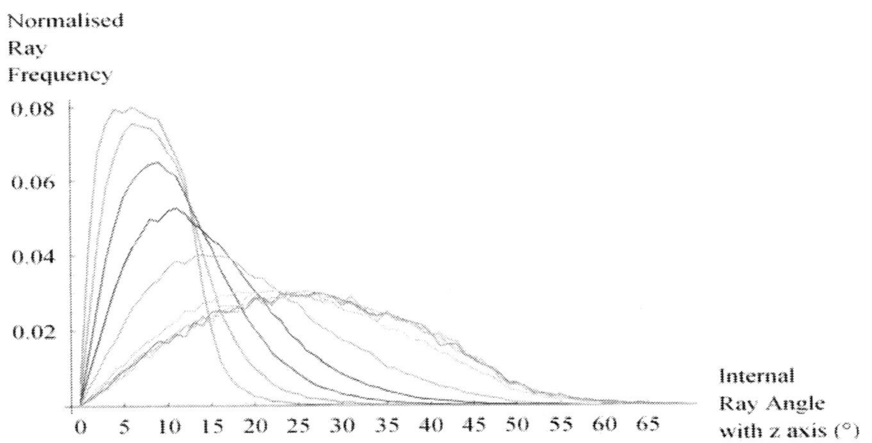

Figure 14. Evolution of internal beam spread with transport distance along a TRIMM doped optical fiber. The first five normalised curves are for transport distances 1, 10, 25, 50, and 100 cm. The remaining curves show a constant beam profile for various distances ranging from 2.5–100 m. Increasing doping simply compresses the distance scale.

A plot showing output profiles for different distances travelled is in fig. 14. The onset of the steady state distribution is clearly seen in this figure. By varying concentration over a wide range, we can achieve energy efficient side lighting over fiber lengths from a few cm up to 20–30 meters depending on the optical quality of the undoped fiber. The main challenge with the longer lengths is to achieve good colour balance far from the source, but this seems possible as long as these lengths of clear fiber also do not lead to faster loss of blue components. If sources are used at both ends, length limits will be dictated either by the source intensity or by the onset of yellowing of output near the middle.

Because the exit rays are being predominantly forward scattered, the output profile is also forward biased, but the low index jacket helps to make it more uniform. With sources at both ends or a reflector at one end, a more uniform output occurs, but it is still weakest when viewed along any line normal to the guide walls. Many applications require light only from one side of the guide, so a diffuse reflector is applied to re-direct rear exiting rays. It is applied either to the outside surface of the low index jacket or mounted externally. The net forward output is then usually quite homogeneous.

This ability to easily and efficiently make use of light going away from useful directions is a major advantage with fiber optic side lighting as opposed to the conventional elongated tubular powered source or fluorescent lamps. After reflection, rear-going light passes straight back through the optical fiber. In contrast, in a tube lamp mounted in a reflective luminare, much light can be blocked or lost due to the lamp tube itself. Thus, we again gain on energy efficiency. Related to this is the ability to engineer the light after it exits the guide without any impact on the internal transport performance of the guide. The coating can be on or flush with the outer surface of the low index layer, and still does not affect transport. With undoped fiber, it is difficult to simply control output with a coating directly applied to the bare guide walls. So, complex screen printing patterning is needed, and is only used in practice over limited lengths otherwise the output would appear too "spotty". A continuous coating on bare guide walls would severely limit achievable transport lengths, and hence, applications.

Multiple small lamps such as LEDs can be used simultaneously as fiber input sources, and are rapidly mixed within "supersidelight" or weakly doped end light guides, so that a wide variety of uniform new colours can be output, and also, changed in a dynamic fashion. If different coloured lamps are input from opposite ends or smaller diameters twisted together, even more varieties of colour and colour dynamics as a function of position along the guide can be achieved with very low loss. This mixer approach to arbitrary coloured lighting is clearly more energy efficient than using colour filters. An undoped clear guide using external means to get light out its side, and diffuse it, can easily yield inhomogeneities. Individual caustics, such as those reported by Swift *et al* [65] for clear polymer rods used for mirror skylights (as in fig. 1), can cause quite unpleasant visual effects. Light mixing in doped guides occurs very soon after the input end, and requires very low doping levels. Mixing for end light without side loss is thus possible, and is discussed next.

4.3. Energy Efficient Light Mixing and Homogenisation

The applications here are concerned with using end light from optical fibers or edge light from sheet light guides. There are several challenges when using single core fibers or other light guiding systems such as clear flat panels for illumination, when light is to be used from one side or the fiber end. They include efficient input coupling, removal of caustics, for multiple sources having them fully mixed, keeping mixing distance small and keeping energy loss low or luminous efficacy high. It is often possible to achieve mixing and homogenisation by using strongly doped conventional diffusers, but then much brighter sources are needed to compensate for the large energy losses that occur. So, net efficacy is then very poor. Also, as many applications are in compact, portable and hand held devices, longevity of battery power between re-charges is of major concern. Thus, achieving high brightness with low power is of considerable value. As an example, in surgical applications involving hand held devices, for instance, looking into a complex structure such as the mouth or throat, uniform well directed but dispersed and suitably localised illumination makes the task much easier, and reduces risk of errors. However, residual caustics can distort perception and create false highlights, dark spots and errors in depth perception. A lightly doped TRIMM small diameter guide is ideal for such applications. Inspection lighting used in such manufacture can benefit in the same way.

Mixing and homogenisation is aided by both multiple reflections off the light guide walls (effectively creating multiple sources), and by the spread induced from interaction with TRIMM spheres. The combination leads to much better performance than either alone or with the use of the light guiding reflections, meaning that very few collisions with particles are needed along the guide. This is important for energy efficiency since it means most light can reach the output end or edge without exiting the guide walls. As before, backscattering losses from the particles are not important while Fresnel back reflections from the front and exit ends are the most significant losses, as shown schematically in fig. 15. The Fresnel losses reduce net efficiency to around 88%. These end losses are increased by the extra divergence from the sphere interactions, but can be reduced in various ways and to various degrees, sometimes with quite simple procedures. 92% efficiencies (and maybe even 95%) should be achievable at low cost. In experiments with short mixing rods (~10 cm) backed up by computer simulations [47], it is found that exact profiles of the beams from LEDs are essential to get optimum results. In practice, they often have sharp features though many modellers and much supplier data approximate them as having smoothly varying intensity profiles with exit angle and change in direction. Furthermore, in simulation with doped guides, the extended source area, and precise placement of the source relative to the guide centre, must be used. Significant errors occur if it is treated as a point source, when moderately close to the guide entrance. Earlier mixing work [69] had used clear rods which had to be quite long to remove most caustics, but still could not provide homogeneous colour or brightness output without an additional diffuser.

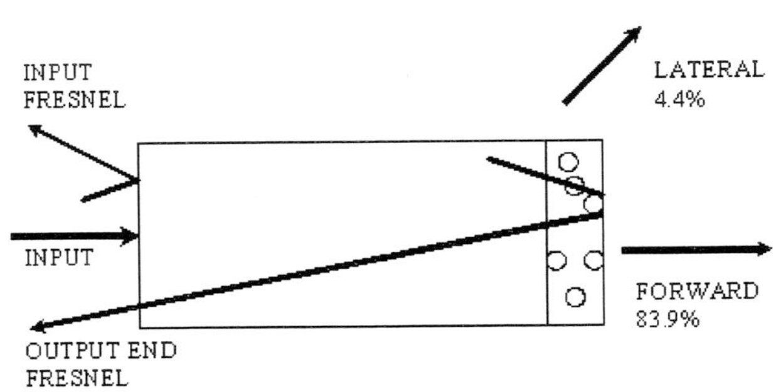

Figure 15. Composition of light mixer losses in a clear light guide terminated with an optically bonded TRIMM sheet. Simple modifications of exit end to eliminate most Fresnel losses can provide > 92% efficiency and combined with LED optical integration at input end up to ~ 96% mixing efficiency [49].

The results for 3×3 mm^2 red, green, and blue LED outputs mixed in a 10 cm long, 12.7 mm diameter PMMA rod, doped with just 0.45% by volume of TRIMM particles show mixing is just complete, and over 93% of lumens, that enter the guide, are output as useful lumens. The average ray has struck only ~ 20 particles. Mixing in even shorter lengths with similar losses is possible with slightly higher concentrations. The central bright output zone is also quite uniform in terms of luminance, then falls off smoothly. In contrast, a clear rod produces both local colour fluctuations, which are easily discerned by eye, and luminance spikes which can be disconcerting. An alternative combination of guide and TRIMM that works well is that shown schematically in fig. 14. It uses a clear rod and a thin TRIMM doped sheet with concentration high enough to get around 20 collisions across the sheet. It is optically glued to the output end [49]. The side loss from this top sheet will generally contribute to desired illumination. This system or the lightly doped rod thus has the potential to be the nucleus of the compact, energy efficient, low cost white light "lamp" based on RGB LED combinations, that has been sought as an alternative to phosphor based LED white light systems. Other mixing options such as micro-optic arrays on polymer are much more expensive.

The decorative, display and sign options with these types of lamps or the longer doped guides in terms of colour options, dynamic colour shifting, light distribution and visual interest and excitement are numerous, and remain to be exploited.

5. Roof, Car and Wall Cooling Based on Spectrally Selective Coatings and Pigments

5.1. Coloured Solar Reflective Paints

Until recently, solar reflection required white or pale coloured paints on buildings or cars. This is often not attractive to many people, limits decorative variety in warm climates where it is used in most buildings, and can cause substantial exterior glare. However, by using pigments or dyes with narrow absorption bands in combination with appropriate substrates, it is possible to achieve a variety of colours, even black, with strong to medium solar reflection. In actual outdoor tests, we have found that interior spaces can be kept several degrees cooler in some spaces using such paints compared with paints that are visually almost identical, but do not reflect the solar NIR component [5,6].

There are two underlying mechanisms that have been identified and tested, and both rely on using metal to reflect the near IR wavelengths. In one case, the latter is the substrate, and in the other, it is the core of the pigment particle. Figure 16 shows reflection off a binder layer on a copper substrate, with and without various dyes which make it black to the eye. This is compared with traditional black paint in the same figure. This normal black has a flat spectral response, and absorbs 95% of incident solar radiation. The binder is partly transparent in the NIR so that the substrate reflects a lot of incident NIR. An even more NIR transparent binder would allow a black with even higher R_{sol} than the one shown. The best solar reflective "black" in fig. 16 has R_{vis} = 5% and R_{sol} = 30% compared with the standard high quality black with R_{sol} and R_{vis} both about 5%. A solar reflective black reflecting up to 40% of incident solar energy may be possible.

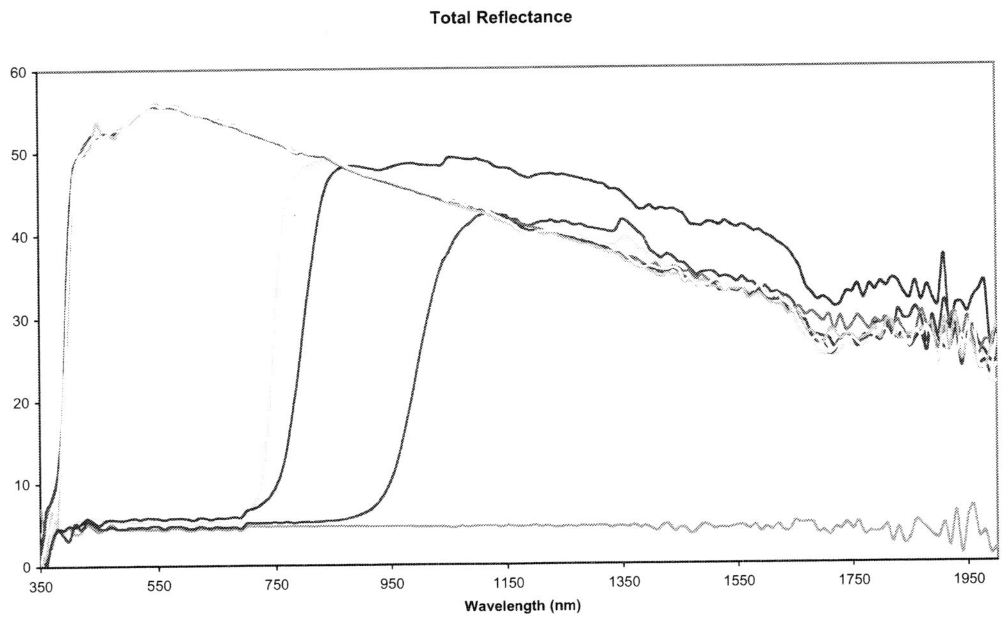

Figure 16. Spectral reflectance of a NIR reflecting painted surface on metal. It is visibly black, and is compared with a standard black (bottom curve) and the binder only with no additive. A binder which is more NIR transmitting would allow the same degree of "blackness" and solar reflectance over 40%.

The production of colours on large area metal sheets so as to also reflect the NIR has been known for some time. It relies on simple interference layers, usually oxides, where the thin film absorption maximum is located at appropriate visible wavelengths for the desired colour. The position of this maximum is tuned simply by varying oxide thickness, and with an appropriate metal substrate, it can be a strong NIR reflector. To achieve this capability in the convenience of paint, the pigment itself is a metallic flake, such as aluminium, and is coated with a thin oxide layer for interference on each flake. The layer thickness must be identical on all flakes for a uniform colour. Coatings with such precision on flake batches to over 200 kgs of metal can be achieved on an industrial scale using either sol-gel or hot fluidised bed chemical vapour deposition. Coating thicknesses are typically between 20–150 nm.

A detailed treatment of this type of paint for energy efficiency applications shows a number of attractive colours with good solar reflectance, which are possible together with using iron oxide as the interference layer on an Al flake [5,6]. These papers also include a schematic of the paint layers, showing how the flakes are oriented. The flake layering details have a strong visual and thermal impact. Colours currently available include blue, grey, silver, orange, red, gold, and purple. Solar absorptance varies between 0.7–0.4. These values were higher than that expected from a simple thin film theoretical analysis of individual flakes in the binder where A_{sol} values below 0.4 were predicted for most colours. The production paints still gave much superior thermal rejection than conventional paints of the same colour in outdoor tests. The discrepancy between theory and experiment was traced to the distribution, gaps and orientation in the flakes, and associated multiple reflections and light trapping. Improvements may be possible if a simple way of improving paint application to limit the extent of flake miss-orientation could be found, but this slight miss-orientation can add visual appeal via slight texturing. A final point on these systems concerns the enhanced decorative options which retain solar reflectance when a double layer coating is put on each flake. A study was also presented of a thin SiO_2 film as the first layer, over-coated with a thin Fe_2O_3 layer. Whereas the single layer flakes have little variation in perceived colour with viewing angle, the double layer flakes give a distinct colour shift as viewing angle varies relative to the plane of the coating. Various colour shifts with viewing direction are possible, for instance, from pink to green and aqua. Paints are also available that give such colour shifts, but do not reflect the NIR. These are also flakes with thin film coatings, but since the flakes themselves are not metallic (for instance mica), the desired thermal benefiting NIR effect is lost.

5.2. Radiative Cooling Paint and Pigmented Plastic Layers

The final example of spectral selectivity, we treat in detail, is of interest for cooling of buildings and for harvesting water from the atmosphere. These are surfaces which have high solar reflectivity and low thermal IR reflectivity – the latter one means high thermal emittance. In particular, they must have low reflectivity in the wavelength range of the atmospheric window between 8–13 μm. Moderately low reflectivity across the thermal IR is also usually common in most paint systems, though ideally the black body radiation outside the atmospheric window should be reflected or transmitted by the pigment, and reflected by the substrate [40,41].

A paint, which is specially designed for roofs containing several additives to provide both the special texture (which enhances strong solar reflection) and high thermal emittance, has recently been tested in our laboratories, and used on several Australian buildings. It contains TiO_2, other inorganic particles and binders. Work on related coatings, which contain TiO_2 as the main ingredient, goes back many years [39]. Related work on ZnS pigments for use in polymers has also been reported [40,41] and trialled for atmospheric water collection by radiative cooling. The coating we have studied gave a number of intriguing and useful results in outdoor tests [70] and some quite astounding reductions in HVAC cooling energy requirements, when used on some Australian supermarkets. Such systems work in the day by minimising solar gain through the roof while simultaneously radiating to the sky. They are ideal for locations where clear sky is common, and humidity is not excessive. Heat from inside can build up under the roof even if gains from incident sun are small, and can then be partly radiated away. The paint we studied had a hemispherical solar reflectance of 0.85 and a hemispherical thermal emittance of (0.95±0.02). At night, large cooling is possible with these coatings, the benefits of which can carry over well into the next day, especially if daytime solar gains are low.

The key feature we found was that on a clear sunny day it was still possible to achieve a small but net cooling from the roof, while at night cooling to ~2 °C below ambient was achieved in the space under the small test roof space. Identical spaces covered with more conventional roof integrated whitish coatings, and sold as "helping to cool", performed much worse. Both systems had a low level of insulation under the metal roof ($R = 1.5\ W^{-1}\ m^2\ °C$). The special coating under a solar flux of 1062 Wm^{-2} reflected 894 Wm^{-2}, the roof temperature equilibrated at 32 °C, and the space under the roof at 27.5 °C. In comparison, the standard whitish metal roof reflected 542 Wm^{-2}, the roof equilibrated at around 60 °C and the air under the roof at around 40 °C.

The actual supermarket cooling energy savings exceeded 50% over one year and considerably exceeded the savings predicted by a rigorous computer simulation, which at first made us to think our materials optical and emittance data were wrong. We believe now that the ability of the coating to keep the roof cool also impacted strongly on the performance of the roof mounted heat exchangers, since they were drawing in cooler air. Hence the coefficient of performance of the whole HVAC system was elevated considerably. Another, at first glance surprising result, came when we compared this coating in the daytime on an uninsulated roof in a small test unit, with one with R = 1.5 insulation underneath. In the insulated case, the roof was slightly cooler. Normally, over an insulation layer, solar irradiated metal roofs get much hotter, so they can send more heat up than down. This unexpected result occurred only under a clear sky. Given it is unusual, it was checked a number of times with different sensors, which were also recalibrated. The observation was confirmed. This is only possible when the net flow of heat from the roof is larger than that absorbed in both cases (insulated and not insulated). The uninsulated roof can extract more heat from below, and hence, stays warmer.

At night, the interior space under the roof cooled to a lower temperature than ambient by between 0.5–1 °C with insulation, and between 2–3 °C without uninsulation. This ability can be adapted to collect water from the atmosphere by giving much longer periods for the material below the dew point. Even better cooling performance can be achieved with attention to other thermal flows from the environment. ZnS pigments in polymer have been successfully tested for water collection. Ultimately we need pigments or materials that selectively radiate at the atmospheric window while being transparent or reflective over the rest of the thermal IR [42]. Then very substantial cooling and water collection can be possible.

Conclusion

It is becoming increasingly recognised that the visual appearance of energy saving technologies is often just as important as the amount of energy they save. Pleasing aesthetics is essential in achieving major market impacts. The core goal of energy savings follows if that capability co-exists. Daylighting and the popularity of skylights are longstanding examples of the appeal factor, but until recently, most skylights gave thermal energy problems. As pointed out by Kurtz recently [71], off-putting visual impact is a common weakness when new energy saving technologies first come to market, both in lighting and in solar energy. Solar energy technology has often been held back by poor aesthetics, although major improvements in recent years are helping its growth. Paints now exist that provide solar gain or rejection as required, while preserving full decorative options, and in some cases, with the ability to cool or collect water.

Attention to both design and to how new optical science can help to achieve energy savings in an attractive package is now overcoming these obstacles, as demonstrated in the examples given in this discussion. Lighting has additional problems. Glare and better light distribution, without further energy loss and reduced appeal, is the key challenge. A scientific approach to control glare without energy loss has been demonstrated using special TRIMM additives in clear polymer. This, in turn, led to novel decorative options integrated with the energy savings, such as colour control, dynamics, and luminance homogeneity. It also gave a new approach to efficient white light lamps using LEDs. Appearance and light distribution control, combined with energy saving, involves complex optical and materials science. In this chapter, many examples have been given where both issues are helped by new materials and optical developments over recent years. Advances in polymer and nanostructure optics underpin almost all of these advances, and promise many more such opportunities in the future, given the large current activity in both fields.

A final example of optical multi-functionality for energy efficiency and aesthetics: as an example, how far we might go in this quest to make the most of our every day building elements and their inter-play with the resources in the environment, consider a window which we engineer to do all of the following at once: block UV, photocatalytically using solar UV clean intake air or organic deposits, self clean with rain or light spray, provide a view, provide plenty of well distributed and comfortable daylight, give good interior thermal control, appear visually interesting from the outside, and finally use the solar NIR radiation to do one of (i) provide PV electricity, (ii) provide useful heat within a suitable fluid, and (iii) desalinate water. Having an optically clear, clean window is also an important aesthetic issue. This ambitious goal is quite feasible using known physics and chemistry, especially with further developments in nano-systems, to create the requisite materials, surfaces, and systems.

Acknowledgements

My colleague Jim Franklin and recent research students Christine Deller, Alan Earp, Stefan Schelm, Angus Gentle, and Swedish (Uppsala) student Jacob Jonsson, contributed much to the work described. Skilled technical and imaging support from Geoff McCredie and Ric Wuhrer is much appreciated. A number of companies supported or had input into different aspects of this work including BASF (Ludwigshafen), Poly Optics Pty Ltd (Burleigh Heads, Qld), Skydome Industries Pty Ltd (Sydney), Solutia (Springfield, USA), Lehmann Pacific (Sydney) and Roehm Degussa (Darmstadt) plus CSIRO Industrial Physics labs. Much of the research described was supported in part by various Australian Research Council grants.

References

[1] Balcomb, D., *Proc. PLEA '98*, 1998, 33–37.
[2] Fontoynont, M. *Solar Energy* 2002, 73, 83–94.
[3] Mardaljevic, J.; Rylatt M. *Energy and Buildings* 2003, 35, 27–35.
[4] Ward Larson, G.; Shakespeare, R. Rendering with radiance: the art and science of lighting visualisation, Morgan Kaufmann, San Francisco, CA, 1998.
[5] Smith, G.B.; Gentle, A.; Swift, P.; Earp, A.; Mronga, N. *J. Solar Energy Materials and Solar Cells* 2003, 79, 163–177.
[6] Smith, G.B.; Gentle, A.; Swift, P.; Earp, A.; Mronga, N. *J. Solar Energy Mat. and Solar Cells* 2003, 79, 179–197.
[7] Granqvist, C.G.; Energy-efficient windows: present and forthcoming technology, *In* Material Science for Solar Energy Conversion Systems, Pergamon, Oxford, UK, 1991.

[8] Smith, G.B.; Dligatch, S.; Sullivan, R.;. Hutchins, M.G. *Solar Energy* 1998, 62, 229–244.

[9] Weber, M.F.; Stover, C.A.; Gilbert, L.R.; Nevitt, T.J.; Ouderkirk, A.J. *Science* 2000, 287, 2451–2456.

[10] Granqvist, C.G.; Handbook of Inorganic Electrochromic Materials; Elsevier, Amsterdam, The Netherlands, 1995.

[11] Wittwer, V.; Datz, M.; Ell, J.; Georg, A.; Graf, W.; Walze, G. *J. Solar Energy Mat. and Solar Cells* 2004, 84, 1–4, 305–314.

[12] Takabayashi, T.; Ozeki, Y.; Tanabe S. *J. Soc. Automotive Engg.* 2003, 2791, 520–525.

[13] Roos, A.; Polato, P.; Van Nijnatten, P.A.; Hutchins, M.G.; Olive, F.; Anderson, C. *Solar Energy* 2000, 69 (Supplement), 15–26.

[14] Karlsson, J.; Rubin, M.; Roos, A. *Solar Energy* 2001, 71, 23–32.

[15] Apian-Bennewitz, P.; *Proc. SPIE* 1999, 2255, 697–706.

[16] Andersen, A.; Rubin, M.; Scartezzini, J.-L. *Solar Energy* 2003, 74, 157–173.

[17] Eriksson, T.S.; Granqvist, C.G. *J. Appl. Phys.* 1986, 60, 2081–2091.

[18] Hamberg, I.; Granqvist, C.G. *J. Appl. Phys.* 1986, 60, R123–159.

[19] Cochrane, G.; Zheng, Z.; Smith, G.B. *Proc. SPIE* 1993, 2017, 68–74.

[20] Smith, G.B.; Deller, C.A.; Swift, P.D.; Gentle, A.; Garrett, P.D.; Fisher, W.K. *J. Nanoparticle Res.* 2002, 4, 157–165.

[21] Schelm, S.; Smith, G.B. *Appl. Phys. Lett.* 2003, 82, 4346–4348.

[22] Schelm, S.; Smith, G.B.; Fisher, W.K.; Garrett, P.D. *J. Appl. Phys.* 2005, 97, 124314-1–8.

[23] Alfrey, T.; Gurnee, E.F.; Schrenk, W.J. *Polymer Engg. Sci.* 1969, 9, 400.

[24] Smith, G.B.; Swift, P.D.; Bendavid, A. *App. Phys. Lett.* 75, 630–632 (1999).

[25] Andersson, K.E.; Wahlstrom, M.K.; Roos, A. *Thin Solid Films* 1992, 214, 2132–218.

[26] Ginley, D.S.; Bright, C. *MRS Bulletin* 2000, 25, 15–21.

[27] Jahan, F.; Smith, G.B. *Thin Solid Films* 1998, 333, 185–190.

[28] Le Bellac, D.; Niklasson, G.A.; Granqvist, C.G. *J. Appl. Phys.* 1995, 77, 6145–6150.

[29] Gombert, A.; Blasi, B.; Buhler, C.; Nitz, P.; Mick, J.; Hossfeld, W.; Niggemann, M. *Opt. Engg.* 2004, 43, 2525–2533.

[30] Blasi, B.; Buhler, C.; Georg, A.; Gombert, A.; Hossfeld, W.; Nitz, P.; Walze, G.; Wittwer, V. *Microstructured surfaces in architectural glazings* (preprint).

[31] Reppel, J.; Edmonds, I.R. *Solar Energy* 1998, 62, 245–253.

[32] Hui. S.C.M.; Muller, H.F.O. *Architectural Sci. Rev.* 2001, 44, 221–226.

[33] Zhang, Q.C.; Mills, D.R. *App. Phys. Lett.* 1992, 60, 545–547.

[34] Zajac, G.; Smith, G.B.; Ignatiev, A. *J. Appl. Phys.* 1980, 51, 5544–5554.

[35] Andersson, A.; Hunderi, O.; Granqvist, C.G. *J. Appl. Phys.* 1980, 51, 754–763.

[36] Smith, G.B.; Sabine, T.M. *J. Australian Ceram. Soc.* 1978, 14, 4–8.

[37] Orel, Z.C.; Gunde, M.K. *Solar Energy Mat. and Solar Cells* 2000, 61, 445–450.

[38] Orel, Z.C.; Gunde, M.K. *Solar Energy Mat. and Solar Cells* 2001, 68 3–4, 337–353.

[39] Addeo, A.; Nicolais, L.; Romeo, G.; Bartoli, B.; Coluzzi, B.; Silverstrini, V. *Solar Energy* 1980, 24, 93–98.

[40] Nilsson, T.M.J.; Niklasson, G.A.; Granqvist, C.G. *Solar Energy Mat. and Solar Cells* 1992, 28, 175–193.

[41] Nilsson T.M.J., Niklasson G.A. *Solar Energy Mat. and Solar Cells* 1995, 37, 93–118.

[42] Eriksson, T.S.; Lushiku, E.M.; Granqvist, C.G. *Proc. SPIE* 1983, 428, 105–111.

[43] Smith, G.B.; Jonsson, J.; Franklin, J. *Appl. Opt.* 2003, 42, 3981–3991.
[44] Jonsson, J.C.; Smith, G.B.; Niklasson, G.A. *Opt. Commun.* 2004, 240, 9–17.
[45] Horibe, A.; Baba, M.; Nihei, E.; Koike, Y. *IEICE Trans. Electron.* 1998, E81-C, 1697–702.
[46] Edmonds, I.R.; Moore, G.I.; Smith, G.B.; Swift, P.D. *Lighting Res.Tech.* 1995, 27, 27–35.
[47] Deller, C.; Smith, G.; Franklin, J. *Opt. Exp.* 2004, 15, 3327–3333.
[48] Deller, C.; Smith, G.B.; Franklin, J.; Joseph, E. *Proc. Aust. Institute of Physics National Congress* 2002, 307–309 (CD Causal Productions, Adelaide). Ph.D. dissertation Deller, C. (UTS library Online), 2005.
[49] Deller, C.A.; Smith, G.B.; Franklin, J.B. *Proc. SPIE* 2004, 5530, 231–240.
[50] Earp, A.A.; Smith, G.B.; Swift, P.D.; Franklin, J. *Solar Energy* 2004, 76, 655–667.
[51] Earp, A.A.; Smith, G.B.; Franklin, J.; Swift, P.D. *Solar Energy Mat. and Solar Cells* 2004, 84, 411–426.
[52] Stuart, H.R.; Hall, D.G. *Appl. Phys. Lett.* 1996, 69, 2327–2329.
[53] Schelm, S.; Smith, G.B. *J. Opt. Soc. Am.* A. 2005, 22, 1288–1292.
[54] Maxwell Garnett, J.C. *Philos. Trans. R.. Soc.* 1904, 203, 385–420 and 1906, 205, 237–288.
[55] McPhedran, R.C.; McKenzie, D.R.; *Proc. Roy. Soc. Lon.*, 1978, 359, 45–63.
[56] Reuben, A.J.; Smith, G.B. *Phys. Rev. E* 1998, 58, 1101–1111.
[57] Ung, T.; Liz-Marzan, L.M.; Mulvaney, P. *Colloids and Surfaces A: Physicochemical and Engineering Aspects* 2002, 202, 119–126.
[58] Prodan, E.; Radloff, C.; Halas, N.J.; Nordlander, P. *Science* 2003, 302, 419–422.
[59] Schelm, S.; Smith, G.B. *J. Physical Chem. B* 2005, 109, 1689–1694.
[60] Schelm, S.; Smith, G.B.; Wei, G.; Vella, A.; Wiecorek, L.; Muller, K.H.; Raguse, B. *Nano Lett.* 2004, 4, 335–339.
[61] Smith, G.B. *Appl. Phys. Lett.* 1979, 35, 668–670.
[62] Vargas, W.E. *J. Opt. Soc. Am. A* 1999, 16, 1362–1372.
[63] Nitz, P.; Ferber, J.; Stangl, R.; Wilson, H.R.; Wittwer, V. *Solar Energy Mat. and Solar Cells* 1998, 54, 297–307.
[64] Deller, C. Ph.D. Dissertation 2005 (Available Online UTS library).
[65] Swift, P.D.; Smith, G.B; Franklin, *J. Lighting Res. and Tech.* 2006, 38, 19–31.
[66] Deller, C. ; Smith, G.B. ; Franklin, J.; Joseph, E., *Proc. Aust. Institute of Physics National Cong.* 2002, 307–309.
[67] Berthet, R.; Brun, G.; Zerrouchi, A.; Druetta, M. *Proc. POF* 2001, 361–367.
[68] Jonsson, J.C.; Smith, G.B.; Deller, C.; Roos, A. *Appl. Opt.* 2005, 44, 2745–2753.
[69] Zhao, F.; Narendran, N.; Van Derlofske, J. *Proc. SPIE*, 2002, 4776, 206–214.
[70] Bell, J.M.; Smith, G.B.; Lehman, R. *Proc. SASBE*, 2003 (Paper T606) .
[71] Kurtz, S.; *Proc. SPIE* 2004, 5530, 316–325.

Geoff Smith is a Professor of Applied Physics at the University of Technology (Sydney, Australia), and currently leads research programs in Physics at the Institute of Nanoscale Technology. He has worked on solar energy and energy efficiency using advanced optical materials for twenty eight years. His recent focus has been on windows, skylights, lighting, daylighting and thermal control, using novel polymer composites, polymer light guides, and nanostructured thin films. Recent technologies he has helped to develop include solar driven fluorescence for light piping to interiors, nanoparticle doping for transparent solar control glazing, energy efficient RGB light mixers, and one-step polymer fibers which either integrate efficient transport and controlled side emission, or produce uniform caustic free output from one end. Theoretical and modelling studies of complex composite media have featured strongly in this work. He is a Renewable Energy Pioneer (World Renewable Energy Network), has a Doctoral degree (honoris causa) from the University of Uppsala (Sweden), and is a Fellow of the Australian Institute of Physics. He has over 160 reviewed publications, 4 book chapters and a number of patents.

In: Frontiers in Optical Technology: Materials and Devices
Editors: P.K. Choudhury and O.N. Singh, pp. 159-190
ISBN 1-60021-084-8

Chapter 6

HIGH-SPEED OPTICAL MODULATORS AND PHOTONIC SIDEBAND MANAGEMENT

Tetsuya Kawanishi
National Institute of Information and Communications Technology,
Tokyo, Japan

Abstract

This chapter shows mathematical expressions for various types of electro-optic modulators, such as, intensity modulators, single-sideband modulators, frequency-shift-keying modulators, phase-shift-keying modulators, etc, based on optical phase modulation. We also discuss electrode structures for high-speed operation. Resonant electrodes can enhance modulation efficiency in a particular band. Design schemes for two types of resonant electrodes: asymmetric resonant structure and double-stub structure, are given. Most important application of optical modulators is conversion of electric signals into lightwave signals. This is indispensable to high-speed optical links. However, recently, we need new functions such as optical label processing at nodes, signal timing control to prevent packet collision, etc. In addition to basic function of optical modulators, this chapter describes the photonic sideband management techniques which can manipulate optical or electrical signals in time or frequency domain, by using sideband generation at the modulators.

1. Introduction

In lightwave telecommunication systems, we have to use some interactions with unmodulated lightwave, in order to obtain optical signals having some information. There are two types of interactions. One is the interaction between electric and optical signals. The other is between two or more lightwaves. Recently, many types of all-optical systems using interaction among lightwaves were proposed to enhance processing performance at nodes. However, it does not decrease the importance of the interaction between electric and optical signals, because most of information which can be accessed by human beings is described by electric signals, at this moment. We have to use the interaction between electrical and optical signals at terminals of the lightwave telecommunications system, while all-optical interaction can be used at the nodes. This chapter describes recent results in high-speed optical modulation and its applications. Optical intensity modulators, which are most commonly used in commercial systems, can be constructed by using electro-optic (EO) or electro-absorption (EA) effect. However, recently, other modulation formats, for example, differential phase-shift-keying (DPSK) or frequency-shift-keying (FSK), are used for long-haul system, optical labeling technique, etc. In this chapter, we focus on the EO modulator, which can control the phase and frequency of lightwave, as well as the intensity. In a material having the EO effect, the refractive index (RI) depends on the voltage applied on the material. Thus, in photonic sideband management (PSBM) techniques, we can obtain an optical phase modulator comprised of an optical waveguide and an electrode. The other types of modulators, such as intensity modulator, FSK modulator, can be constructed by optical phase modulators.

Firstly, some mathematical expressions for optical modulators using the EO effect are given. We also discuss performance of various types of modulators. Secondly, we describe the electrode structures for high-frequency operations [1, 2, 3, 4], and the design scheme for resonant electrodes [5, 6, 7]. In this chapter, we also review a new concept of optical sideband modulation techniques, i.e., PSBM, where optical or electrical signals can be manipulated in time or frequency domain by using sideband generation in optical modulators. A couple of PSBM techniques, such as, optical label swapping (OLS) using double-sideband (DSB) modulation [8], reciprocating optical modulation (ROM) for high-order harmonic generation [9, 10, 11, 12, 13, 14], and optical tunable delay [15], are described, where the frequency and time domain profile of the input lightwave is controlled by an electric signal. PSBM techniques enable us to manipulate optical signals by electric control signals, while most of modulators are just to convert the electric signal to the optical one. In addition, basically, PSBM does not use optical oscillation or non-linearities, so that there are no chaotic phenomena, such as mode-hopping, mode-competitions. Thus, we can obtain stable and agile manipulation of the optical signals in frequency and time domains.

2. Optical Modulators Based on Phase Modulation

Mathematical expressions for optical phase modulators are given to describe the basic principle of modulators using the EO effect. We also describe various types of optical modulators consisting of phase modulators: 1) intensity modulator, 2) single-sideband (SSB) modulator [16, 17, 18], 3) FSK modulator [19, 20], and 4) PSK modulator.

2.1. Phase Modulator

We can electronically control optical phase by using EO materials, such as lithium niobate (LN), lithium tantalate (LT), gallium arsenide (GaAs), etc. Fig. 1 shows a schematic of an optical modulator using EO effect. An electrode is placed along with an optical waveguide. By applying electric voltage on the electrode, we can change the RI in the waveguide, so that the phase of the lightwave can be controlled. For simplicity, the input lightwave is assumed to be monochromatic, and be described by $\mathrm{e}^{2\pi\mathrm{i}f_0 t}$. The output lightwave R can be expressed by

$$R = A^{\mathrm{LW}}\mathrm{e}^{2\pi\mathrm{i}f_0 t+\mathrm{i}f(t)} \quad (1)$$

$$f(t) \equiv KV(t) \tag{2}$$

where $V(t)$ and K, respectively, denote the electric voltage on the electrode and the coupling coefficient between the electric signal and the lightwave signal. A^{LW} is the optical transmittance in the waveguide. When $V(t)$ has high-frequency components, the frequency response of K should be taken into account. The coefficient K can be assumed to be a constant if the wavelengths of the high-frequency components on the electrode are much longer than the length of the electrode. The response of EO materials is much faster than the frequency response of K, which is dominated by that of the electrode and by the phase mismatch between the electric and lightwave signals.

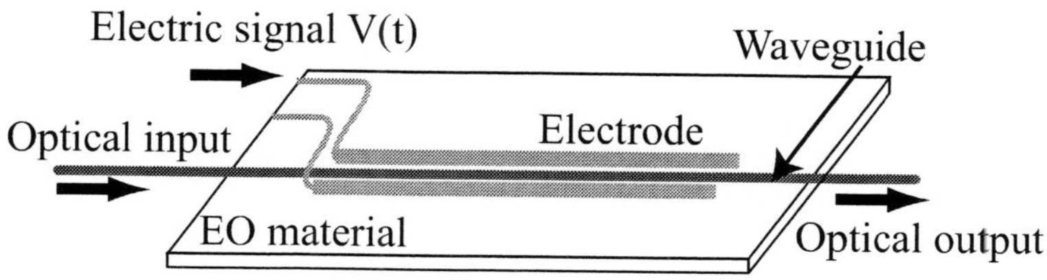

Figure 1. Optical phase modulator.

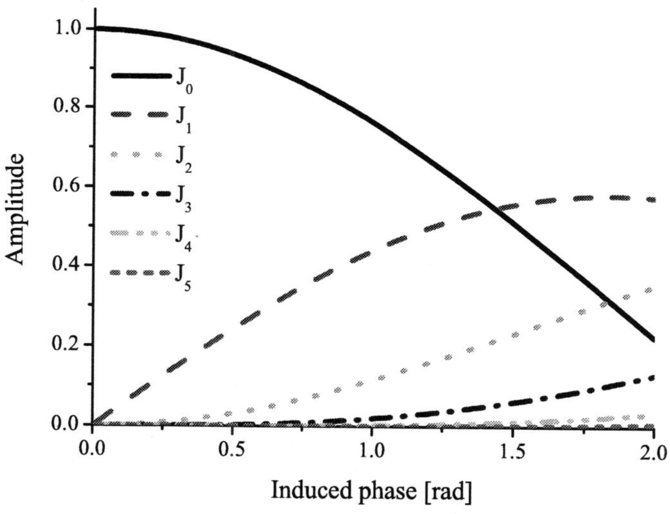

Figure 2. Bessel functions.

Here, we consider that $f(t)$ is a sinusoidal signal described by $A^{\mathrm{RF}} \sin 2\pi f_m t$. The optical output can be expressed by

$$\begin{aligned} R &= A^{\mathrm{LW}} \exp \mathrm{i}[f_0 t + A^{\mathrm{RF}} \sin f_m t] \\ &= A^{\mathrm{LW}} \sum_{n=-\infty}^{\infty} J_n(A^{\mathrm{RF}}) \exp 2\pi \mathrm{i}[f_0 t + n f_m t], \end{aligned} \tag{3}$$

where $J_n(A)$ is the first-kind n-th order Bessel function. A^{RF} is an index for optical phase deviation induced by the electric signal, so that it is called induced phase. When $f(t) = A^{\mathrm{RF}} \cos f_m t$, the output is expressed by

$$\begin{aligned} R &= A^{\mathrm{LW}} \exp \mathrm{i}[2\pi f_0 t + A^{\mathrm{RF}} \cos 2\pi f_m t] \\ &= A^{\mathrm{LW}} \sum_{n=-\infty}^{\infty} \mathrm{i}^n J_n(A^{\mathrm{RF}}) \exp 2\pi \mathrm{i}[f_0 t + n f_m t]. \end{aligned} \tag{4}$$

The spectrum component of $\mathrm{e}^{2\pi\mathrm{i}(f_0+nf_m)t}$, so-called the n-th order sideband, can be expressed by the n-th order Bessel function, which are shown in fig. 2. The first order function $J_1(A)$ has the maximum point which can be defined by

$$J_1'(A_m) = 0, \quad \left(J_1'(A) \equiv \frac{\mathrm{d}J_1(A)}{\mathrm{d}A}, \quad 0 \leq A \leq \pi \right), \tag{5}$$

where $J_1(A_m) = 0.583, \quad A_m = 1.841$. In most applications of optical modulation techniques, the desired component is the first order sidebands, so that the induced phase A should be less than A_m to achieve effective modulation without generation of undesired high-order sideband components.

2.2. Intensity Modulator

Mach-Zehnder (MZ) structures are commonly used to achieve intensity modulation by EO effects. As shown in fig. 3 the MZ structure consists of two optical phase modulators. We can control the output optical intensity by applying a pair of electric signals to the two phase modulators. Balanced push-pull scheme is ideal for intensity modulation, where the induced phase of the modulator A is $g(t)/2$, and that of the modulator B is $-g(t)/2$. $g(t)$ describes the optical phase difference between the two arms of the MZ structure. We can obtain the balanced push-pull operation by using an x-cut LN substrate and a coplanar waveguide (CPW) structure for electric signal guiding, as shown in fig. 4, where an electric signal is fed to the CPW. Fig. 5 shows the principle of operation. When the lightwaves in the two arms are in phase $g(t) = 2n\pi \quad n = \cdots, -1, 0, +1, \cdots$, the modulator is in "on" state. At the junction of the output, lightwaves interfere constructively. On the other hand, when lightwaves are in opposite phase $g(t) = (2n+1)\pi$, they are converted into high-order radiative mode waves which cannot propagate through the waveguide for output. Thus, we turn the modulator "on" and "off" by changing the voltage on the electrode. The intensity can be expressed by

$$\begin{aligned} |R| &= \frac{1}{2} \left| \left(\mathrm{e}^{\mathrm{i}g(t)/2} + \mathrm{e}^{-\mathrm{i}g(t)/2} \right) \mathrm{e}^{2\pi\mathrm{i}f_0 t} \right| && (6) \\ &= \left| \cos\left[g(t)/2 \right] \right|. && (7) \end{aligned}$$

We call the MZ structure for intensity modulation as the MZ modulator (MZM). The output intensity ratio between the "on"-state and the "off"-state, called the extinction ratio, is an important index of the MZM. In the ideal case, described by eq. (7), the extinction ratio goes to infinity. However, there is some residual lightwave even in the "off" state, due to amplitude imbalance in the arms, and crosstalk between the high-order and waveguide modes, as shown in fig. 6. Typical extinction ratios of commercially available LN modulators are 20-30 dB. Recently, we proposed a novel technique for compensation of the amplitude imbalance due to fabrication errors [21]. By using the trimmers, we can compensate the amplitude imbalance due to fabrication errors. Fig. 7 shows the on-off switching performance of the LN MZM with trimmers. The extinction ratio was 50 dB, while that of conventional technique was 37 dB.

When $g(t)$ is a sinusoidal signal described by $2A^{\mathrm{RF}} \sin 2\pi f_m t + \phi_{\mathrm{B}}$, the optical output can be expressed by

$$R = \frac{1}{2} A^{\mathrm{LW}} \mathrm{e}^{2\pi \mathrm{i} f_0 t} \left[\mathrm{e}^{\mathrm{i}(A^{\mathrm{RF}} \sin 2\pi f_m t + \phi_{\mathrm{B}}/2)} + \mathrm{e}^{-\mathrm{i}(A^{\mathrm{RF}} \sin 2\pi f_m t + \phi_{\mathrm{B}}/2)} \right] \quad (8)$$

$$= \frac{1}{2} A^{\mathrm{LW}} \mathrm{e}^{2\pi \mathrm{i} f_0 t} \sum_{n=-\infty}^{\infty} J_n(A^{\mathrm{RF}}) \mathrm{e}^{2\pi \mathrm{i}[f_0 t + n f_m t]} \left[\mathrm{e}^{\mathrm{i}\phi_{\mathrm{B}}/2} + (-1)^n \mathrm{e}^{-\mathrm{i}\phi_{\mathrm{B}}/2} \right] \quad (9)$$

$$= A^{\mathrm{LW}} \mathrm{e}^{2\pi \mathrm{i} f_0 t} \left[\cos \frac{\phi_{\mathrm{B}}}{2} \sum_{n=-\infty}^{\infty} J_{2n}(A^{\mathrm{RF}}) \mathrm{e}^{2\pi \mathrm{i}[f_0 t + 2n f_m t]} \right.$$

$$\left. + \mathrm{i} \sin \frac{\phi_{\mathrm{B}}}{2} \sum_{n=-\infty}^{\infty} J_{2n+1}(A^{\mathrm{RF}}) \mathrm{e}^{2\pi \mathrm{i}[f_0 t + (2n+1) f_m t]} \right]. \quad (10)$$

The output optical intensity $|R|^2$, which can be detected by a high-speed photodetector, is expressed by

$$|R|^2 \simeq |A^{\mathrm{LW}}|^2 \left[J_0^2(A^{\mathrm{RF}}) \cos^2 \frac{\phi_{\mathrm{B}}}{2} + 2 J_1^2(A^{\mathrm{RF}}) \sin^2 \frac{\phi_{\mathrm{B}}}{2} \right.$$
$$- 4 J_0(A^{\mathrm{RF}}) J_1(A^{\mathrm{RF}}) \sin \frac{\phi_{\mathrm{B}}}{2} \cos \frac{\phi_{\mathrm{B}}}{2} \sin 2\pi f_m t$$
$$\left. + 2 \left\{ 2 J_0(A^{\mathrm{RF}}) J_2(A^{\mathrm{RF}}) \cos^2 \frac{\phi_{\mathrm{B}}}{2} - J_1^2(A^{\mathrm{RF}}) \sin^2 \frac{\phi_{\mathrm{B}}}{2} \right\} \cos(2 \cdot 2\pi f_m t) \right], \quad (11)$$

where we assumed that $A^{\mathrm{RF}} << 1$, and the high-order components are neglected. By using Taylor's expansion of Bessel function, we get a simple equation,

$$|R|^2 / |A^{\mathrm{LW}}|^2 = \frac{1}{2} + \frac{1 - |A^{\mathrm{RF}}|^2}{2} \cos \phi_{\mathrm{B}} - A^{\mathrm{RF}} \sin \phi_{\mathrm{B}} \sin 2\pi f_m t$$
$$+ \frac{1}{2} |A^{\mathrm{RF}}|^2 \cos \phi_{\mathrm{B}} \cos(2 \cdot 2\pi f_m t) \quad (12)$$

which is useful to discuss performance of optical systems for rf-signal transmission where the photodetector generates rf-signals. The intensities of the fundamental component $\sin(2\pi f_m t)$ and the second order harmonic $\cos(2 \cdot 2\pi f_m t)$ can be controlled by the dc-bias ϕ_{B}. The fundamental and second order components are proportional to $\sin \phi_{\mathrm{B}}$ and $\cos \phi_{\mathrm{B}}$, respectively. The ratio between the average power and the rf-signal component dominates conversion efficiency from lightwaves to rf-signals at the photodetector. The ratios for the fundamental and second order components are, respectively,

$$D_1 = \left| \frac{2A^{\mathrm{RF}} \sin \phi_{\mathrm{B}}}{1 + (1 - |A^{\mathrm{RF}}|^2) \cos \phi_{\mathrm{B}}} \right| \quad (13)$$

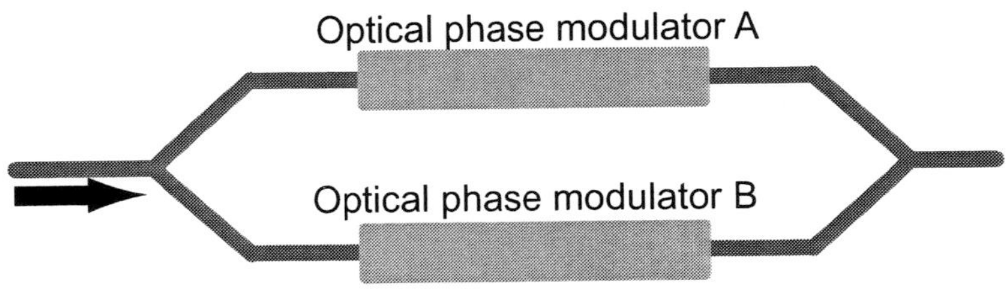

Figure 3. MZ structure

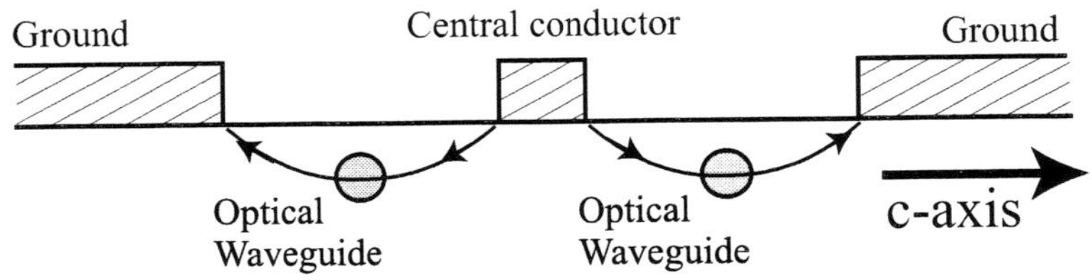

Figure 4. Cross section of balanced MZ structure for push-pull operation.

$$D_2 = \left| \frac{|A^{\mathrm{RF}}|^2 \cos\phi_{\mathrm{B}}}{1+(1-|A^{\mathrm{RF}}|^2)\cos\phi_{\mathrm{B}}} \right|. \tag{14}$$

In the case of $\phi_{\mathrm{B}} = \pi$, the even order components in the output R, including the carrier component $\mathrm{e}^{2\pi \mathrm{i} f_0 t}$, become zero and the average power $|R|^2$ goes to a minimum of $\frac{|A^{\mathrm{RF}}|^2}{2}$, where the dominant components are the first order upper and lower sidebands. This modulation scheme, called double-sideband suppressed carrier (DSB-SC) modulation, is suitable for rf-signal transmission. This is because undesired dispersion effects in optical fibers can be suppressed, and the ratio between the average power and the second order rf-signal D_2 becomes a maximum. In addition, DSB-SC has an impotant role in PSBM to manipulate lightwave signals in frequency domain, as described in later part of this chapter. The carrier suppression ratio depends on the extinction ratio of the MZM, so that the carrier can be also suppressed largely by using the intensity trimmers for high-extinction ratio [21]. Fig. 8 shows optical spectra of DSB-SC signals with and without imbalance compensation in the LN MZM with the trimmers, where the modulation frequency f_m was 10.5 GHz.

2.3. SSB Modulator

Fig. 9 shows a schematic of an optical SSB modulator consisting of parallel four optical phase modulators [16, 17]. The SSB modulator has two electrodes $\mathrm{RF_A}$ and $\mathrm{RF_B}$ for high-speed signals, where a pair of rf-signals of the same frequency f_m are fed to the electrodes. The electric field of the output can be expressed by

$$R = \frac{\mathrm{e}^{2\pi \mathrm{i} f_0 t}}{4} \sum_{j=1}^{4} \sum_{n=-\infty}^{\infty} J_n(A_j^{\mathrm{RF}}) \mathrm{e}^{\mathrm{i}n(2\pi f_m t + \phi_j^{\mathrm{RF}})} \times A_j^{\mathrm{LW}} \mathrm{e}^{\mathrm{i}\phi_j^{\mathrm{LW}}} \tag{15}$$

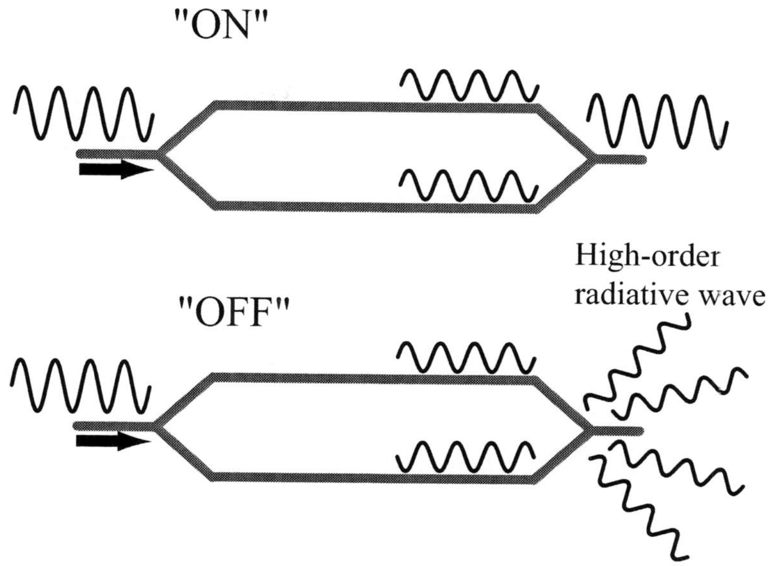

Figure 5. Principle of intensity modulation.

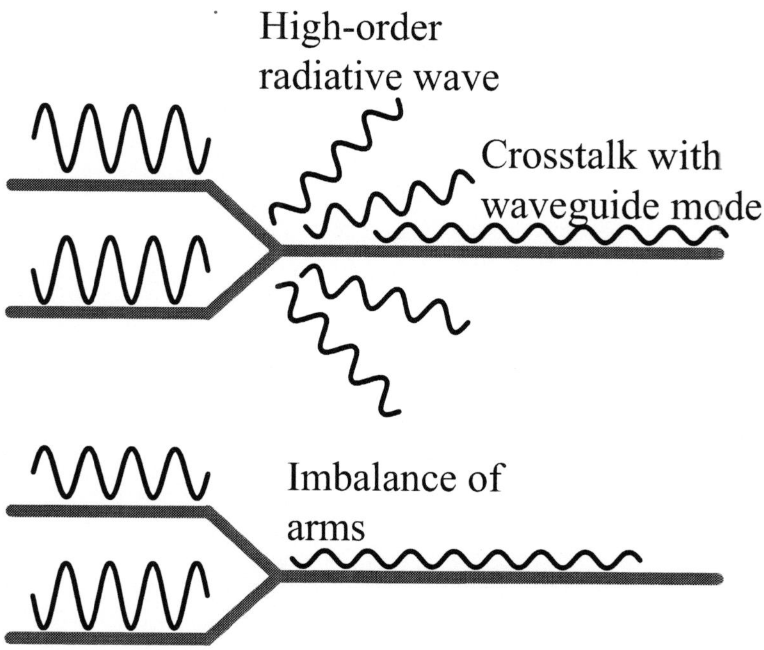

Figure 6. Degradiation of extinction ratio.

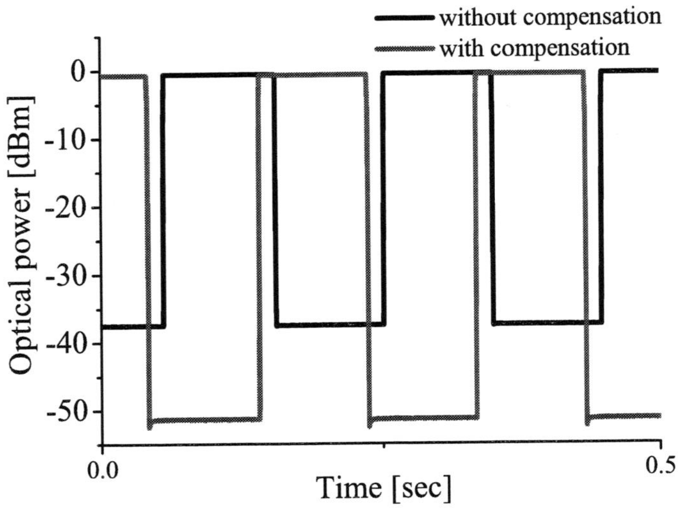

Figure 7. High-extinction ratio on-off switching with compensation of imbalance in MZM using intensity trimmers.

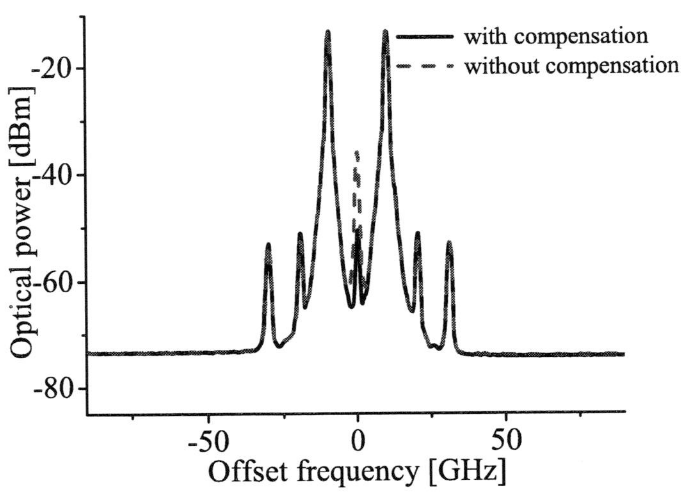

Figure 8. High carrier suppression in DSB-SC using an MZM with intensity trimmers.

$$= \frac{e^{2\pi i f_0 t}}{4} \sum_{n=-\infty}^{\infty} e^{2\pi i n f_m t} \sum_{j=1}^{4} J_n(A_j^{\mathrm{RF}}) P_{n,j} A_j^{\mathrm{LW}}, \tag{16}$$

where

$$P_{n,j} \equiv \exp i \left[(T - Sn) j \frac{\pi}{2} + \Delta\phi_j^{\mathrm{LW}} + n \Delta\phi_j^{\mathrm{RF}} \right] \tag{17}$$

$$\phi_j^{\mathrm{LW}} = T \frac{j\pi}{2} + \Delta\phi_j^{\mathrm{LW}} \tag{18}$$

$$\phi_j^{\mathrm{RF}} = -S \frac{j\pi}{2} + \Delta\phi_j^{\mathrm{RF}} \tag{19}$$

$$S = \pm 1, \quad T = \pm 1. \tag{20}$$

ϕ_j^{RF} and ϕ_j^{LW}, respectively, denote the phases of rf-signal and lightwave in path j. $\Delta\phi_j^{\mathrm{RF}}$ and $\Delta\phi_j^{\mathrm{LW}}$ are the deviations of the phases from the ideal condition for the SSB modulation. A_j^{RF} is the induced optical phase due to the rf-signal in path j, and is proportional to the amplitude of the rf-signal applied to the electrode. A_j^{LW} denotes the amplitude of lightwave in path j. f_0 and f_m are the frequencies of input lightwave and rf-signal, respectively. When $\Delta\phi_j^{\mathrm{RF}} = \Delta\phi_j^{\mathrm{LW}} = 0$, the phases are 0, $\pi/2$, π and $3\pi/2$ (0°, 90°, 180° and 270°). The SSB modulator has a pair of Mach-Zehnder structures on an x-cut LN substrate with a CPW, so that we can apply rf-signals of 0°, 90°, 180° and 270° by feeding a pair of rf-signals with 90° phase difference at two rf-ports ($\mathrm{RF_A}$, $\mathrm{RF_B}$) [17]. The optical phase differences are also set to be 90°, by using dc-bias ports ($\mathrm{DC_A}$, $\mathrm{DC_A}$, $\mathrm{DC_A}$). The output optical spectrum, defined by eq. (16), can be expressed by

$$R = A^{\mathrm{LW}} e^{2\pi i f_0 t} \sum_{n=0}^{\infty} J_N\left(A^{\mathrm{RF}}\right) e^{2\pi i N f_m t} \tag{21}$$

$$\begin{aligned} &= A^{\mathrm{LW}} U e^{2\pi i f_0 t} \Big[J_1(A^{\mathrm{RF}}) e^{2\pi i U f_m t} \\ &\quad - J_3(A^{\mathrm{RF}}) e^{-3\cdot 2\pi i U f_m t} + J_5(A^{\mathrm{RF}}) e^{5\cdot 2\pi i U f_m t} \\ &\quad - J_7(A^{\mathrm{RF}}) e^{-7\cdot 2\pi i U f_m t} + \cdots \Big], \end{aligned} \tag{22}$$

where N and U are defined by

$$N \equiv U \times (2n+1) \times (-1)^n \tag{23}$$

$$U \equiv S \times T \tag{24}$$

When the intensity of the electric field is so small that we can neglect high-order harmonic generation at the optical phase modulation, the output for $U = 1$ can be approximately expressed by

$$R \simeq A^{\mathrm{LW}} e^{2\pi i f_0 t} \left[J_1(A^{\mathrm{RF}}) e^{2\pi i f_m t} - J_3(A^{\mathrm{RF}}) e^{-3\cdot 2\pi i f_m t} \right]. \tag{25}$$

Because $J_1 > J_3$, the dominant component in the output is the first order upper sideband (USB), which corresponds to the frequency shifted component. Fig. 10 shows the principle of the SSB modulation. Each sub Mach-Zehnder structure (the pair of paths 1 and 3, or that of paths 2 and 4) is in null-bias point. Thus, the input lightwave component ($e^{2\pi i f_0 t}$) is vanished, and lower and upper sidebands are generated. Due to the $90°$ phase differences in the rf-signal and the lightwave, the polarity of lower sideband (LSB) component at point "P" is the opposite of "Q". The LSB would be vanished at the output port, so that the SSB modulated signal consisting of USB can be obtained. On the other hand, in the case of $U = -1$, LSB can be generated instead of USB. We can easily switch the polarity of T, by changing the dc-bias voltage applied on the port $\mathrm{DC_C}$. The signal-to-noise-ratio (SNR) and conversion efficiency of the frequency shift by the modulator are given by $J_1(A^{\mathrm{RF}})/J_3(A^{\mathrm{RF}})$ and $J_1(A^{\mathrm{RF}})$, respectively (see fig. 11). The conversion efficiency has a maximum of $0.582 \quad (-5.36\,\mathrm{dB})$ when $A^{\mathrm{RF}} = A_m$. As shown in eq. (22), the output lightwave has undesired high-order sidenband components whose orders are $-3, +5, -7, \cdots$ for $U = +1$, and $3, -5, +7, \cdots$ for $U = -1$. Thus, the frequency difference between the generated spectral components equals $4f_m$. We note that the dominant high-order component (J_3) can be suppressed by using a predistortion technique [18].

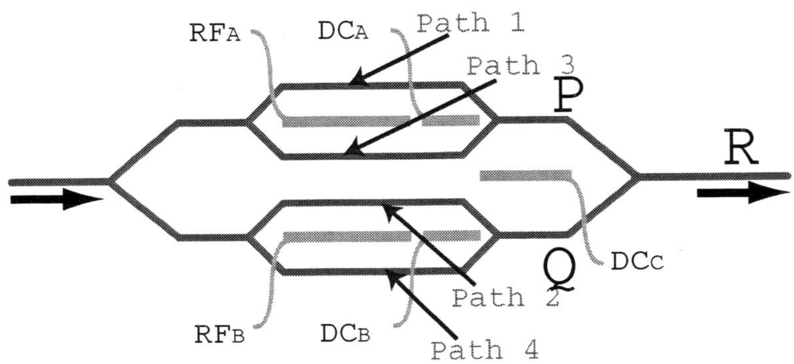

Figure 9. Optical SSB modulator.

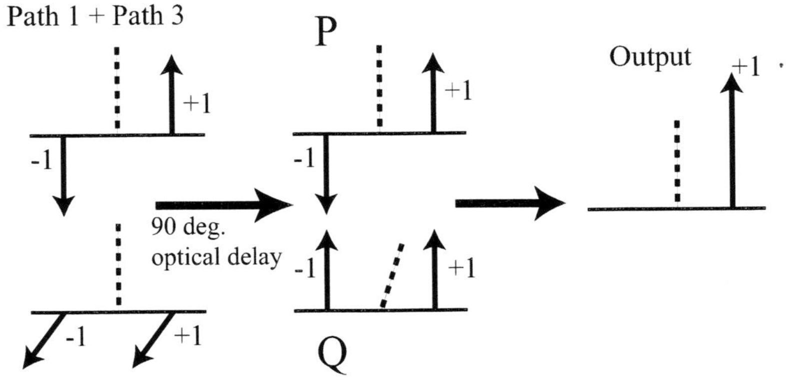

Figure 10. Principle of optical SSB modulation.

We consider the optical frequency shift with the SSB modulator in wavelength-domain-

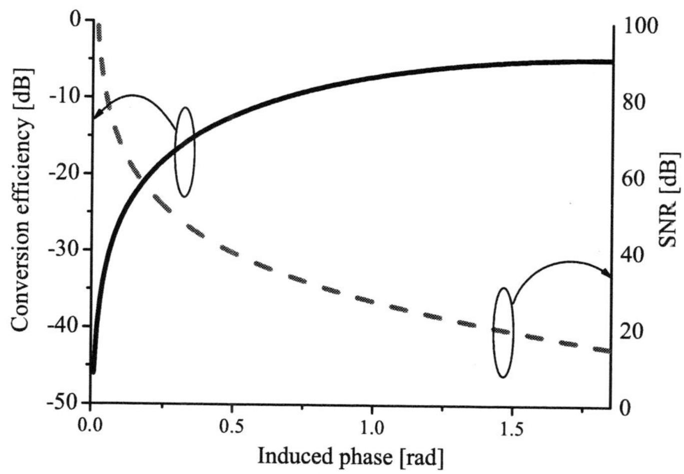

Figure 11. Conversion efficiency and SNR of wavelength shifter using SSB modulation.

multiplexing (WDM) systems with numerical results. Fig. 12 shows a model of a 9-channel WDM system for the simulation with a lightwave network simulator, $Optisystem^{\rm TM}$ 2.2, where the channel spacing was 25 GHz. The bit rate and the optical frequency of the n-th channel were, respectively, $2.5 \times [(n-5) \times (r/100+1)]$ Gbps and $193.5 + (n-5) \times 0.025$ THz, where $r = 2.0$. Thus, the bit rate of the fifth channel was 2.5 Gbps. Those of neighboring channels were 2% higher or lower than that of the fifth, in order to suppress unphysical cross-talks which are due to the finite time windows in this simulation, where the sampling rate and time window in the simulator were 160 Gbps and 40.96 ns, respectively. The non-return-to-zero (NRZ) signals were generated by an LN optical intensity modulator. We used a model for the SSB modulator consisting of four optical phase modulators as shown in fig. 9. The lightwave signals were fed to the SSB modulator via a WDM multiplexer. In order to compensate the loss at the wavelength shift, an EDFA was put at the output lightwave port of the modulator, where the output power was controlled to be 10 dBm by tuning the pump laser power. The noise figure of the EDFA was assumed to be 4 dB. We calculated the optical spectra at the input and the output ports of the SSB modulator with the resolution of 0.01 nm. As shown in fig. 13, optical frequency shift (25 GHz) of WDM channels was successfully demonstrated in this simulation. The undesired third order sideband components were also generated in the output spectra. The frequency difference between the desired frequency shifted component (the first order USB) and the undesired sideband was $4f_m$, so that there were four peaks of them in the spectrum. The other third order components were overlapped with the desired components. In the case of $A^{\rm RF} = 0.92$, the fifth order sidebands were also generated, in addition to the third order sidebands which were dominant in undesired components. Fig. 14 shows Q-factors of the fifth channel at the electric signal monitor in fig. 12, and the total optical power at the output port of the SSB modulator. The electric output of the photo detector was fed to the monitor via a transimpedance amplifier whose impedance was 600Ω and noise figure was 6 dB. Q-factor of the reference signal, where an attenuator was placed at the output port of the WDM multiplexer instead of the SSB modulator, is also shown in fig. 14. The output power was an increasing function of the induced phase A^{RF}. However, the SNR of the output signal is a decreasing function as shown in fig. 11. Due to this trade-off relations, the Q-factor had a peak around $A^{\mathrm{RF}} = 0.5$.

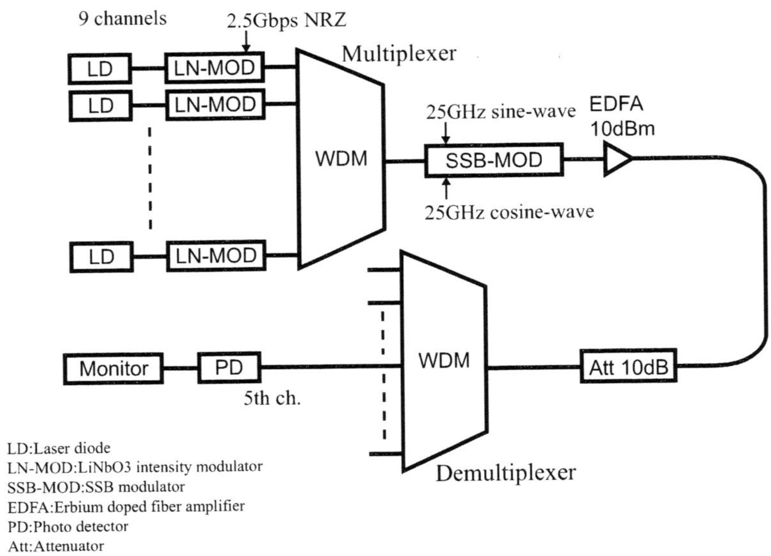

Figure 12. Model for a 2.5 Gbps WDM system with a wavelength shifter using SSB modulation.

2.4. FSK Modulator

The FSK modulator consists of parallel four optical phase modulators as shown in fig. 15 [19, 20]. The device structure is almost the same in the SSB modulator, but the FSK modulator has an electrode ($\mathrm{RF_C}$) for high-speed FSK signal, instead of a dc-bias electrode ($\mathrm{DC_C}$) in the SSB modulator. When we apply a pair of rf-signals, which are of the same frequency f_m and have a 90° phase difference, to the electrodes $\mathrm{RF_A}$ and $\mathrm{RF_B}$, frequency shifted lightwave can be generated at the output port of the modulator. A sub Mach-Zehnder structure of paths 1 and 3 should be in null-bias point (lightwave signals in the paths have 180° phase difference), where the dc-bias can be controlled by $\mathrm{RF_A}$. The other sub Mach-Zehnder structure of paths 2 and 4 are also set to be in null-bias point by using $\mathrm{RF_B}$. To eliminate USB or LSB, the lightwave signal in each path also should have 90° phase difference with each other, as described in Section 2.3.. When the phase difference induced by $\mathrm{RF_C}$ is $\pm 90°$, we can get carrier-suppressed SSB modulation comprising one of the sideband components (USB or LSB). Thus, the optical frequency of output lightwave can be switched by changing the induced phase at $\mathrm{RF_C}$. In the SSB modulator, the electrode for optical phase control $\mathrm{DC_C}$ was not designed for high-speed operation, so that the switching time was limited by the response of the electrode. On the other hand, the FSK modulator has the electrode $\mathrm{RF_C}$ for high-speed optical phase switch at the junction of a pair of sub Mach-Zehnder structures. The amplitudes of USB and LSB are, respectively, described by

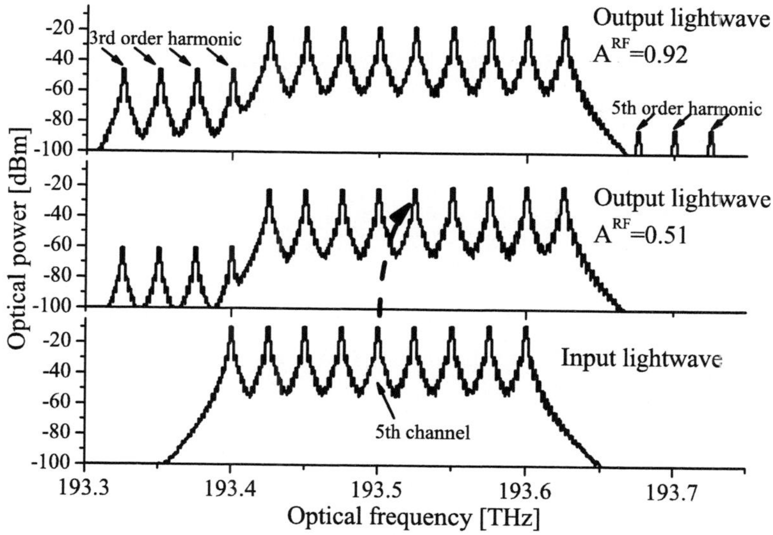

Figure 13. Input and output optical spectra of the SSB modulator.

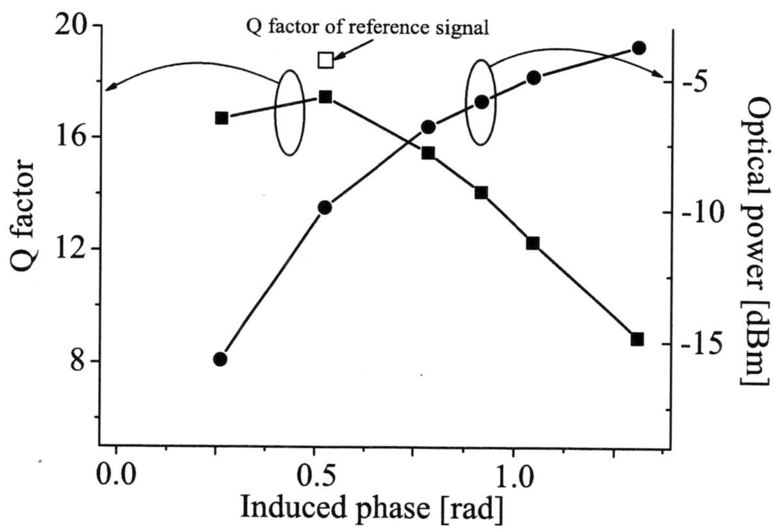

Figure 14. Q-factor of the fifth channel measured by an electric signal monitor, and the total optical power at the output port of the SSB modulator. Q-factor of the reference signal is 18.8, where an attenuator is placed at the output port of the WDM multiplexer instead of the SSB modulator.

$[1 + \mathrm{i}\exp(\mathrm{i}\phi_{\mathrm{FSK}})]/2$ and $[-1 + \mathrm{i}\exp(\mathrm{i}\phi_{\mathrm{FSK}})]/2$, where ϕ_{FSK} is the induced phase difference at $\mathrm{RF_C}$, and $\phi_{\mathrm{FSK}} = -90^\circ$ corresponds to an optimal condition for USB generation ($U = 1$). Thus, by feeding an NRZ signal, whose zero and mark levels correspond to $\phi_{\mathrm{FSK}} = -90, +90^\circ$, to $\mathrm{RF_C}$, we can generate an optical FSK signal, without any parasitic intensity modulation. The bandwidth for the FSK signal should be smaller than the rf-frequency f_m, when the FSK signal is demodulated by optical filters.

An optical FSK modulator having an LN lightwave circuit with traveling wave electrodes to obtain high-speed response, was reported in Ref. [8], where the rise time was about 100 ps, which was limited by driver and receiver circuits, and 3 dB bandwidths of the electrodes ($\mathrm{RF_A}$, $\mathrm{RF_B}$ and $\mathrm{RF_C}$) were 18 GHz, approximately. Thus, the frequency switching time of the optical FSK modulator would be less than 100 ps. We also measured the FSK response by feeding a sinusoidal signal to $\mathrm{RF_C}$. The FSK signal was demodulated into an intensity modulated (IM) signal by an optical filter. We also measured bit-error-ratio (BER) performance of FSK transmission. As shown in fig. 2.4., a 9.95 Gbps NRZ ($2^{31} - 1$ pseudorandom binary sequence:PRBS) signal was applied to $\mathrm{RF_C}$ of the FSK modulator, where a fiber Bragg grating (FBG) was used to convert an FSK signal into an IM signal for demodulation. Fig. 17 shows BER curves and eye-diagrams of 9.95 Gbps FSK for back-to-back, and after transmission through a 60 km single-mode fiber (SMF). The rf frequency f_m was 12.5 GHz. The results show that the eyes are clearly open and that error-free transmission of 10 Gbps FSK-60 km SMF is possible. Power penalties of 60 km transmission with reference to back-to-back at $-\log(BER) = 9$ was -0.8 dB. We deduce that the negative penalty was due to the dispersion of the FBG. In addition, by using balanced detection scheme, we can achieve long-distance transmission and enhanced receiver sensitivity. 10 Gbps FSK transmission over a 130 km SMF was demonstrated by balanced receiver with group delay compensation between USB and LSB components [22].

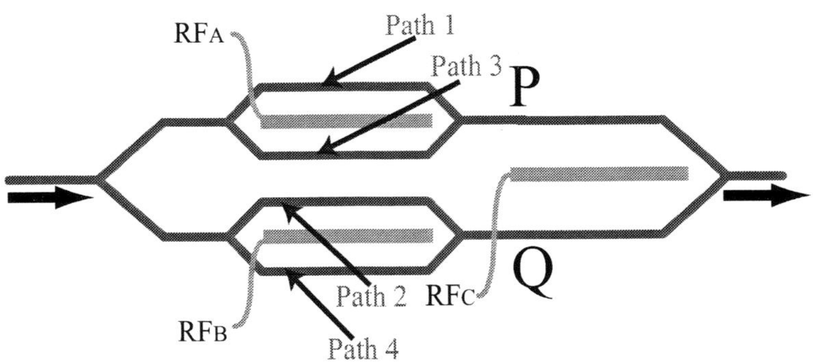

Figure 15. Optical FSK modulator.

2.5. PSK Modulator

By using the SSB modulator designed for optical frequency shift, we can control both phase and amplitude of the output lightwave simultaneously. We apply a pair of baseband signals $g_1(t)$ and $g_2(t)$ to the electrodes $\mathrm{RF_A}$ and $\mathrm{RF_B}$, respectively, where the dc-bias condition

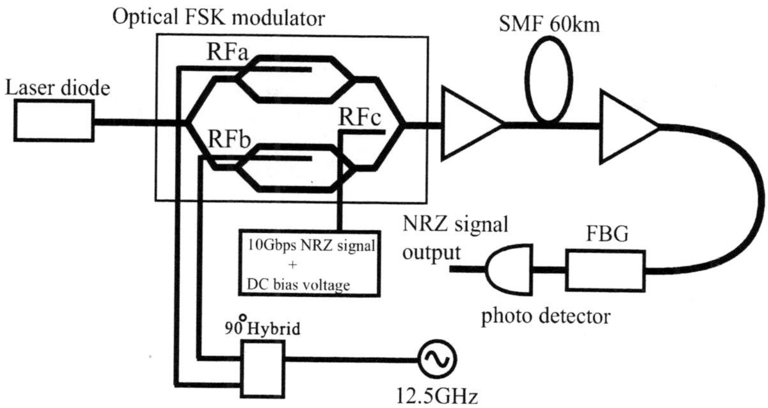

Figure 16. Setup for FSK transmission.

is identical to that of SSB modulation. The output can be expressed by

$$R = \frac{A^{\mathrm{LW}} \mathrm{e}^{\mathrm{i}\pi/4}}{2} \mathrm{e}^{2\pi \mathrm{i} f_0 t} \left[\cos[g_1(t)/2] + \mathrm{i}\cos[g_2(t)/2]\right] \tag{26}$$

$$= \frac{A^{\mathrm{LW}} \mathrm{e}^{\mathrm{i}\pi/4}}{2} \sqrt{\cos^2[g_1(t)/2] + \cos^2[g_2(t)/2]} \times \exp \mathrm{i}\left[2\pi f_0 t + \tan^{-1}(\cos[g_2(t)/2]/\cos[g_1(t)/2])\right] \tag{27}$$

where U is assumed to be -1 ($\phi_{\mathrm{FSK}} = \pi/2$). By using four symbols of $(g_1, g_2) = (0, 0)$, $(2\pi, 0)$, $(0, 2\pi)$, $(2\pi, 2\pi)$, we can generate a quadrature PSK signal, where the phases of the output, $\tan^{-1}(\cos[g_2(t)/2]/\cos[g_1(t)/2])$, are $0, \pi/2, \pi, 3\pi/2$. Similarly, binary PSK can be achieved by an MZM with symbols of $g = 0, 2\pi$.

3. Electrode Design for High-Speed Signals

The EO effects of LN, LT, GaAs, etc. are so small that the length of the modulator should be a few centimeters to obtain effective modulation. When the frequency of the electric signal on the electrode is high, the wavelength of the electric signal would be less than the electrode length. In addition, there is phase delay due to lightwave propagation along the waveguide. Thus, we have to use a transmission line theory to design the electrode for high-frequency operation. The moving coordinate system also should be used to investigate the effect of the phase delay due to the propagation. There are two types of electrode for high-frequency operation: traveling-wave and resonant electrodes. The traveling-wave electrode gives broadband response from dc to 50 GHz. Thus, the modulator with the traveling-electrode is suitable for high-speed digital transmission system. On the other hand, the resonant electrode can be used for band operation, for example, 25-26 GHz band [23, 24]. The merits of the resonant electrodes are 1) the length of the electrode is much shorter that that of the traveling-wave, and 2) the modulation efficiency is large at the particular frequency. The modulator cannot be used for broadband digital systems, but it is useful for some special applications, such as, clock generation, sideband generation with a fixed

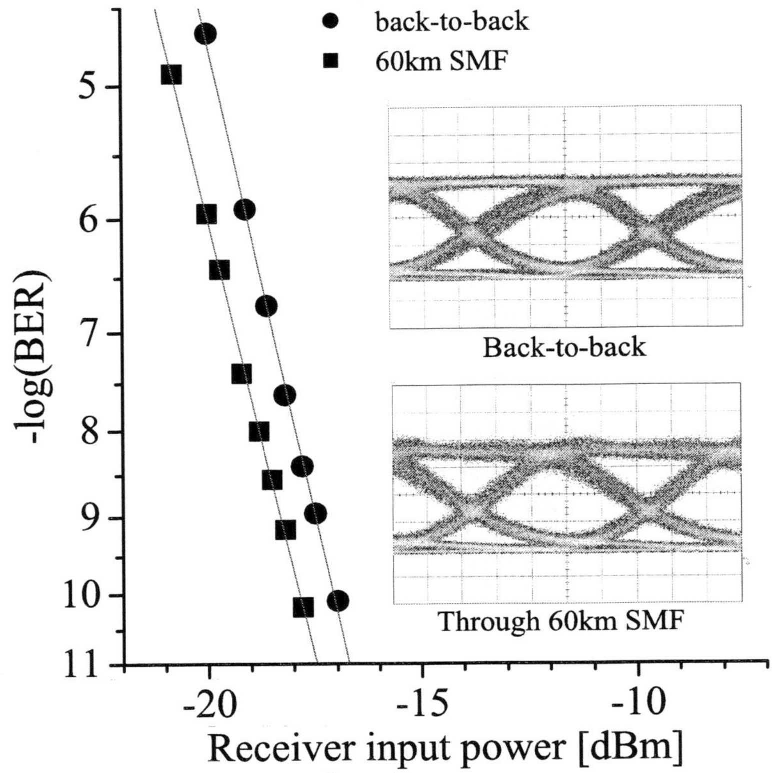

Figure 17. Bit-error-ratio curves and eye-diagrams of 9.95 Gbps FSK.

frequency separation and radio-on-fiber systems. The modulator size is much smaller than the conventional one, so that this is suitable for integrated modulators which have several functions. In this section, we investigated two types of resonant structures: an asymmetric resonant structure and a double-stub structure, to obtain compact and effective optical modulators for band-type operation. Combination of series and parallel resonance was used in these structures. The impedance at the feeding point can be increased by parallel resonance between a modulating electrode and a stub.

3.1. Asymmetric Resonant Structure [5]

As shown in fig. 18, the modulating electrode has two arms that are asymmetric coplanar waveguides (ACPWs). The ground plane, modulating electrode, and feeding line consist of gold whose thickness is t_{EL}, where s is the width of the gap between the ground plane and the modulating electrode, and w is the width of the modulating electrode. The feeding line, which is connected at the junction of the two arms, is a CPW. A buffer layer, whose thickness is t_{BUF}, is there to reduce the loss of the lightwave propagating in the optical waveguides under the metal electrodes. The short arm (arm I), with length denoted by L_1, is short-ended, while the long arm (arm II), with length denoted by L_2, is open-ended. The two arms differ in length and termination, so that we termed this configuration the asymmetric resonant structure. The equivalent circuit is shown in fig. 18 (c), where the

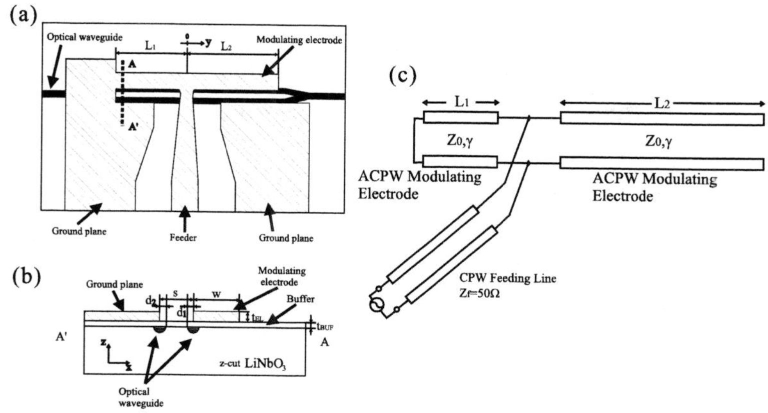

Figure 18. Asymmetric resonant structure. (a) top-view, (b) cross-section, (c) equivalent circuit.

impedance of arm I (Z_1) and that of arm II (Z_2) are expressed by

$$Z_1 = Z_0 \tanh \gamma L_1 \tag{28}$$

$$Z_2 = Z_0 \coth \gamma L_2. \tag{29}$$

Z_0 is the characteristic impedance of ACPW. $\gamma(= \alpha + i\beta)$ denotes the propagation coefficient. The total impedance of the asymmetric resonant structure is given by

$$Z_L = Z_0 \frac{\tanh \gamma L_1 \coth \gamma L_2}{\tanh \gamma L_1 + \coth \gamma L_2}. \tag{30}$$

The voltage on the modulating electrode is given by

$$V(y,t) = T_r G(y) V_{in} e^{2i\pi f_m (t-t_0)}, \tag{31}$$

where $V_{in} e^{2i\pi f_m (t-t_0)}$ is the input voltage whose frequency is f_m. The voltage transmittance at the junction is defined by

$$T_r \equiv \frac{2Z_L}{Z_L + Z_f}, \tag{32}$$

where Z_f is the characteristic impedance of the feeding line. $|T_r|$ is an increasing function of $|Z_L|$, and the range is from 0 ($|Z_L| = 0$) to 2 ($|Z_L| = \infty$). $G(y)$ expresses the distribution of the standing voltage wave on the electrode, and is defined by

$$G(y) \equiv \begin{cases} \frac{\cosh \gamma(L_2 - y)}{\cosh \gamma L_2} & (y > 0) \\ \frac{\sinh \gamma(L_1 + y)}{\sinh \gamma L_1} & (y < 0) \end{cases}. \tag{33}$$

The induced phase at each optical waveguide is the sum of Pockels effect with respect to the coordinate system moving along with the lightwave propagating in the optical waveguide. Thus, the difference of the induced phases of the two optical waveguides of the Mach-Zehnder structure can be expressed by

$$\phi = \frac{\pi}{\lambda_0} n_0^3 r_{33} \frac{\Gamma}{s} L \Phi V_{\text{in}} \quad (34)$$

$$\Phi \equiv \frac{1}{L V_{\text{in}}} \int_{-L_1}^{L_2} V(y, \frac{y}{c} n_0 + t) \mathrm{d}y \quad (35)$$

$$\Gamma = \Gamma_1 + \Gamma_2 \quad (36)$$

where c is the speed of light. λ_0 and n_0 are the wavelength and RI of the lightwave, respectively. $L \equiv L_1 + L_2$ is the total length of the modulating electrodes. Γ_1 is the overlap integral between the field induced by the electrode and the field of the lightwave in the waveguide under the edge of the modulating electrode. Γ_2 is that of the ground plane. We consider a z-cut LN substrate where the c-axis is in z-direction. Pockels effect between the z-direction component of the electric field induced by the modulating electrode and lightwave whose electric field is polarized in the z-direction is considered, because the electrooptic coefficient of the effect mentioned above (r_{33}) is the largest one. The factor of $\frac{y}{c} n_0 + t$ means the phase difference due to the propagation delay of the lightwave. The integral of the voltage on the electrode defined by eq. (35), Φ is termed normalized induced phase, and shows the effect of the resonant structure. The normalized induced phase for an optical modulator with a resonant structure can be larger than unity, while the normalized induced phase for a lossless perfectly velocity-matched traveling wave modulator equals unity. The half-wave voltage of the modulator V_π is given by $\pi/|\phi|$. Similarly, $V_\pi L$, defined by $\pi L/|\phi|$ ($\propto |\Phi|^{-1}$), is the voltage-length product which shows the degree of the modulation efficiency and the scale of the modulator.

To obtain compact and effective optical modulators for band-operation, we designed them to have large normalized induced phases. We consider an ACPW with the following: $t_{BUF} = 0.55$ μm, $s = 27$ μm, $w = 50$ μm, and $t_{EL} = 2$ μm. γ and Z_0 at 10 GHz were $11.24 + \text{i}891.8\, m^{-1}$ and $34.71\ \Omega$, which were obtained by using an electro-magnetic field calculator with the finite element method (HP-HFSS5.4). Thus, the wavelength of a 10 GHz voltage wave on the ACPW was 7.03 mm. The following parameters were used for numerical calculations: $\varepsilon_{\text{SiO}_2} = 4.0$, $\varepsilon_{\text{LN[xy]}} = 43.0$, $\varepsilon_{\text{LN[z]}} = 28.0$ and $\sigma_{Au} = 4.3 \times 10^7 (\Omega\text{m})^{-1}$. $\varepsilon_{\text{LN[xy]}}$ and $\varepsilon_{\text{LN[z]}}$ denote permittivities of LN in the xy-plane and the z-direction, respectively. $\varepsilon_{\text{SiO}_2}$ and σ_{Au} are permittivity of the buffer layer (SiO_2) and conductivity of the electrodes (Au), respectively. The central inductor of the feeding line was 50 μm wide and the characteristic impedance was 50 Ω. By using eqs. (34) and (35), we obtained a combination of L_1 and L_2, which gave a maximum of Φ. When $L_1 = 0.03\lambda$ and $L_2 = 0.22\lambda$, Φ had a maximum of 3.7 at 10 GHz. The total length of the modulating electrodes L was 1.76 mm. Fig. 19 shows the normalized induced phase as a function of the frequency f_m, and the voltage wave distribution on the modulating electrode. In addition to a peak at 10 GHz, there are peaks due to high-order resonance at 30 and 50 GHz. In fig. 20, the voltage reflectivity at the junction and the total impedance of these two modulating electrodes were shown as functions of f_m. The denominator of eq. (30) goes to a minimum at resonance, which corresponds to parallel resonance of the two arms, so that the impedance has peaks at the resonant frequencies (10, 30, 50, and 70 GHz). As shown in fig. 20(a), the reflectivity has dips, where the impedance has peaks. The peaks of the optical response were caused

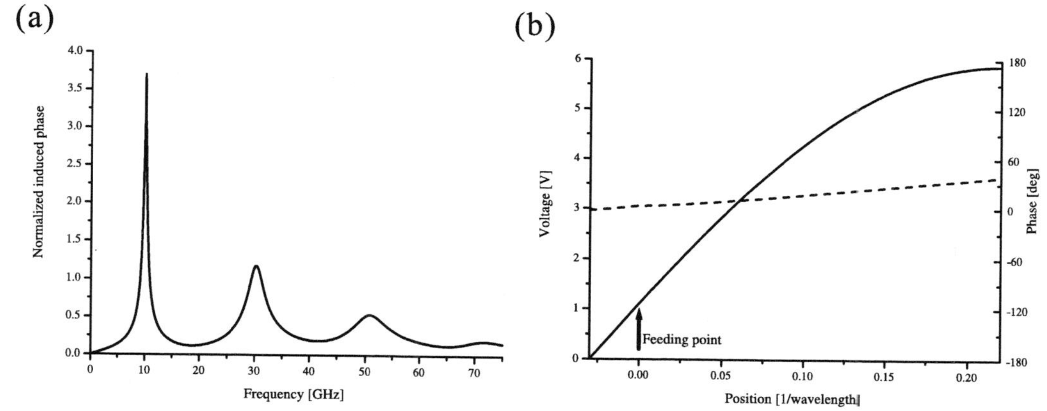

Figure 19. (a) Optical response of the asymmetric resonant structure, and (b) voltage wave distribution on modulating electrodes, with respect to the coordinate system moving along with the lightwave propagating in the optical waveguide $V(y, \frac{y}{c}n_0 + t)$. Solid line and dashed line denote the amplitude and the phase of the voltage wave, respectively.

by enhancement of T_r. Actually, the impedance Z_L is not equal to Z_f at the first-order resonance (10 GHz), as shown in fig. 20(b). The impedance matching performed at the feeding point is not necessary to obtain a large optical response, because Pockels effect is caused by the electric field induced by the electrodes regardless of the power input from the feeding line.

If both arms were terminated by the same impedance (e.g. open or short), the polarity of the standing-wave voltage profile would change on the electrodes and the length L_1 or L_2 should be longer to obtain the resonance. In an asymmetric resonant structure, however, both L_1 and L_2 are smaller than a quarter of the wavelength λ. As shown in fig. 19(b), the polarity does not change, which is why an asymmetric structure can reduce the half-wave voltage and the electrode length. At the end of arm II, the voltage amplitude is approximately six times as large as the input voltage. The length of arm II is slightly shorter than $\lambda/4$, so that arm II is nearly in resonance and that the impedance of arm II is small. But, the total impedance Z_L is large owing to arm I which has two functions: as a stub to enhance the impedance and as a modulating electrode. Arms I and II are in parallel resonance, where arm I is capacitive and arm II is inductive.

3.2. Dual-Stub Structure [6]

As shown in fig. 21 (a), the electrode structure consists of a modulating electrode and two stubs. The modulating electrode has two arms of the same length, and is an ACPW. A feeding line, which is a CPW, is connected at the center of the modulating electrode. Two stubs, which are also CPWs of the same length, are also connected at the junction of the modulating electrode and the feeding line, so that we call it the double-stub structure. The equivalent circuit is shown in fig. 21 (b). The total impedance is given by

$$Z_L = \frac{Z_{01}Z_{02}\coth\gamma_1 L_1 \tanh\gamma_2 L_2}{2(Z_{01}\coth\gamma_1 L_1 + \tanh\gamma_2 L_2)}, \tag{37}$$

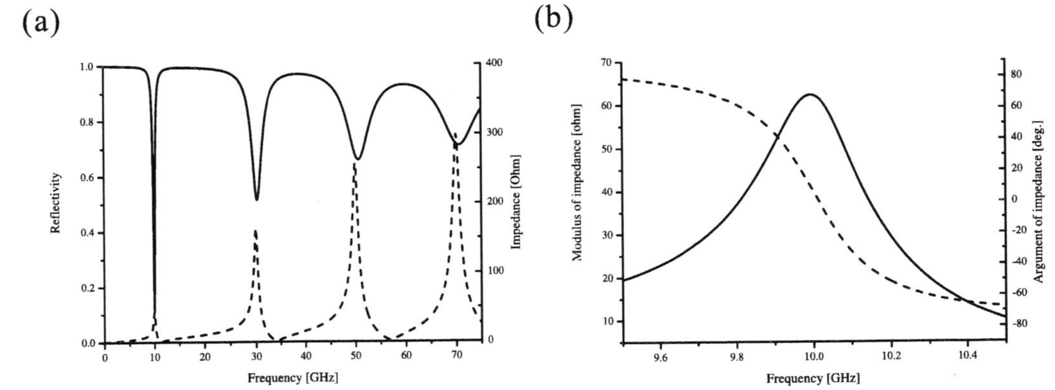

Figure 20. Electric response of the asymmetric resonant structure (a) solid line and dashed line denote voltage reflectivety at the junction and impedance of the optimized structure, respectively. (b) impedance of the modulator near the peak at 10 GHz. Solid line and dashed line denote modulus and argument of the impedance.

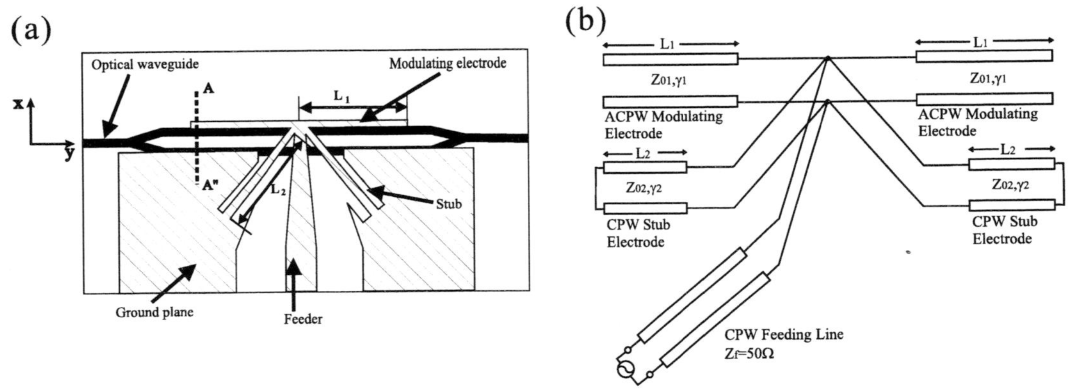

Figure 21. Double-stub structure (a) top-view, (b) equivalent circuit.

when we consider the case in which the modulating electrode is open-ended and the stub is short-ended. L_i, γ_i and Z_{0i} denote the lengths, the propagation coefficients and the characteristic impedances of the modulating electrode $(i = 1)$, and of the stub $(i = 2)$. The voltage on the modulating electrode is given by eq. (31), where the distribution of the standing wave voltage wave on the modulating electrode of the double-stub structure is expressed by

$$G(y) = \frac{\cosh \gamma_1 (L_1 - |y|)}{\cosh \gamma_1 L_1}. \tag{38}$$

The difference of the induced phases of the two optical waveguides of the Mach-Zehnder structure can be expressed by eq. (34), but the normalized induced phase for the double-stub structure is defined by

$$\Phi \equiv \frac{1}{V_{\text{in}} L} \int_{-L_1}^{L_1} V(y, \frac{y}{c} n_0 + t) \mathrm{d}y, \tag{39}$$

where $L \equiv 2L_1$ is the total length of the modulating electrodes.

We considered the electrode structure with the following parameters. The gap between the electrodes of ACPWs and CPWs is 27 μm. The width of the modulating electrode is 5 μm, while that of the stub is 50 μm to reduce loss at the stubs. The characteristic coefficients of the modulating electrode (ACPW) are $\gamma_1 = 25.92 + \mathrm{i}735.2\ m^{-1}$, $Z_{01} = 70.1\Omega$. Those for the stub (CPW) are $\gamma_2 = 14.29 + \mathrm{i}863.5\ m^{-1}$, $Z_{02} = 28.7\Omega$. By using eq. (35), we obtained a combination of L_1 and L_2, which gave a maximum of Φ. When $L_1 = 0.19\lambda_1$ and $L_2 = 0.12\lambda_2$, Φ had a maximum of 2.44 at 10 GHz, where λ_1 and λ_2 are, respectively, the wavelength on the modulating electrode and the stub. The total length of the modulating electrodes L was 3.25 mm. As shown in fig. 22, the optical response and

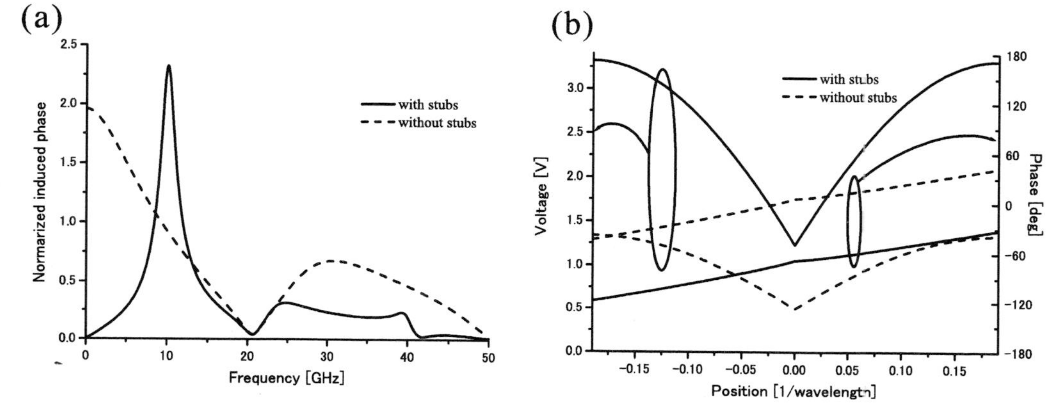

Figure 22. (a) Optical response of the double-stub structure, and (b) voltage wave distribution on modulating electrodes, with respect to the coordinate system moving along with the lightwave propagating in the optical waveguide $V(y, \frac{y}{c}n_0 + t)$. Solid line and dashed line correspond to the structures with stubs and without stubs, respectively.

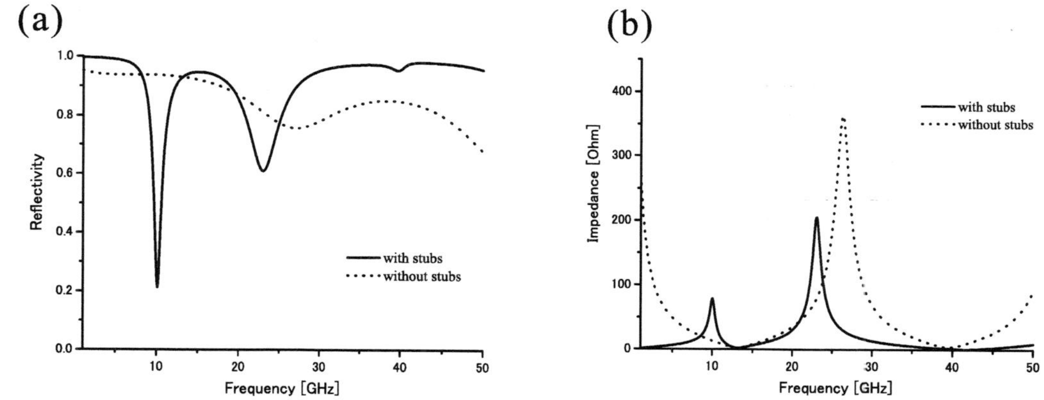

Figure 23. Electric response of the double-stub structure. (a) voltage reflectivity at the junction. (b) total impedance of the resonant structure. Solid line and dashed line correspond to the structures with stubs and without stubs, respectively.

the voltage profile at 10 GHz can be enhanced by using the stubs. Fig. 23 shows the electric response of the designed structure. The impedance had a peak at the resonant frequency, due to the parallel resonance between the modulating electrode and the stub, in the same way as the asymmetric resonant structure. Even in the case of the structure without stubs, there was a peak in the impedance, but it does not correspond to the enhancement of the optical response.

3.3. Numerical and Experimental Results

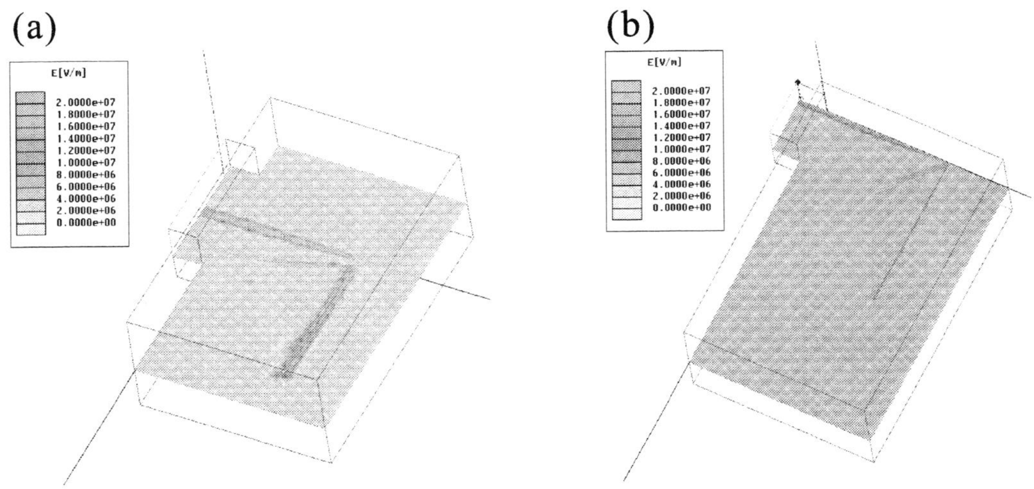

Figure 24. Electric field distribution on the resonant structures, calculated by FEM. Input power and frequency were 1W and 10GHz, respectively. (a) asymmetric resonant structure. (b) double-stub structure.

By using the finite element method (FEM), we investigated electro-magnetic field in the resonant structures. Fig. 24 shows the electric field distribution on the resonant structures. The double-stub structure is symmetric with respect to xz-plane, so that a half of the structure $(y > 0)$ was used in FEM, as shown in fig. 24 (b). The electric field on the modulating electrodes were enhanced both in the asymmetric resonant structure and in the double-stub structure.

We also measured optical and electric responses of fabricated modulators. The Mach-Zehnder interferometer consisted of two optical waveguides formed by the titanium diffusion technique on a z-cut LN substrate. The RI n_0 was assumed to be 2.2 for $\lambda_0 = 1.55\ \mu$m. The effective width and depth of the optical waveguides were 10 μm and 8 μm, respectively. The positions of the waveguides were as follows: $d_1 = 2\ \mu$m and $d_2 = 3\ \mu$m. The overlap integral Γ, calculated by TEM approximation [25], was 0.92, where the thickness of the electrode t_{EL} was neglected. The profile of the mode guided in the optical waveguide was elliptic and the field distribution was assumed to be uniform. The modulator also had an open-ended dc-bias electrode to set the Mach-Zehnder switch function to the quadrature point. The normalized induced phase can be measured as the ratio of the optical response of a resonant electrode at a particular frequency to that of a non-resonant electrode at dc, where both electrodes were of the same length. Γ obtained by the response measured at dc was 0.82, which was 89% of the calculated one. This discrepancy was due to the effect of the finite electrode thickness or deformation of the lightwave field pattern. Fig. 25 (a)

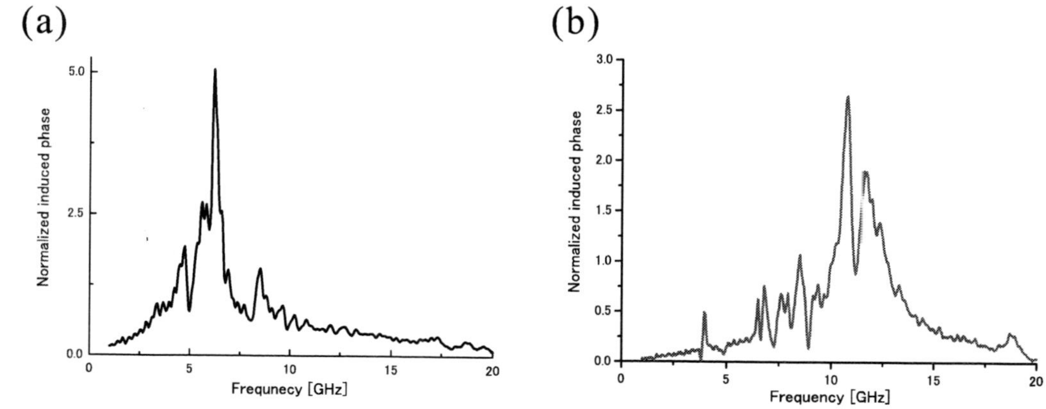

Figure 25. Measured optical responses (normalized induced phase) of fabricated modulators. (a) asymmetric resonant structure. (b) double-stub structure.

shows the optical responses of the asymmetric resonant structure for the 1.55-μm region, obtained from the responses of the resonant and dc-bias electrodes. The responses had a peak of 4.94 at 6.2 GHz, while the designed resonant frequency was 10 GHz. We deduce that this difference is due to the asymmetry of the resonant structure. The electric field of the fundamental CPW mode is symmetric, so that the mismatch of the field pattern occurs at the junction of the modulating electrode, as shown in fig. 24 (a). On the other hand, the double-stub structure is symmetric with respect to the central electrode of the feeding line, so that the electric field is also symmetric. The optical response of the double-stub structure had a peak of 2.65 at 10.75 GHz, which is in good agreement with the result of the equivalent circuit. We also measured the responses of the double-stub structures whose electrode lengths L_1 and L_2 were slightly changed from the designed values [6]. V_π and Φ were sensitive to the stub length L_2. The normalized induced phase of a modulator whose stub length was 90% of the designed one, and had a peak of 3.41 at 10.55 GHz. The half-wave voltage was 13.7 V at the peak, while the total length L was 3.25 mm.

Two types of resonant structures for band-operation optical modulators, the asymmetric resonant structure and the double-stub structure, were investigated, by using the equivalent circuits, the finite element method technique, and conventional measurements for electric and optical responses. The normalized induced phase of the fabricated modulator using the asymmetric resonant structure was 4.94. The half-wave voltage was 17.9 V in the 1.55 μm wavelength range, in spite of the short electrode length of 1.76 mm. However, due to the asymmetry of the structure, the resonant frequency was shifted from the frequency designed by the equivalent circuit. On the other hand, the optical response of the double-stub structure had a resonant peak near the designed frequency, and the half-wave voltage was 13.7V with the electrode length of 3.25 mm.

Resonant-type modulators can acts both on forward and backward lighwaves, while traveling-wave electrodes cannot on backward lightwaves. Thus, the lightwave reflection in the resonant-type modulator can double the modulation efficiency [7]. The halfwave voltages of a double-stub structure with lightwave reflection was 2.9 V at 7.0 GHz.

4. Photonic Sideband Management (PSBM) Techniques

PSBM techniques, can manipulate optical or electrical signals in time or frequency domain by using sideband generation at optical modulators. A couple of PSBM techniques, such as, reciprocating optical modulation (ROM), optical label swapping (OLS) using double-sideband (DSB) modulation and optical tunable delay, are described, where the frequency and time domain profiles of the input lightwave are controlled by an electric signal. ROM can generate lightwave modulated by a high-frequency signal whose frequency is an integer multiple of the electric signal fed to the modulator. Thus, we can easily obtain a lightwave signal having a millimeter-wave from a microwave signal. OLS with DSB modulation can be used for optical packet switching systems. This technique can be applied for a bundle of wavelength-domain-multiplexing (WDM) channels, so that we can construct simple nodes. The tunable delay line can also be used in optical packet systems, to prevent the packets from colliding at the nodes.

4.1. IM/FSK Modulation and OLS Using Double-Sideband Modulation [8]

A combination of IM and FSK is a promising technique for optical label switching in optical packet systems [26, 27]. In this technique, payload signals are in IM format, while label information is written by FSK signal. The merit of this FSK labeling is that an FSK transmitter generates the label information on the optical carrier frequency without affecting its intensity. Simultaneously modulated IM and FSK signals are independently demodulated by using an optical filter, so that the label information can be extracted without affecting the payload signal. OLS is a promising technique for implementing packet routing and forwarding functions over optical packet networks. Previously reported OLS systems used cross modulations between optical spectral components, such as cross phase modulation in an optical amplifier, in order to generate a new label signal. Thus, the OLS system should have additional light sources for the cross modulation. In an OLS system having wavelength-domain-multiplexing (WDM) channels, we have to use a pumping light source and a cross modulation device for each WDM channel, so that the setup for the OLS with WDM becomes extremely complicated. In this section, we describe an OLS technique using DSB modulation technique. As shown in fig. 26, the setup is very simple where pumping light sources are not necessary. IM/FSK signal has a pair of optical carriers whose frequencies are $f_0 - f_m$ and $f_0 + f_m$, where f_0 is the optical carrier frequency at the output port of the light source. f_m denotes the frequency of the modulating signal fed to the electric input ports $\mathrm{RF_A}$ and $\mathrm{RF_B}$ of the FSK modulator. When the FSK switching signal fed to $\mathrm{RF_C}$ is in “1” state where the modulator generates USB ($\phi_{\mathrm{FSK}} = -90°$), the output carrier frequency is $f_0 + f_m$. That of “0” state, where the modulator generates LSB ($\phi_{\mathrm{FSK}} = +90°$), is $f_0 - f_m$. The IM/FSK signal is fed to an intensity modulator which is in null-bias point to obtain DSB-SC modulation. The modulating signal is a sinusoidal wave whose frequency is f_m. The output has three optical spectral components whose carriers are $f_0 - 2f_m$, f_m and $f_0 + 2f_m$. When the FSK switching signal is in “1” state, the output of the intensity modulator has two carriers of f_0 and $f_0 + 2f_m$. On the other hand, the output of “0” has two carriers of f_0 and $f_0 - 2f_m$. The output always has the f_0 component regardless of the FSK switching signal. Thus, by using an optical bandpass filter whose center frequency is f_0, we can remove the FSK signal from the IM/FSK signal, and restore a pure IM signal which does not have an FSK signal. The output of the DSB-SC modulation has only one optical carrier of f_0, so that we can add a new FSK label by using another FSK modulator. As shown in fig. 27, the FSK modulator can be used for a WDM signal which has many IM channels. We can simultaneously shift the carrier frequencies of the channels by one FSK modulator, as described in Section 2.3.. For an FSK-WDM signal, an optical filter which has a periodic optical frequency response, such as an interleaver, an arrayed-waveguide, and so on, can be used to demodulate the FSK signal. The proposed label swapping technique is also applicable to WDM systems by using an optical filter having the periodic response.

An experimental setup for IM/FSK modulation and our proposed OLS system is shown in fig. 28, where the bit rates of FSK and IM were, respectively, 1 Gbps and 9.95 Gbps (NRZ, PRBS $2^{31} - 1$). The frequency deviation of FSK (f_m) was 12.5 GHz. The extinction ratio of the IM modulation was 3.9 dB. Residual IM in the FSK modulation was so small that the IM output which can be generated just by feeding the IM/FSK signal to a high-speed photodetector has a clearly opened eye-diagram. The FSK signal was demodulated by using an optical filter which can discriminate between USB and LSB. Fig. 29 shows BER curves and eye-diagrams of the FSK and IM signals, where error-free transmissions for FSK and IM were obtained. By using DSB-SC modulation, we demonstrated the restoration of a pure IM signal which has one optical carrier. Fig. 30 shows the optical spectra at the output port of the DSB modulator and at the output port of the optical bandpass filter. The spectral components of $f_0 - 2f_m$, f_0 and $f_0 + 2f_m$ were generated at the DSB modulator. The restored pure IM signal having the f_0 component was extracted by the bandpass filter. We also obtained clear eye-opening and error-free transmissions for the restored IM, as shown in fig. 31. An IM/FSK signal with a new FSK label can be generated by another FSK modulator placed at the output port of the optical filter, so that we can construct a simple OLS system consisting of a DSB-SC and an FSK modulators.

4.2. Reciprocating Optical Modulation [14]

An ROM, consisting of a pair of optical filters and an optical modulator, can generate high-order sideband components of EO modulation, as shown in fig. 32, where the FBGs were fixed in V-grooves on SiO_2 substrates and directly attached to the LN modulator chip. One of the optical filters is placed at the optical input port (input filter), and the other is at the output port (output filter). By using an ROM, we can generate a lightwave modulated by an rf signal whose frequency is an integer multiple of an electric rf signal applied to the modulator. The phase-shifted FBG has a narrow pass band of 1.5 GHz in the reflection band. The output filter was a conventional FBG with a heater to let the center wavelength equal to that of the input filter. When the input wavelength was in the pass band of the input filter, both USB and LSB reciprocate between the input and output filters, so that high-order DSB components can be effectively generated. The resonant electrode enhances modulation efficiency for forward and backward lightwaves, while a traveling-wave electrodes acts only on forward propagating lighwaves. The resonant frequency of the electrode was designed to be equal to inverse of delay in a reciprocation process, so that the successive modulation process can be in phase. Fig. 33 shows the output spectrum of the fabricated ROM and that

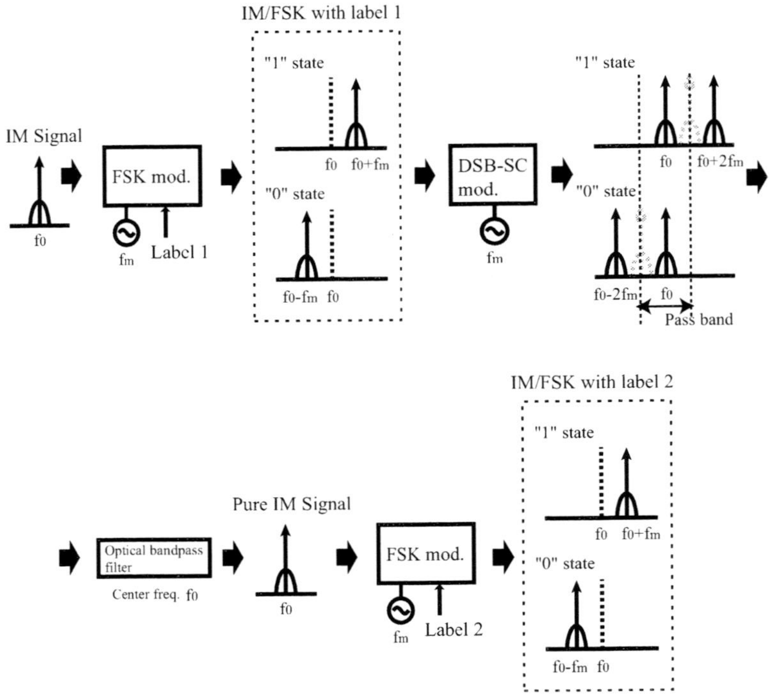

Figure 26. Principle of OLS of IM/FSK signal.

of the conventional optical phase modulation, where the input rf-power and frequency were, respectively, 19.8 dBm and 5 GHz. Optical power and frequency of the input lightwave were, respectively, 4.2 dBm and 192.956 THz. Sideband components lower than 5th-order were in the reflection band, so that 5th or 6th order harmonic components were generated effectively.

4.3. Optical Tunable Delay Line Using SSB Modulation [15]

Optical buffers that are composed of delay lines are key components in optical packet switching systems. To prevent packets from colliding, it is necessary to vary the delay at the buffer. There are two types of configurations for tunable optical delay lines. One is comprised of an optical switch and a fiber loop, where the operation of the switch must be synchronized with the propagation cycle of the packet in the loop. The other consists of an array of several delay lines of various lengths connected to an optical switch by which we can select one from the array delay line. Each line has an optical amplifier and an optical switch, which results in a complicated setup. In this section, we describe an optical tunable delay line consisting of an optical single-sideband (SSB) modulator, a fiber Bragg grating, and an optical fiber loop. The number of times, a lightwave circulates in the loop, depends on the frequency of the electric signal applied to the modulator. Thus, we can control the delay by switching the frequency without using delay lines of various lengths. In addition, it is not necessary to synchronize the switching timing with the cycle of the packet propagation.

As shown in fig. 34(a), the tunable delay line has an optical port consisting of two optical

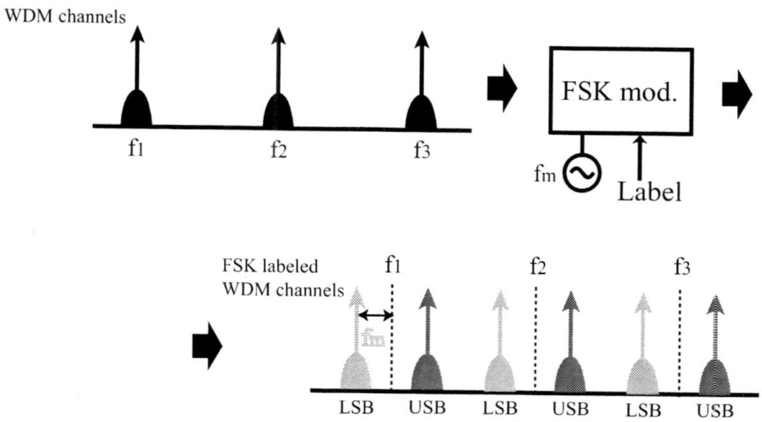

Figure 27. FSK labeling on bundled WDM channels.

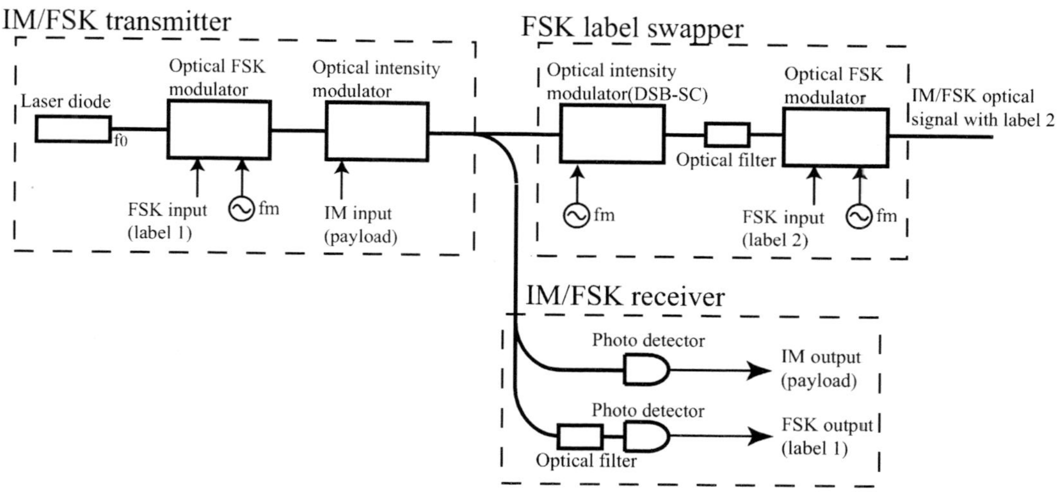

Figure 28. Setup for IM/FSK modulation and OLS using DSB-SC.

circulators and an FBG placed between them. A lightwave whose optical frequency is out of the reflection band of the FBG can pass through the port, while a lightwave in the band cannot. Thus, an input lightwave whose frequency is out of the reflection band comes into the loop. When the modulator is not in operation, the lightwave goes through the loop once, and then it exits from the port. Consider that the SSB modulator is in operation, and that the input optical frequency (f_0) is slightly lower than the edge of the reflection band, as shown in fig. 34(b), where f_m denotes the frequency of the electric signal fed to the modulator, and f_r is the reflection band width of the FBG. The input optical frequency can be shifted at the SSB modulator. We assume that the dc-bias voltages for the modulator are tuned to generate the upper sideband. Of course, we can use the lower sideband instead of the upper sideband when f_0 is slightly higher than the upper edge of the reflection band. The f_0 lightwave, coming into the loop through the port, is input to the SSB modulator, and then its frequency is shifted into the reflection band (f_0+f_m). Thus, the lightwave is reflected by the FBG, and circulates in the loop again. During the successive circulation steps, the optical frequency is shifted at the modulator. After n times circulation, the frequency of the lightwave becomes

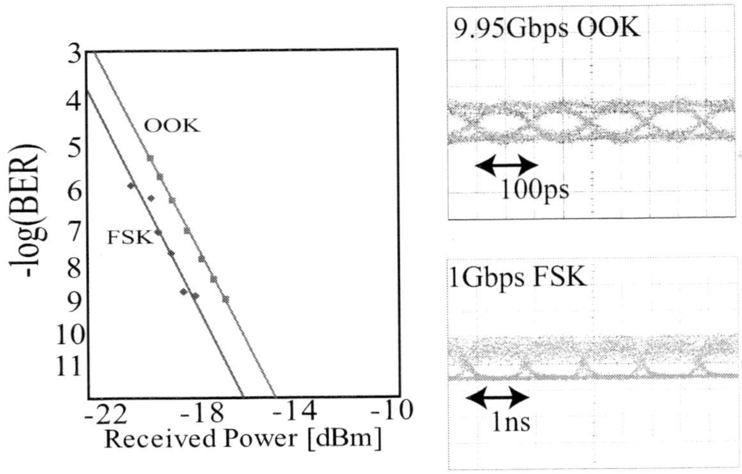

Figure 29. Bit-error-ratio curves and eye-diagrams for FSK and IM signals.

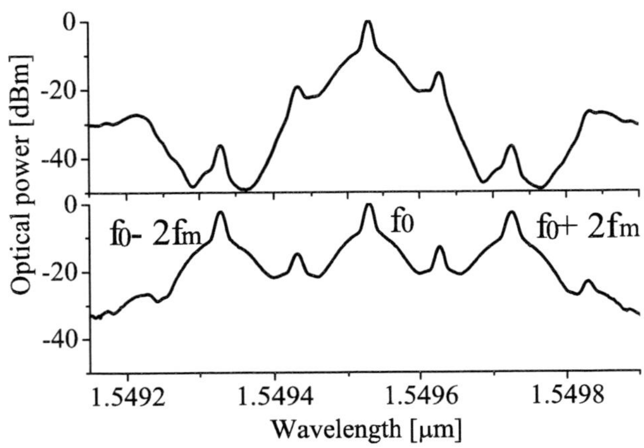

Figure 30. Optical spectra at the output port of the DSB modulator (lower) and at the output port of the optical bandpass filter (upper).

$f_0 + nf_m$. We call such a spectral component an n-th order channel. After circulating several times, the lightwave exits the reflection band, where the number of circulation steps p is given by $pf_m > f_r > (p-1)f_m$. In this section, we call the spectral component of $f_0 + pf_m$ the prime channel, which is the lowest order channel in the channels whose frequency is higher than the reflection band. The output lightwave contains some other components resulting from the nonlinearity of the modulator and residual transmittance in the reflection band. However, the prime channel would be dominant when we can consider that f_m is larger than the edge slope width of the reflection band, and that any high-order harmonic generation at the modulator is negligible. The propagation delay of a circulation step is given by $\tau = l/c$, where l denotes the effective optical length of the loop, and c is the speed of light. The delay of the prime channel at the loop is dependent on f_m, as follows: $p\tau = \tau \times Int(f_r/f_m)$, where $Int(x)$ gives the smallest integer larger than x. Thus, we can change the delay $p\tau$ by switching the frequency of the electric signal f_m. The frequency

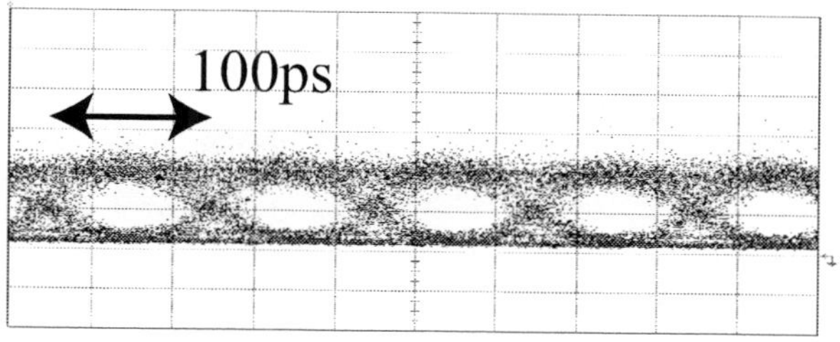

Figure 31. Eye-diagram of restored pure IM signal at the output port of the optical band pass filter.

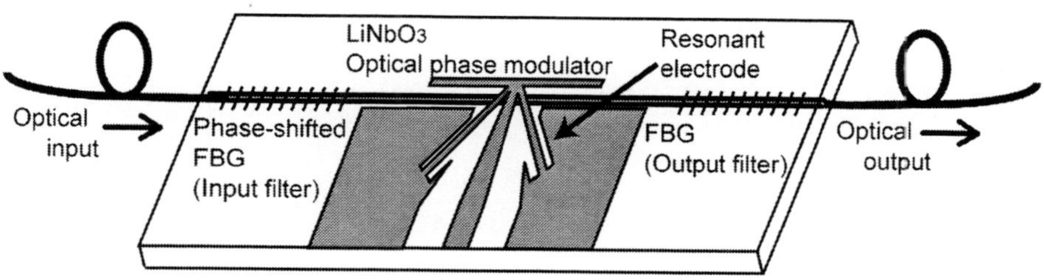

Figure 32. Reciprocating optical modulator with resonant electrode.

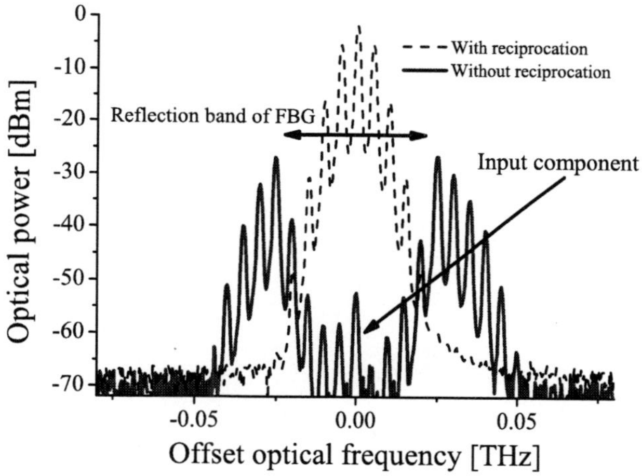

Figure 33. Output spectra modulated by ROM and by conventional phase modulation.

can be switched at any time except when a packet is passing through the SSB modulator. Thus, it is not necessary to synchronize the switching timing with the propagation cycle of the packet. In addition, we note that pulses longer than the optical length l can pass through the delay loop without any collision owing to frequency shift at the modulator.

Fig. 35 shows the time domain envelopes and the frequency domain spectra of the output lightwaves. The input lightwave was intensity modulated by a train of pulses whose period and duty ratio were, respectively, 4 μs and 20%, as shown in the plot at the bottom of fig. 35 (b). An FBG was placed at the optical output port to eliminate the spectral component of the input lightwave reflected at the FBG located between the two circulators. The reflection band width f_r was 22.5 GHz (FWHM), and the edge slope width Δf_r was 6.0 GHz. Δf_r was defined by the width of the slope where the reflectivety is from

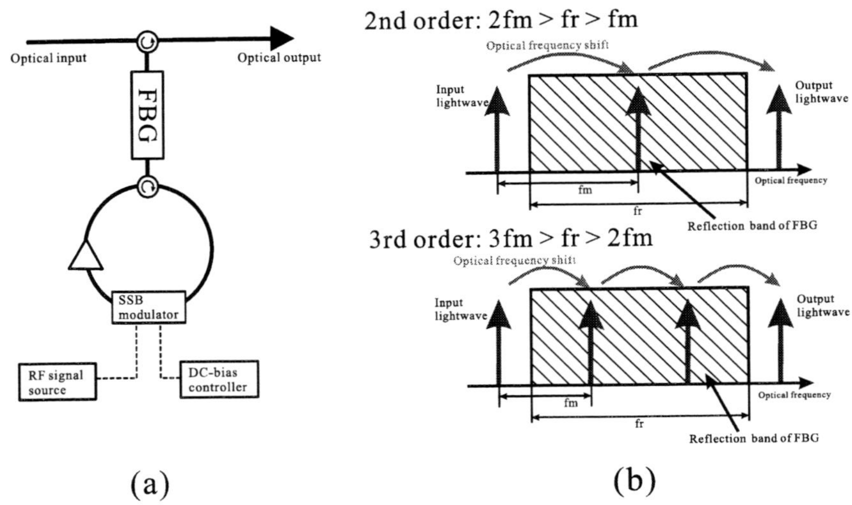

Figure 34. Tunable delay using SSB modulation.

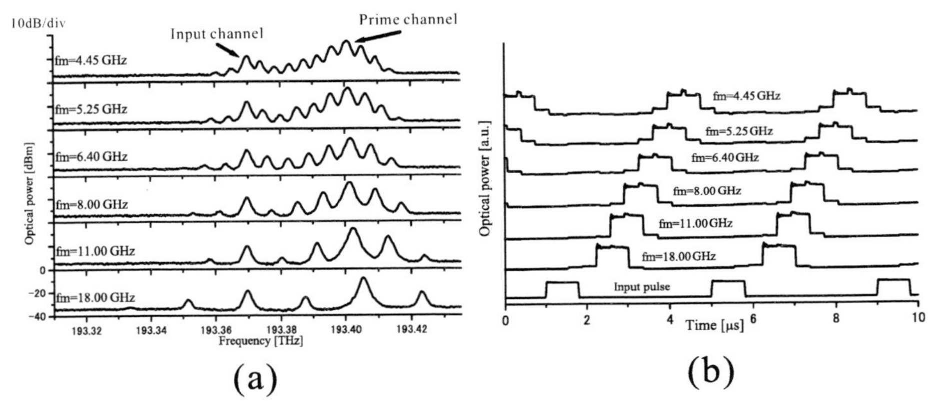

Figure 35. Output lightwaves: (a) Optical spectra (b) Time domain envelopes.

10% to 90% of the maximum. We used an x-cut LN optical SSB modulator, as the optical frequency shifter, where the carrier-suppression ratio was over 25 dB, and the frequency conversion loss was 23 dB. The dc-bias voltages were adjusted to maximize the first order upper sideband in the output of the SSB modulator. An optical amplifier was used, as shown in Fig. 34(a), to compensate for loss at the modulator. When $f_m = 18.00$ GHz, the 2nd-order channel is the primary channel in the output. Similarly, when $f_m = 11.00, 8.00, 6.40, 5.25$, and 4.45 GHz, the primary channels were the 3rd, 4th, 5th, 6th, and 7th channel, respectively. In the case of $f_m = 18.00$ GHz, the peak of the primary channel exists at 193.406 THz, where the optical frequency of the input lightwave is 193.370 THz. As shown in fig. 35(b), the delay can be controlled by changing the frequency of the electric signal fed to the SSB modulator. The delay due to one step of the circulation (τ) corresponding to the time difference between adjoining channels, was 350 ns, while the pulse width is 800 ns. In the cases of $f_m = 5.25$ GHz and 4.45 GHz, undesired spectral components in the outputs were not negligible, as shown in fig. 35(a), because f_m is not larger than Δf_r. The time domain profiles shown in the figure are deformed due to overlapping of the adjacent channels onto the primary one.

References

[1] Noguchi, K.; Mitomi, O.; Miyazawa, H. *J. Lightwave Technol.* 1998 16, 615–619.

[2] Zhou, Y.; Izutsu, M.; Sueta, T. *J. Lightwave Technol.* 1991 9, 750–753.

[3] Yoshida, K.; Nomura, A.; Kanda, Y. *IEICE Trans.* 1994 E77-C, 1181–1184.

[4] Izutsu, M.; Mizuochi, Y.; Sueta, T. *IEICE Trans.* 1995 E78-C, 55–60.

[5] Kawanishi, T.; Oikawa, S.; Higuma, K.; Sasaki, M.; Izutsu, M. *IEICE Trans. Electron.* 2002 E85-C, 150–155.

[6] Kawanishi, T.; Oikawa, S.; Higuma, K.; Matsuo, Y.; Izutsu, M. *Electron. Lett.* 2001 37, 1244–1246.

[7] Kawanishi, T.; Oikawa, S.; Higuma, K.; Matsuo,Y.; Izutsu, M. *Electron. Lett.* 2002 38, 1204–1205.

[8] Kawanishi, T.; Higuma, K.; Fujita, T.; Ichikawa, J.; Sakamoto, T.; Shinada, S.; Izutsu, M. *J. Lightwave Technol.* 2005 23, 87–94.

[9] Kawanishi, T.; Sasaki, M.; Shimotsu, S.; Oikawa,S.; Izutsu, M. *IEEE Photonics Tech. Lett* 2001 13, 854–856.

[10] Kawanishi, T.; Oikawa, S.; Izutsu, M. *J. Lightwave Technol.* 2002 20, 1408–1415.

[11] Kawanishi, T.; Yoshiara, K.; Oikawa, S.; Shinada, S.; Sakamoto, T.; Izutsu, M. *Jpn. J. Appl. Phys.* 2004 43, 5791–5794.

[12] Kawanishi, T.; Sakamoto, T.; Shinada, S.; Izutsu, M. *IEEE Microwave and Wireless Components Lett.* 2004 14, 566–568.

[13] Kawanishi, T.; Oikawa, S.; Yoshiara, K.; Sakamoto, T.; Shinada, S.; Izutsu, M. *IEEE Photon. Tech. Lett.* 2005 17, 669–671.

[14] Kawanishi, T.; Shinada, S.; Sakamoto, T.; Oikawa, S.; Yoshiara, K.; Izutsu, M. *Electron. Lett.* 2005 41, 69–70.

[15] Kawanishi, T.; Oikawa, S.; Higuma, K.; Izutsu, M. *IEEE Photon. Tech. Lett.* 2002 14, 1454–1456.

[16] Shimotsu, S.; Oikawa, S.; Saitou, T.; Mitsugi, N.; Kubodera, K.; Kawanishi, T.; Izutsu, M. *IEEE Photon. Tech. Lett.* 2001 13, 364–366.

[17] Higuma, K.; Oikawa, S.; Hashimoto, Y.; Nagata, H.; Izutsu, M. *Electron. Lett.* 2001 37, 515–516.

[18] Kawanishi, T.; Izutsu, M. *IEEE Photon. Tech. Lett.* 2004 16, 1534–1536.

[19] Kawanishi, T.; Higuma, K.; Fujita, T.; Ichikawa, J.; Shinada, S.; Sakamoto, T.; Izutsu, M. *IEICE Electron. Express* 2004 1, 69–72.

[20] Kawanishi, T.; Higuma, K.; Fujita, T.; Ichikawa, J.; Sakamoto, T.; Shinada, S.; Izutsu, M. *Electron. Lett.* 2004 40, 691–692.

[21] Kawanishi, T.; Sakamoto, T.; Tsuchiya, M.; Izutsu, M., "High Carrier Suppression Double Sideband Modulation Using an Integrated $LiNbO_3$ Optical Modulator," *Microwave Photonics Conference 2005*

[22] Kawanishi, T.; Higuma, K.; Fujita, T.; Ichikawa, J.; Sakamoto, T.; Izutsu, M., *IEICE Electron. Express* 2005 2, 333–337

[23] Lim, C.; Nirmalathas, A.; Novak, D.; Waterhouse, R. *Electron. Lett.* 2000 36, 442–443.

[24] Kuri, T.; Kitayama, K.; Takahashi, Y. *IEEE Photonics Tech. Lett.* 2000 12, 419–421.

[25] Collin, E. R., Field Theory of Guided Waves, IEEE Press, NY 1991, 247–328.

[26] Vegas Olmos, J.J.; Tafur Monroy, I.; Koon, J. M. A. *Optics Express* 2003 11, 3136–3140

[27] Vlachos, K.; Zhang, J.; Cheyns, J.; Sulur; Chi, N.; Van Breusegem, E.; Monroy, T. I.; Jennen, L. G. J.; Holm-Nielsen, V. P.; Peucheret, C.; O'Dowd, R.; Demeester, P.; Koonen, J. M. A. *J. Lightwave Technol.* 2004 21, 2617–2628.

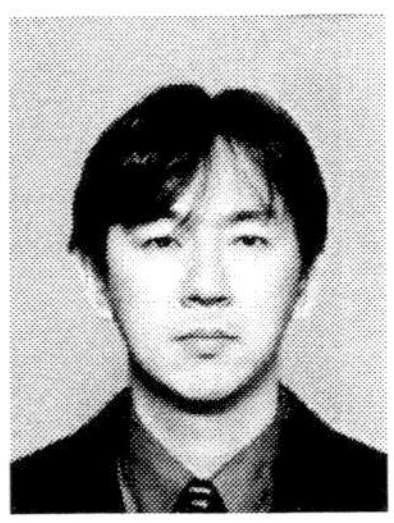

Tetsuya Kawanishi received the B.E., M.E. and Ph.D. degrees in electronics from Kyoto University, in 1992, 1994 and 1997, respectively. From 1994 to 1995, he worked for Production Engineering Laboratory of Matsushita Electric Industrial (Panasonic) Co., Ltd. In 1997, he was with Venture Business Laboratory of Kyoto University, where he had been engaged in research on electromagnetic scattering and on near-field optics. He joined the Communications Research Laboratory, Ministry of Posts and Telecommunications (from April 1, 2004, National Institute of Information and Communications Technology), Koganei, Tokyo, in 1998. He was a Visiting Scholar at the Department of Electrical Computer Engineering, University of California at San Diego, USA, in 2004. He is now a Senior Researcher of National Institute of Information and Communications Technology, and is currently working on high-speed optical modulators and on RF photonics. He received URSI Young Scientists Award in 1999.

In: Frontiers in Optical Technology: Materials and Devices ISBN: 1-60021-084-8
Editors: P.K. Choudhury and O.N. Singh, pp. 191-212

Chapter 7

LATERAL BIASING IN P-N JUNCTION LIGHT EMITTERS: HELLISH DEVICES

***J.Y. Wah*[1] *and N. Balkan*[2]**
[1] Faculty of Engineering, Multimedia University
63100 Cyberjaya, MALAYSIA
[2] Department of Electronic Systems Engineering
University of Essex, Colchester CO4 3SQ, UK

Abstract

Conventional vertical biasing on *p-n* junction has induced several problems when carriers are injected through the distributed Bragg reflectors (DBRs). In this chapter, we introduce a series of lateral biasing devices, namely HELLISH (hot electron light emission and lasing in semiconductor heterostructure). It is a novel hot electron surface emitter, however, carriers are injected parallel to the junction plane, and hence, bypasses the DBRs. The injection of hot electron-hole pairs into the quantum well (QW) is achieved via tunnelling and thermionic emission processes. The structure has been then evolved and developed to Ultra-Bright light emitting diode (LED), vertical cavity surface emitting laser (VCSEL), vertical cavity semiconductor optical amplifier (VCSOA), wavelength converter and semiconductor saturable absorber mirror (SESAM)-added device. One of the four-point HELLISH structures is capable to provide emission and absorption regions simultaneously, and hence, gives rise to wavelength conversion functionality as well as potential pulsation devices. These novel structures are very easy to fabricate, and their functionalities are discussed in detail.

1. Introduction

Hot electron light emission and lasing in semiconductor heterostructure (HELLISH) devices are surface emitters based on longitudinal transport. It is a novel hot electron surface emitter consisting of a GaAs quantum well (QW) on the *n-* side of a $Ga_{1-x}Al_xAs$ *p-n* junction. This has been currently adapted to GaAs/GaInNAs material structure for 1.3–1.6 μm communication window. These structures utilise hot electron transport parallel to the junction plane in high applied electric fields. The injection of hot electron-hole pairs into the QW is achieved via tunnelling and thermionic emission processes. However, at low applied electric fields, the carrier transport mechanism would be appropriately interpreted using "quasi-flat-band condition". This will be discussed in one of the sections.

The first type of HELLISH devices, HELLISH-1 [1], was a bare *p-n* junction with single QW on *n*-side of the junction. Due to continuous interest in the operations of the device, it was then added with a bottom distributed Bragg reflector (DBR) to enhance the surface light emission. It constructed a quasi-cavity and gives superb performance in term of light intensity; it was given a nick name as Ultra-Bright HELLISH or UB-HELLISH [2,3]. Although light emission was improved, the measured spectrum shows that it operates as a LED. Nevertheless, the bandwidth of the emission spectra was further narrowed, and demonstrated an output power of 5.5 mW by incorporating top and bottom DBRs to form a vertical cavity surface emitting laser (VCSEL) cavity or VCSEL-HELLISH [4,5]. The concept was applied to vertical cavity semiconductor optical amplifier (VCSOA) as well [6].

These devices are able to emit via the top surface, and therefore, 2D arrays of light emitters, monolithic fabrication and integration can be achieved easily and cheaply. The HELLISH contacts arrangement is very simple. To contact the active layers, Au-Ge-Ni contacts are directly evaporated onto the *n*-layer, and subsequently diffused through the active layers. External bias of either polarity is applied in the pulsed or CW mode between contacts of a simple bar shaped device, as shown in fig. 1. Due to several unique behaviours of HELLISH devices, and to explore the possible influences of the contacts, the *p*-layer contacts were separated and formed a 4-contact device. This device has a top-hat shape where the *n*-layers were etched till the *p*-layer, as shown in fig. 2. It was then given a nickname as Top-Hat HELLISH (THH) [7]. There are few unexpected interesting functionalities provided by this Top-Hat Structure (THS), such as emission and absorption regions at one piece of device, wavelength conversion and optical modulation [7].

The current line of this work is to adapt the semiconductor saturable absorber mirror (SESAM) integrated to the VCSEL-HELLISH where the active layers will be placed with GaInNAs material to match the optical communication windows. This integration will be expected to produce cost effective high frequency mode-locked pulse trains for the use of optical clock.

2. Hellish Operation

2.1. Hot Electrons

The advent of novel crystal growing techniques and advanced fabrication techniques have resulted in the production of semiconductor devices of dimensions well under micrometer. Therefore, the application of a few volts is able to produce large electric fields E across the sample, where free electrons are no longer in equilibrium with the crystal lattice. Furthermore, the mobility of electrons in heterostructure devices can be significantly improved by modulation doping, which was first proposed in 1969 [8] and demonstrated in 1978 [9]. In modulation doping structures, the ionised impurity scattering is essentially eliminated due to the spatial separation between the mobile electrons and the fixed ionised parent atoms, and hence, a 2D electron gas with very high mobility is achieved. This technique has produced several high-speed devices such as the high electron mobility transistor (HEMT) [10].

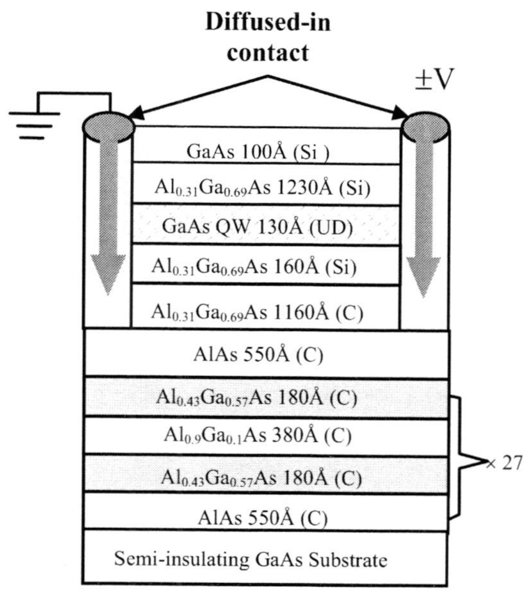

Figure 1. UB-HELLISH, its contacts and biasing arrangements.

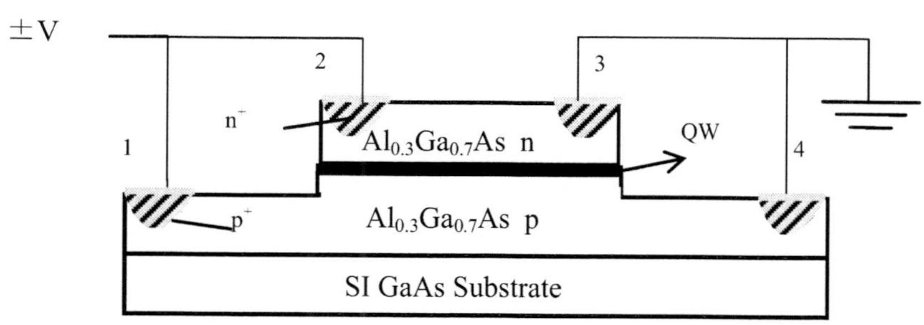

Figure 2. Schematic of the THS and contact configuration of the device.

Modulation doping coupled with small device sizes makes it possible to observe hot electron effects at much reduced fields in these low dimension semiconductors. In the hot electron regime, the *I-V* characteristic becomes nonlinear due to enhanced optical phonon scattering [11,12]. Experimentally, information on the hot carrier distribution in 2D devices can be obtained by using optical probing techniques including absorption, luminescence and Raman spectroscopy [13–15]. Moreover, information about hot electron scattering mechanisms with phonons and real space transfer (RST) scattering can be obtained experimentally by transport [16], time of flight [17] and hot electron photoluminescence measurements [18]. The term RST is used for the massive thermionic emission of hot electrons over heterojunction interfaces [19], where the heating of electrons is caused by electric fields parallel to the interface, and the perpendicular momentum is gained through scattering events [20]. One of the effects of RST of hot electrons in parallel fields is negative differential resistance (NDR) in the *I-V* characteristics. An important hot electron device based on RST was demonstrated in 1984, and referred in the literature as a CHINT (charge injection transistor) [21,22] or NERFET (negative resistance field-effect transistor). A light emitting version of the CHINT is also demonstrated [23].

HELLISH devices are hot electron surface emitters. Several types of light emitters have been demonstrated including HELLISH-1 [1], HELLISH-2 [24], Ultra-Bright HELLISH [2,3] and HELLISH-VCSEL [4]. All HELLISH devices utilise longitudinal transport, and hence, electrical fields are applied parallel to the layers which cause the carriers to heat in their respective channels. Therefore, hot carrier effects, thermionic emission, hot carrier tunnelling, and RST are the physical concepts behind the operation at the devices.

2.2. The Epitaxial Structures of HELLISH-1 and UB-HELLISH

The HELLISH-1 and UB-HELLISH structures are shown in Table 1. The wafer of the HELLISH-1 was grown at Sheffield EPSRC III-V Semiconductor Growth Facility by MOVPE on semi-insulating GaAs substrates. The device (HELLISH-1) consists of a 130 Å GaAs QW situated on the *n*-side of an $Al_{0.31}Ga_{0.69}As$ *p-n* junction within the depletion region. The *p-n* junction is formed by doping $Al_{0.31}Ga_{0.69}As$ with Si and C using a doping density of $N_D = 2\times10^{18}$ cm^{-3} and $N_A = 1.24\times10^{18}$ cm^{-3}, respectively. The GaAs capping layer of 200 Å is used to protect the $Al_{0.31}Ga_{0.69}As$ layer from oxidising while providing a small bandgap energy material at the surface. Ohmic contacts are formed by diffusing Au-Ge-Ni to all the layers.

Table 1. Structural diagram of the HELLISH-1 and UB-HELLISH.

Repeats	Material	Layer Thickness	Dopant	Concentration /cm^3	Layer Thickness	Dopant	Concentration /cm^3
		UB-HELLISH			HELLISH-1		
1	GaAs	100Å	Si	2E18	200 Å	Si	2E18
1	Al0.31Ga0.69As	1230 Å	Si	2E18	710 Å	Si	2E18
1	GaAs	130 Å	Ud		130 Å	Ud	
1	Al0.31Ga0.69As	160 Å	Si	2E18	164 Å	Si	2E18
1	Al0.31Ga0.69As	1160 Å	C	1.24E18	18000 Å	C	1.24E18
1	AlAs	550 Å	Bottom DBR Doped C		--		
27	Al0.43Ga0.57As	180 Å					
27	Al0.1Ga0.9As	380 Å					
27	Al0.43Ga0.57As	180 Å					
27	AlAs	550 Å					
Semi-insulating GaAs substrate							

UB-HELLISH is a super radiant structure grown by MOVPE with the addition of a bottom DBR (fig. 2). The DBR provides over 99% reflectivity so that the emission, previously lost from the substrate, is reflected back through the device. The GaAs-air interface has a reflectivity of approximately $R = [(3.6-1)/(3.6+1)]2 \sim 30\%$, due to the discontinuities in the refractive index, and forms a "quasi-cavity". Therefore, it is able to operate as a super radiant structure, so called UB-HELLISH.

2.3. High Field Hellish Operation

HELLISH devices simply consist of two diffused-in contacts which are separated by a length l. In thermal equilibrium, the band profile is uniformly formed along l, as shown in fig. 3. The 3D energy band profile shows how the conduction and valence bands are distributed in the direction of growth. Fermi level under thermal equilibrium is also shown in the fig. 3.

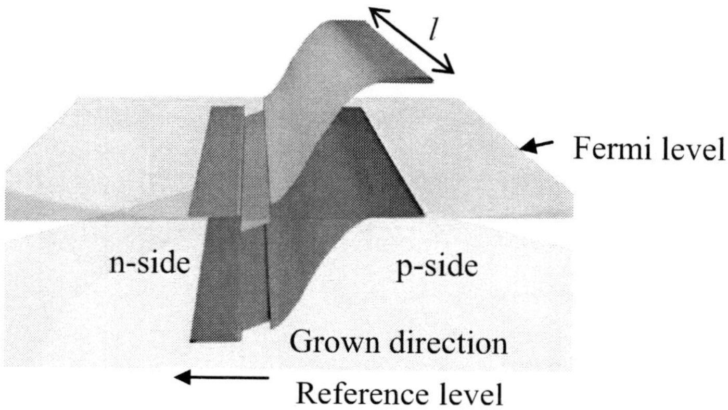

Figure 3. The 3D energy band profiles of HELLISH devices in thermal equilibrium condition. The top and bottom bands are the conduction and valence bands, respectively. Also shows the position of Fermi level and the reference level.

By placing both contacts on the top surface and diffusing through all layers, it is possible to apply longitudinal electric fields and enable mobile carriers to flow in both *n*- and *p*-channels. When the fields are on, energy bands tilt up according to the polarity of the applied voltage. The sequence of illustrations in figs. 4 shows how the bands are being tilted by the longitudinal electric fields. As the applied voltage increases, the degree of tilting of the band increases proportionally as a result of stronger electric fields. However, the built-in field in the growth direction is not affected by the application of the longitudinal bias.

High electric field carrier dynamics of HELLISH devices have also been developed (fig. 5). Briefly, it involves electric field induction heating of mobile carriers, i.e., hot electrons and holes, in their respective channels. Hot electrons are injected into the quantum well via tunneling and thermionic emission. The well acts as a giant trap for the hot electrons diffusing or drifting into the depletion region.

Electrons can gain energy from an applied longitudinal electric field; therefore, when a strong electric field is applied, the electron temperature can go well above the lattice temperature (T_L). The electron temperature (T_e) within the quantum well can be obtained experimentally by investigating the high energy tail of electroluminescence (EL) spectra as a function of applied electric fields at 77 °K. The lattice temperature is set at 77 °K because there are less non-radiative centers and reduced thermal broadening, and therefore, the EL high energy tail can be investigated with high accuracy. The high energy tail of the EL spectrum can be described by a Maxwellian distribution [25]:

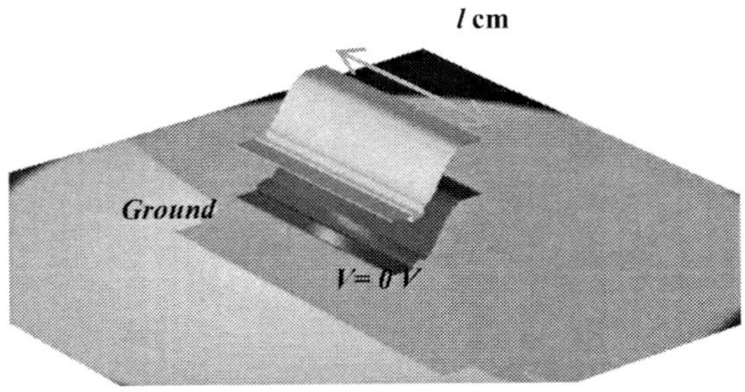

(a) $V = 0$; the band profiles without external bias. No existing longitudinal field. The length of the device is l cm.

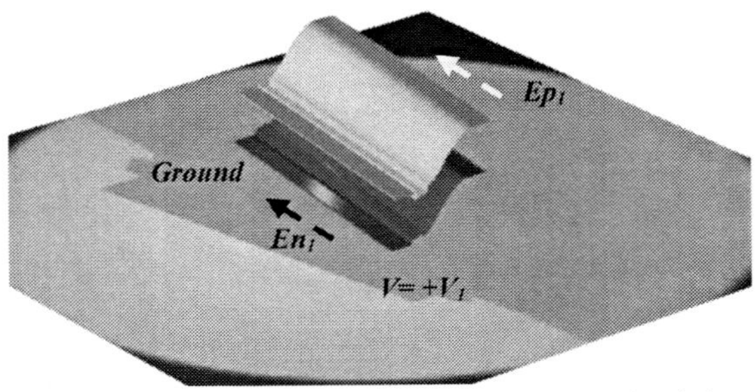

(b) $V = +V_1$; the tilted band profiles when the device is biased with V_1 at the right contact while left contact is grounded. E_{n1} and E_{p1} are the longitudinal electric fields. $E_{n1} = E_{p1} = V_1/l$. It is worth noting that HELLISH devices can be biased either negatively or positively.

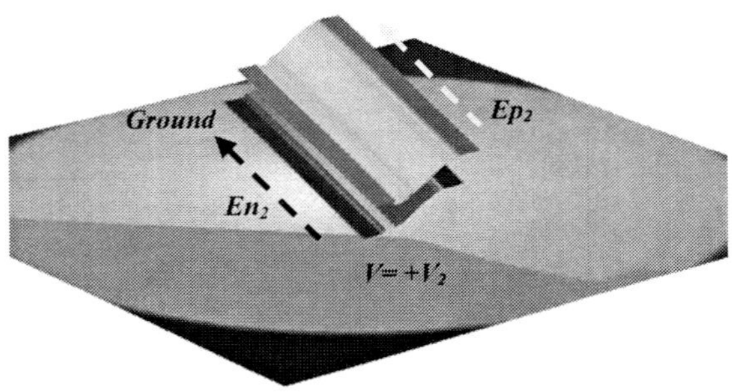

(c) $V = +V_2$, $(V_2 > V_1)$; the more tilted band profiles when the device is biased with V_2, where $V_2 > V_1$. E_{n2} and E_{p2} are the longitudinal electric fields. $E_{n2} = E_{p2} = V_2/l$.

Figures 4 (a), (b), and (c). A sequence of pictures shows how the energy bands of HELLISH are being tilted by longitudinal electric fields.

$$I_{EL} \propto \exp\left(\frac{h\omega}{2\pi kT_e}\right) \tag{1}$$

where I_{EL} is the EL intensity, $\hbar\omega$ is the emitted photon energy, and k is Boltzmann's constant. The logarithm of the high energy tail could be plotted against photon energy at $T_L = 77$ °K to give a linear dependence with the slope $1/kT_e$. From the gradient of the linear region, it was possible to extract the temperature of electrons, using eq. (1). Figure 6(a) shows a set of EL spectra with high energy tails for several applied electric fields at 77 °K. It is clear that, at high applied electric field, electrons occupy higher energies and have higher temperatures than the lattice temperature, as shown in fig. 6(b). The electron temperature is determined by using the following eq. (1) and the corresponding high energy tail gradients. From the figure, the highest energy of hot electrons is estimated about 22 meV (250 °K) in the bulk *n*-channel, which can be transferred to the QW.

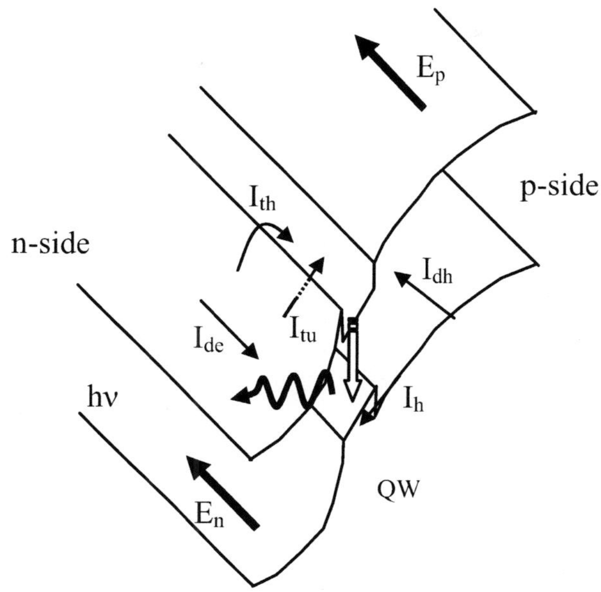

Figure 5. The HELLISH-1 energy band profile with the QW on the *n*-side of the depletion region of $Al_xGa_{1-x}As$ *p-n* junction. $E_{n,p}$ are the electric fields applied parallel to the layers. The injection of non-equilibrium carriers from the 3D $Al_xGa_{1-x}As$ layer into the first sub-band of the GaAs QW is via I_{th} (thermionic) and I_{tu} (tunneling) currents. I_{de} and I_{dh} are the electron and hole drift currents in the QW, respectively. I_h is the hot hole injection current into the well.

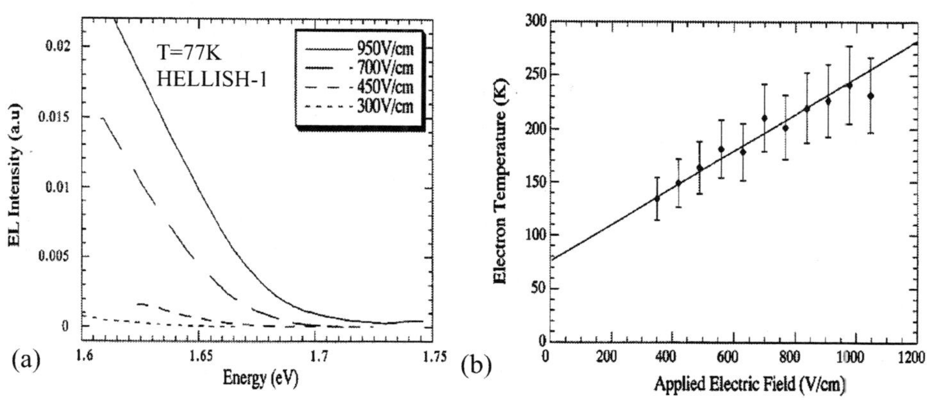

Figure 6. (a) High energy tail of EL spectra for various applied electric fields at 77 °K. (b) QW electron temperature as obtained from the high energy tail as a function of the applied electric field at 77 °K.

2.4. Low Field HELLISH Operation: Quasi Flat-Band Condition

While the high-applied field operation of HELLISH devices can be attributed to hot carrier effects, many of the devices previously reported started to emit light at relatively low applied fields [1–5]. Even a recent GaInAsP/InP VCSEL-HELLISH gives light emission at a voltage of 100 V/cm [26]. At this low field regime, carrier heating is minimal, and the hot carrier model of operation is no longer valid. In this section, we present further results concerning the low-field operation, and attempt to model the operation using a quasi-flat-band model [27].

The 3D model, shown in fig. 7 is a simplified energy band diagram of the standard HELLISH device, where we assumed most of the built-in potential barrier profile is retained, unaffected and parallel to the junction.

When the HELLISH device is externally biased at a low voltage, the energy band profiles will be tilted from the drain to the source region. The quasi-flat-band condition will be established if the bias voltage is such that the tilting establishes a potential at A, which will be equal to the potential at B, as shown in fig. 7. This will be similar to the well-known flat-band condition in a forward biased *p-n* junction, but it formed diagonally. In this situation, the flow of carriers is permitted diagonally across the junction as well as the longitudinal transport in the *p*- and *n*- channels. Because of the existing strong longitudinal electric fields in each of the channels, only a small fraction of carriers are capable of diffusing across the junction plane diagonally, and traversing the active region. The light emission in the low voltage regime is, therefore, expected to be weak compared to conventional vertical biasing.

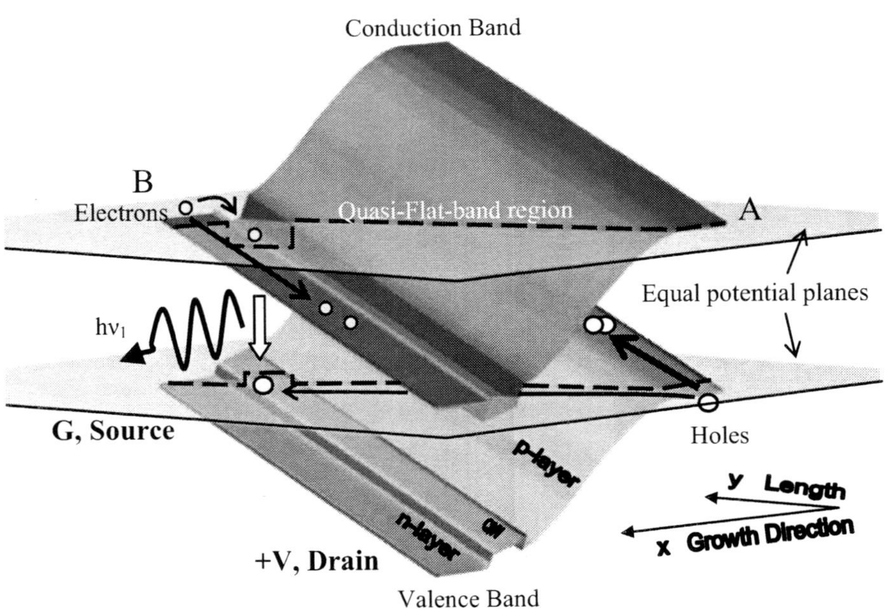

Figure 7. Schematic diagram to illustrate the quasi-flat-band region (broken lines) on the conduction and valence bands. There are two grayscale planes to illustrate the equal potential regions in the conduction and valence bands accordingly. Arrows depict the drift/diffusion of the carriers. The non-filled arrow shows where the radiative recombination takes place at photon energy $h\nu_1$. G = ground.

When the quasi-flat-band condition is established, a number of electrons in the n-channel are able to diffuse into the nearby quantum well. At the same time, most of them will be drifting along the n-channel induced by the electric field. Similarly, a small number of holes will diffuse into the quantum well close to the source region. As depicted in fig. 7, holes take a longer path to the QW (where the recombination occurs) than the electrons, and hence, the efficiency of light emission strongly depends on the number of holes arriving in the quantum well. Therefore, radiative recombination occurs near to the source region of the device. This prediction agrees remarkably well with experimental observations [1–7]. It is interesting to note that, if the external field is increased, the quasi-flat-band region will be established further away from the source. Thus, more excess carriers will reach the active area of the device, enhancing thereby the emission, which cannot be explained by the hot electron model. This growing emission region is also observed in the experiment.

One of the most important features of HELLISH device is the bidirectionality in terms of the polarity of the applied voltage. The emission region can be changed if the polarity of the applied voltage is reversed. This can be understood simply from the band profiles by reversing the tilting/biasing of bands, as shown in fig. 8.

Experimentally, a 2 mm long HELLISH device was fabricated from HELLISH-1 (the length of the device is defined by the separation of the contacts). The device was biased, and the emission captured by infrared film is shown in fig. 9. The figure illustrates the bidirectionality of the HELLISH device. When the device was biased with −10 V (−50 V/cm), light is clearly seen to emerge from the area near the right contact (cathode). When the voltage polarity was reversed (the right hand photograph), the device emits from around the opposite contact. However, in either polarity, the emission is always coming from the source region (i.e. cathode).

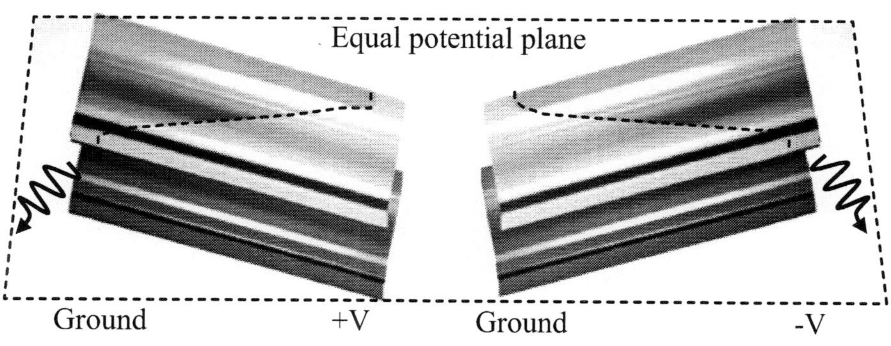

Figure 8. Illustration of bidirectionality of HELLISH devices with different biasing polarities. Left: the device is positively biased and light emission is from the left. Right: the device is negatively biased and light emission is from the right.

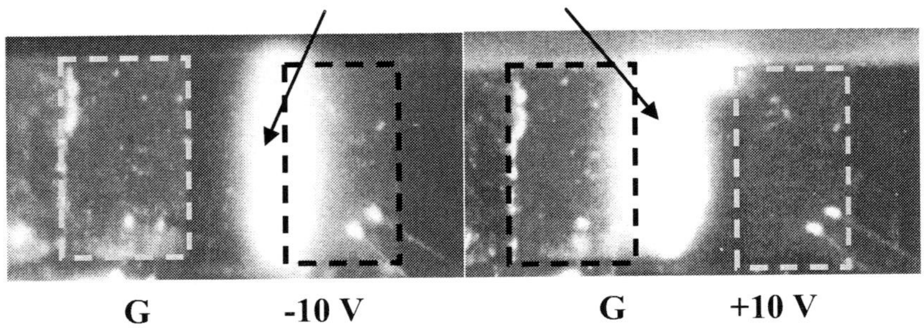

Figure 9. Infrared photographs of light emission from biased HELLISH-1. The contact separation is 2 mm. It is clear that most of the emitted light is from the vicinity of source region. Left: device was biased with −10 V. Right: device was biased with +10 V. Contacts are indicated by broken lines.

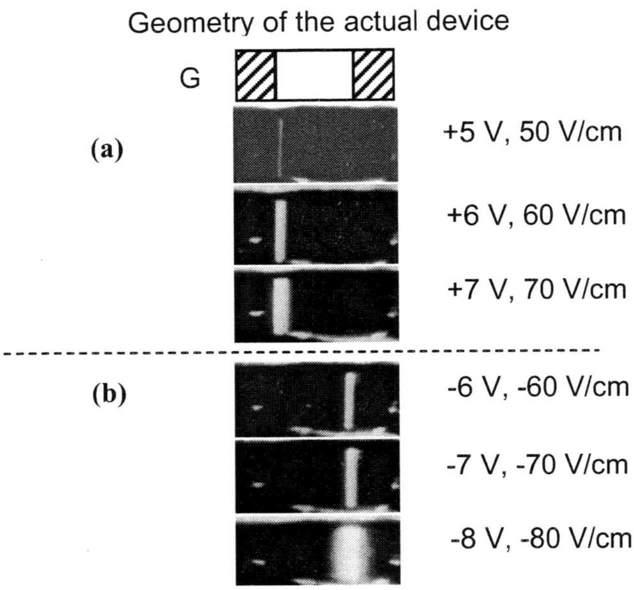

Figure 10. Infrared photographs illustrating the growing emission region of a 1 mm UB-HELLISH device (the schematic is shown on top of the photos) with increasing fields. G = Ground. (a) positive polarity and (b) negative polarity.

Light emission from a 1 mm long UB-HELLISH device was also investigated. The first set of infrared pictures [fig. 10(a)] shows the emission from the device with positive polarity voltages between 5 V and 7 V. It is obvious that the emission region starts from the cathode, growing toward the opposite contact, as the applied voltage increases. The second set of infrared pictures [fig. 10(b)] shows the emission from the device when negative polarity voltages between –6 V and –8 V were applied. Again similar behaviour is observed.

We have shown that the operation of HELLISH devices at low lateral biasing voltages can be explained in terms of a simple model, where the light emission is due to the diffusion of carriers into the quasi-flat-band condition region. At high applied fields, hot electron effects dominate the HELLISH operation.

3. VCSEL-HELLISH

The VCSEL-HELLISH was grown by MOVPE on semi-insulating GaAs substrate. The VCSEL consists of a HELLISH-1 cavity surrounded by upper and lower DBRs which have 17 and 27 periods, respectively, of AlGaAs layers with the Al concentrations indicated. The lower DBR provides a reflectivity in excess of 99%, while the reflectivity of the upper DBR is slightly lower to allow an output from the top surface. The QW is centralized within the cavity so that it is aligned to the anti-node of the confined optical field, and therefore, maximises gain with each round trip. The dimensions of VCSEL-HELLISH varied between 100 μm and 3.5 mm. The upper DBR was selectively removed down to the HELLISH cavity so that the Ohmic Au/Ge/Ni contacts could be evaporated onto the active region, and then diffused throughout the cavity layers. The QW laser displayed the diode characteristics over 77–300 °K while the VCSEL contacts were ohmic at low applied voltages over this temperature range.

The output of HELLISH-VCSEL is determined by both the gain spectra of the HELLISH cavity and reflectivity spectra of the DBRs. EL spectra were obtained in the pulsed mode at temperatures between 77 °K and 300 °K. The reflectivity and EL spectra of VCSEL-HELLISH are shown in figs. 11 and 12, respectively. It can be seen that the cavity resonance wavelength is at 816 nm, the EL spectrum is, therefore, peaked at this wavelength with a FWHM of 1.5 nm.

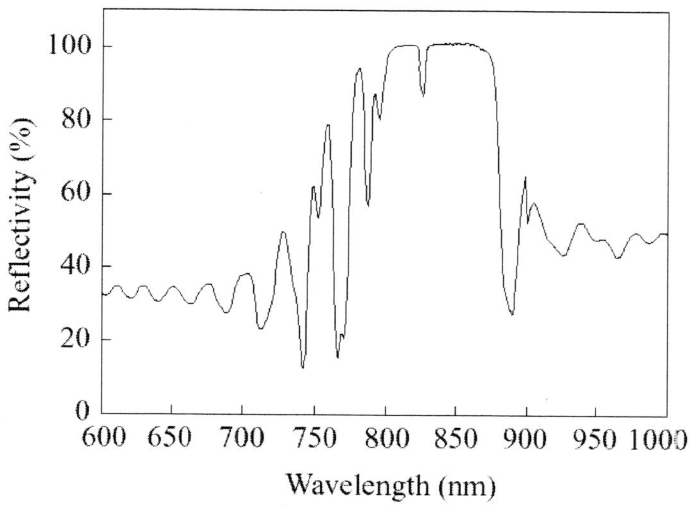

Figure 11. The reflectivity spectrum of VCSEL-HELLISH.

It is essential to consider the output power of VCSEL under HELLISH treatment. This was achieved by measuring the light emission using a power meter, and then normalising with respect to the duty cycle. It can be seen that the VCSEL has an increased output power of 2.6 mW at 1.2 kVcm^{-1}. From fig. 13, there is a distinct VCSEL threshold of 50 Vcm^{-1} after which the light emission begins to increase linearly. After 150 Vcm^{-1} the device exhibits heating effects and the output power is limited all the way up to the maximum fields. It has been shown that VCSEL-HELLISH is thermally stable up to 1.7 $kVcm^{-1}$.

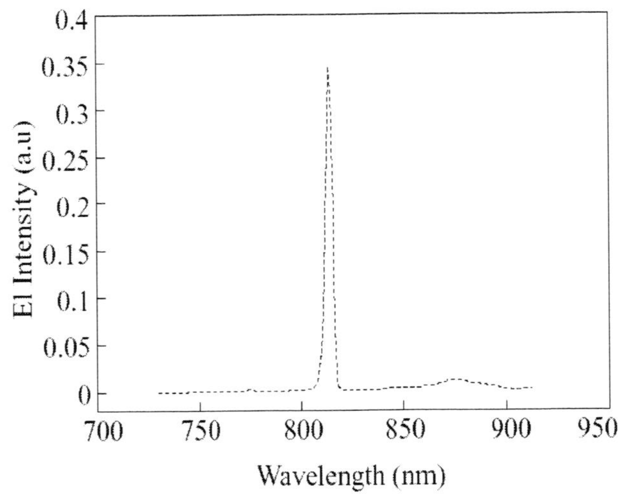

Figure 12. The EL spectrum of VCSEL-HELLISH.

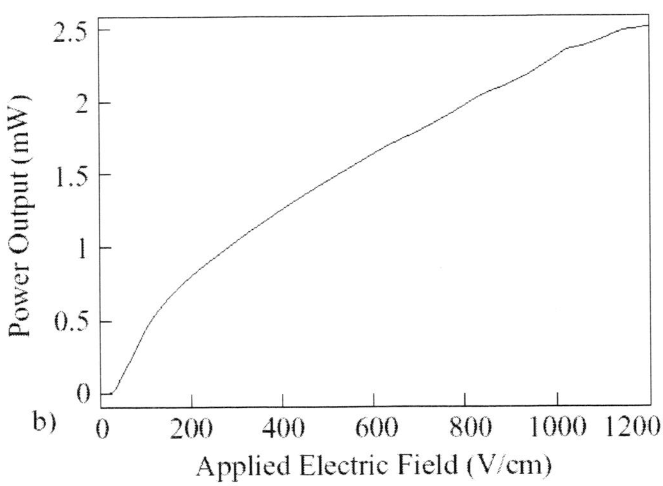

Figure 13. The output power of VCSEL-HELLISH 3 mm × 1 mm bar as a function of applied voltage operating in pulse mode with a constant repetition rate of 10 ms at 300 °K.

We have demonstrated that the VCSEL-HELLISH may be lasing at room temperature. The emission has a peak wavelength of 816 nm, with a FWHM of 1.5 nm at room temperature. One of the benefits of using HELLISH is that the device displays optical XOR logic functions by its bidirectional property.

4. Top-Hat HELLISH (THH) Wavelength Converter

Unlike the standard HELLISH devices, the THH device, as illustrated in fig. 14(a), is fabricated to a shorter *n*-channel than *p*-channel by etching. Ohmic contacts to the two channels are formed using Au-Zn and Au-Ni-Ge systems for the *p*-type and *n*-type contacts, respectively. Under normal operation, the device is biased with positive and negative voltages ($\pm V$) at contacts 1 and 2, while contacts 3 and 4 are grounded. When the device is biased with the positive polarity, it is clear from fig. 14(b) that the potential near to contact 2 (l_2) is higher in the *n*-channel than the *p*-channel ($V_n > V_p$). This results in an effective reverse biased region, which acts as a light absorber. In contrast, the area near to contact 3 (l_3) is effectively forward biased by the field, and therefore, can act as an emission region. In other words, owing to the shorter *n*-channel length, the applied electric field in the *n*-channel is higher than that in the *p*-channel, since both the channels are equally biased. Because of the different electric field strengths within these channels, the energy band profile of the device is altered. The built-in potential barrier of the *p-n* junction is enhanced near the anode (drain) and flattened near the cathode (source), as illustrated in fig. 15. The reverse and forward biased regions of the THS can be interchanged simply by changing the applied field polarity [fig. 14(b)]. It can, therefore, operate as a bi-directional field effect light emitter/detector [7]. Another potential application would be the self-pulsation device, since it consists of both emission and absorption regions.

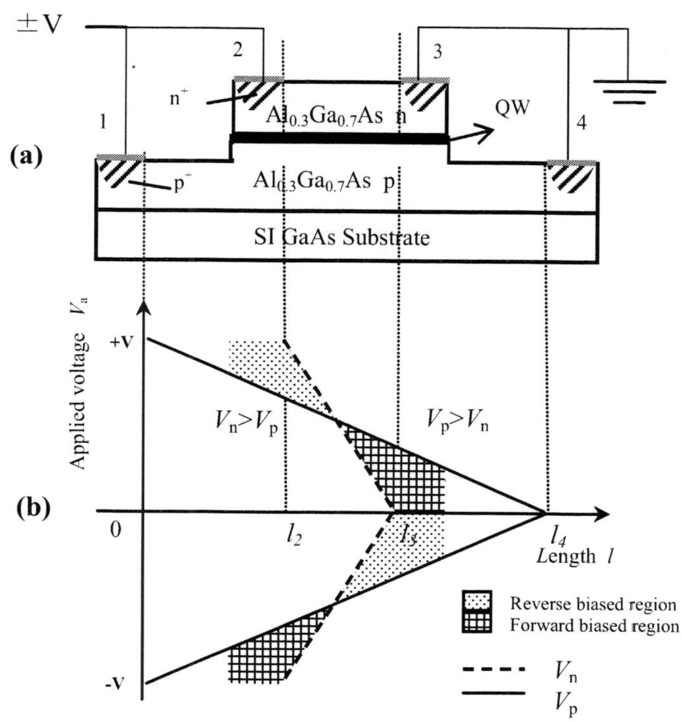

Figure 14. (a) Schematic of the THS device; (b) Illustration of the potential distributions along the *n*- (broken line) and *p*- (solid line) channels of the device. In the region $V_n > V_p$, the device is effectively reverse biased; and in the region $V_p > V_n$, it is effectively forward biased.

The carrier dynamics and wavelength conversion of the THS can be further understood by considering its 3D band profile, as shown in fig. 15. If an incident beam of photons with energies ($h\nu_0 > E_g$ ($Al_{0.3}Ga_{0.7}As$)) impinges on the absorption region, it is absorbed by the material. Photogenerated excess carriers are then created in all the layers within the absorption depth. In the *n*-layer, when an electron-hole pair is created, the electric field in the *n*-channel causes the electron and hole to drift in opposite directions, leading to a photocurrent. The photocurrent is negligible if the electrons and holes recombine before they reach the electrodes. However, if the electron transit time is short compared to the recombination time, then a significant number of photogenerated electrons may reach electrode 2 (drain). Therefore, an increased sample current in the circuit will be observed. A similar process can occur in the *p*-channel, however, as the mobility of the holes is much less than that of electrons and the path of the excess holes to the contact 4 is longer, carriers will tend to recombine before they reach the electrode. The contribution of the excess holes to the photocurrent can, therefore, be ignored.

As well as the longitudinal transport of photogenerated excess carriers, the THS involves vertical transport. In this, the excess carriers within the depletion region and the diffusion length from the depletion edge are rapidly swept across the depletion region by the strong electric field. The electrons and holes are swept into the *n*- and *p*-region, respectively, by the field, and are then collected as the photocurrent by the external circuit. The excess electrons and holes trapped in the intrinsic region can radiatively recombine. Each of these processes results in enhanced radiative recombination near to contact 3 (source) with photon energy of $h\nu_1$. Thus, the impinging optical signal ($h\nu_0$) is absorbed by the device and re-emitted at the conversion wavelength ($h\nu_1$).

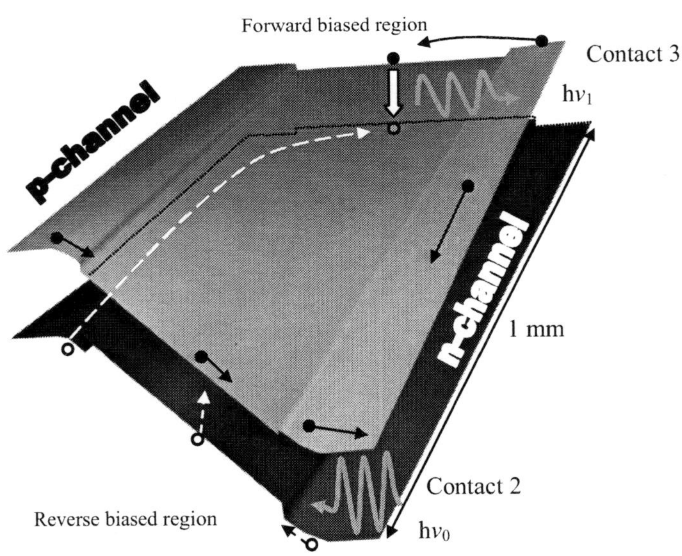

Figure 15. Schematic of the wavelength conversion process in the THH. $h\upsilon_0$ and $h\upsilon_I$ are the incident and converted photon energies, respectively. Electrons and holes are represented by filled and opened circles, respectively, while arrows depict the flow of excess carriers in the band profile of the device. The gradient of the channels corresponds to the strength of the electric fields.

The THH has been demonstrated as an all-optical wavelength converter (down conversion), using the set-up illustrated in fig. 16. The operation of the THH device as a wavelength converter has been reported previously [7,28]. The device was biased slightly below the emission threshold (V_a = 6.5 V), and the 850 nm VCSEL was used as a slave source for the THH, providing a quasi CW light source. The light source can be a tunable laser source as long as the wavelength falls within the gain bandwidth of the device. The input signal (647 nm) was coupled into the absorption end of the device. The 647 nm signal is absorbed by the device and causes a modulation of the 850 nm light, hence acting as a wavelength converter. This is shown in the reflected spectra of the device recorded by the OSA (fig. 17), where the intensity of the 850 nm line is clearly enhanced when the 647 nm signal is turned on. This measurement was carried out using just electrical pumping. Hence the gain could be improved using both optical and electrical pumping. Potentially, multiple wavelength sources may be coupled to the gain region of THH in a fiber-array arrangement; therefore, multiple wavelengths could be modulated simultaneously.

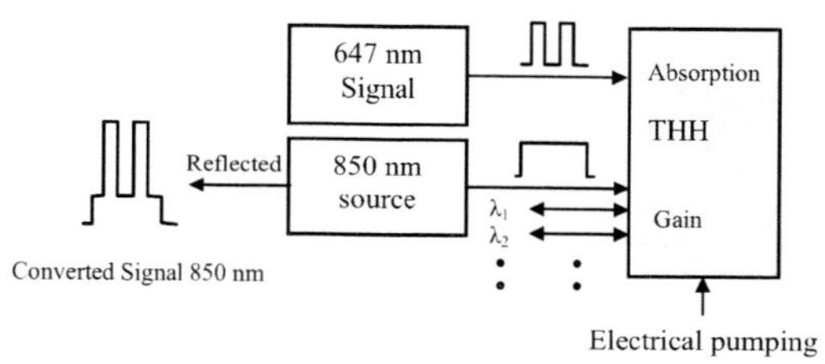

Figure 16. Illustration of THH as a multiple-wavelength converter.

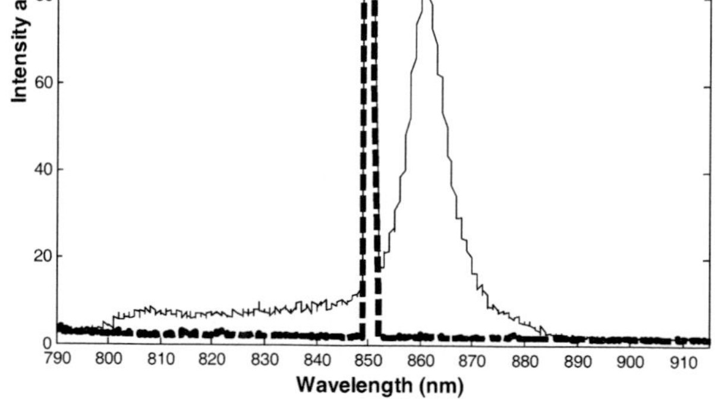

Figure 17. Reflected spectra of the THH operating as a wavelength converter. The 850 nm light is modulated with the 647 nm incoming signal. Applied voltage V_a = 6.5 V, 647 nm laser power is 70 mW.

5. VCSOA Hellish

Semiconductor optical amplifiers (SOAs) have become an important component of optical fiber networks in recent years. Many of the early SOA devices were based on Fabry Perot cavity lasers [29]; since then a number of advanced structures, such as the traveling-wave amplifier (TWA) [30], have been demonstrated leading to improved performance. Similarly, vertical cavity SOAs (VCSOAs), which have attracted considerable interest in recent years, have been demonstrated, and offer a number of potential advantages over conventional in-plane edge-emitting SOA devices. Main advantages are lower cost, inherent polarization insensitivity, higher fiber coupling efficiency, lower noise figure and potential for integration into high-density two-dimensional array architectures [31]. So far, only a limited number of VCSOAs have been reported in the literature. These include both optically [32] and electrically [33] pumped devices operating at 1550 nm as well as an optically pumped 1300 nm device [34]. Each of these devices has a structure similar to that of standard VCSELs. The bottom mirrors of these devices are produced to have reflectivities close to unity; this is achieved using many period of DBR stacks. In contrast, the top mirrors have relatively a few DBR periods. The lower reflectivity of the top mirror can provide enhanced absorption of pump power (optically) in quantum wells, as well as maximizing the gain-bandwidth production of the VCSOA [35]. The low reflectivity of the top DBR designed for operation in reflection mode requires multiple quantum wells in the active region to give high single pass gain.

We have previously demonstrated several novel longitudinal transport surface-emitting HELLISH devices. In this work, we used THS for the VCSOA wafer with a short *n*-channel and a longer *p*-channel. The THS allows different fields to be applied across the *n*- and *p*-channels. Because of these different electric field strengths in the channels, the energy band profile of the THH device can be twisted resulting in an emission region and an absorption region within the same junction plane. It has, therefore, the essential elements required to construct a compact wavelength converter, which has been demonstrated at wavelengths of 850 nm [7,31] and 1.3 μm [36].

In common with conventional VCSOAs, the THH-VCSOA has a high reflectivity lower DBR and a low reflectivity upper DBR [37]. This configuration should provide adequate optical gain to the incoming signal, and hence, provides amplification to the reflected signal. Like other VCSOAs, the device can be both electrically and optically pumped. The most interesting potential application of the THH-VCSOA is a single device which acts as a combined wavelength converter and optical amplifier.

5.1. The Material Structure and Emission Spectrum

The current device (Table 2) was grown by MOVPE on a semi-insulating GaAs substrate. The structure consists of a 7 period top DBR with a reflectivity of around 77.7 %. The bottom mirror is a 27 period DBR, 7 of which are doped with carbon and the remaining 20 are un-doped, giving a reflectivity of around 99.7 %. Although the doping causes significant optical losses, it is necessary to include the 7 doped periods in the lower DBR to allow for electrical pumping as well as giving a conducting channel to the *p*-layer. The active region in the microcavity consists of a simple AlGaAs/GaAs/AlGaAs *p-i-n* structure designed to operate at a resonance wavelength of 850 nm. The 774 Å GaAs intrinsic region, where radiative recombination takes place, is situated at the anti-node of the standing wave pattern in the 1.5λ cavity.

Tabel 2. Epitaxial structure of the THH-VCSOA as designed

Repeat	Material	Layer Thickness	Dopant	Concentration/cm^3
1	GaAs	100 Å	Si	3×10^{18}
1	$Al_{0.2}Ga_{0.8}As$	390 Å		2×10^{18} Top DBR
7	$Al_{0.51}Ga_{0.49}As$	180 Å		
7	AlAs	560 Å		2×10^{18}
7	$Al_{0.51}Ga_{0.49}As$	180 Å		
7	$Al_{0.2}Ga_{0.8}As$	390 Å		
1	AlAs	560 Å		
1	$Al_{0.31}Ga_{0.69}As$	952 Å		
1	GaAs	774 Å	Ud	-
1	$Al_{0.31}Ga_{0.69}As$	952 Å	C	1.24×10^{18}
1	AlAs	560 Å		1.24×10^{18} Bottom DBR
7	$Al_{0.51}Ga_{0.49}As$	180 Å		
7	$Al_{0.2}Ga_{0.8}As$	390 Å		
7	$Al_{0.51}Ga_{0.49}As$	180 Å		
7	AlAs	560 Å		
20	$Al_{0.51}Ga_{0.49}As$	180 Å	Ud	Bottom DBR
20	$Al_{0.2}Ga_{0.8}As$	390 Å		
20	$Al_{0.51}Ga_{0.49}As$	180 Å		
20	AlAs	560 Å		
Semi-insulating GaAs substrate				

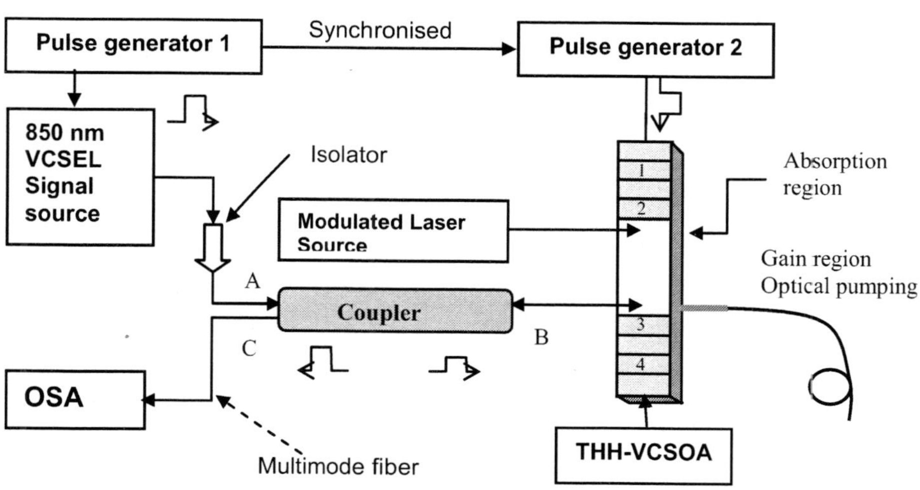

Figure 18. Illustration of the experimental setup of the THH as an optical modulated VCSOA in reflection mode. The numbers on the THH-VCSOA device indicate the contacts.

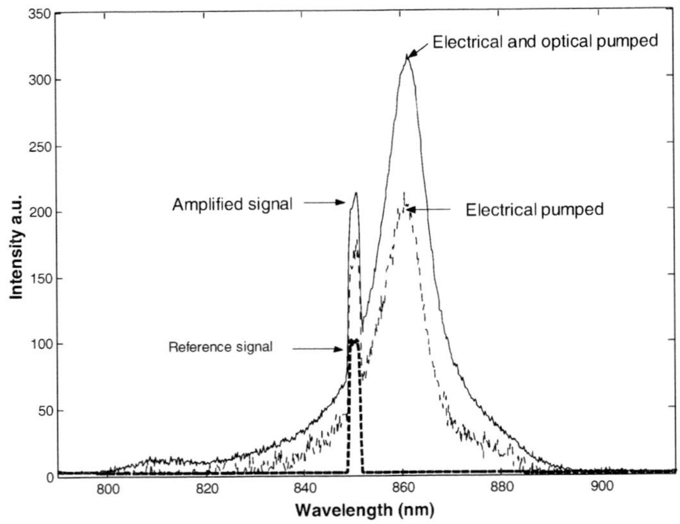

Figure 19. Emission spectra of the THH VCSOA with electrical pumping (dotted line) and with electrical plus optical pumping (solid line). The spectra of the 850 nm signal source as a reference (dashed line). Applied voltage V_a = 13 V, pump power = 85 mW.

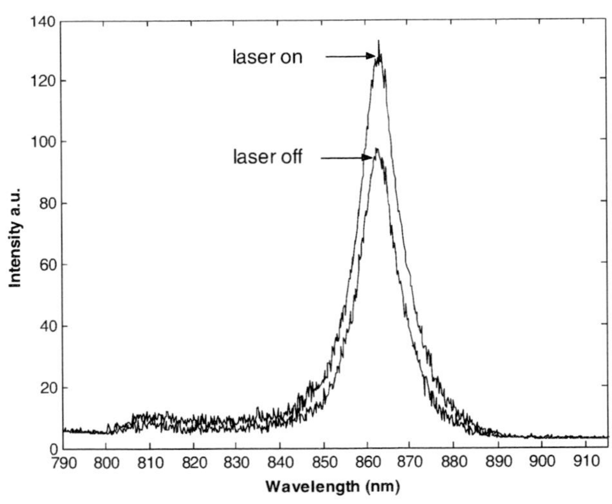

Figure 20. Gain spectrum of the THH device with and without optical modulation. Applied voltage V_a = 7 V, modulation light power = 70 mW.

To investigate the room temperature properties of the THH-VCSOA, we used the experimental setup, as shown in fig. 18. The optical signal to be amplified was generated by a modulated (10 μs pulse width, 1 kHz repetition rate) 850 nm source VCSEL. This was delivered to the gain region of the THH-VCSOA via a three-port (10:90) optical coupler, which also allowed collection of the reflected signal. For comparison, the reference signal generated by the source was measured from the output of the fiber B. The reflected amplified signal and the emission spectrum from the device were investigated with either electrical pumping or electrical and optical pumping. For electrical pumping, 13 V was applied to the test device under THH bias configuration. The required high bias voltage is due to the geometry of the THS device, and can be adjusted by altering the lengths of the channels. The spectra under each of these conditions are shown in fig. 19. This shows that the incident signal is slightly amplified with just electrical pumping (dotted line) and further enhanced (by 3 dB) with both optical and electrical pumping (solid line). The optical pump power used in this measurement was 85 mW. The gain of the device can be further enhanced/modulated by illuminating the absorption region using the Kr-ion laser, as shown in fig. 20, where the gain spectrum intensity increases during illumination. In summary, the signal gain of this device can be modulated by three different methods.

Conclusion

We have demonstrated a number of HELLISH devices using lateral biasing mechanism. Those devices show some unique behaviours, such as bidirectionality, growing emission region, XOR logic function, absorption and emission at once etc. These properties might find useful applications in photonic aspects. The current work is to integrate semiconductor saturable absorber mirror into the VCSEL structure to produce a mode locked laser. GaInNAs has been attractive recently as it is one of the promising materials to allow VCSEL operating in 1.3–1.6 μm optical communication windows. Therefore, it would be an advantage to use GaInNAs as a gain medium, and this has been done in several HELLISH reports [29,38].

References

[1] Straw, A.; Balkan, N.; O'Brien, A.; Cunha, D.A.; Gupta, R. *Superlatt. and Microstr.* 1995, 18, 34–53.

[2] O'Brien, A.; Balkan, N.; Boland-Thoms, A.; Adams, M.; Bel, A.; Serpenquzel, A.; Aydinli, A.; Roberts, J. *Opt. Quantum Electron.* 1999, 31, 183–190.

[3] O'Brien, A.; Balkan, N.; Roberts, J. *Appl. Phys. Lett.* 1997, 70, 366–368.

[4] Balkan, N.; O'Brien, A.; Boland-Thoms, A.; Potter, R.; Poolton, N.; Adams, M.; Masum, J.; Bek, A.; Serpenguzel, A.; Aydinli, A.; Roberts, J. *Proc. SPIE* 1998, 3283, 162–170.

[5] Balkan, N.; Serpenguzel, A.; Sokmen, I.; Hepburn, C.; Potter, R.; Roberts, J.; O'Brien, A.; Adams, M.; *Mat. Sci. & Eng.* 2000, B74, 96–100.

[6] Wah, J.Y.; Boland-Thoms, A.; Balkan, N. *Phys. Stat. Sol. (C)* 2005, 2, 3100–3103.

[7] Wah, J.Y.; Balkan, N., Potter, R.; Roberts, J. *Phys. Stat. Sol. (A)* 2003, 196, 496–503.

[8] Esaki, L.; Tsu, R. *IBM J. of Res. and Dev.* 1970, 14, 61–65.

[9] Dingle, R.; Störmer, H.L.; Gossard, A.C.; Wiegmann, W. *Appl. Phys. Lett.* 1978, 33, 665–667.

[10] Abe, M.; Mimura, T., Kobayashi, N.; Suzuki, M.; Kosugi, M.; Odani, K.; Hanyu, I. *IEEE Trans. Electron. Dev.* 1989, 36, 2021–2031.

[11] Price, P.J. *Solid-State Electron.* 1978, 21, 9–16.

[12] Barker, J.R. *J. Luminescence* 1981, 23, 101–126.
[13] Lyon, S.A. *J. Luminescence* 1986, 35, 121–154.
[14] Shah, J. *IEEE J. Quantum Electron.* 1986, 22, 1728–1743.
[15] Tsang, J.C.; Kash, J.A.; Jha, S.S. *Physica B* 1985, 134, 184–188.
[16] Inoue, M.; Hayahsi, H.; Sasaki, G.; Nakajima, S. *Physica A* 1985, 134, 327–331.
[17] Reggiani, L.; David, K.F.; Barker, J.R.; Jacoboni, C. Physics of nonlinear transport in semiconductors; Plenum Press: New York, 1980.
[18] Kash, K.; Shah, J.; Block, D.; Gossard, A.C.; Wiegmann, W. *Physica B* 1985, 134, 189–198.
[19] Hess, K.; Morkoc, H.; Shichijo. H.; Streetman. B.G. *Appl. Phys. Lett.* 1979, 35, 469–471.
[20] Adachi, S. Properties of Aluminium Gallium Arsenide; IEE INSPEC, 1993; No. 7.
[21] Kastalsky, A.; Luryi, S. *IEEE Trans. Electron Dev. Lett.* 1983, 4, 334–336.
[22] Luryi, S.; Kastalsky, A.; Gossard, A.C.; Hendel, R.H. *IEEE Trans. Electron Dev.* 1984, 31, 832–839.
[23] Mastrapasqua, M.; Capasso, F.; Luryi, S.; Hutchinson, A.L.; Deborah, L.S.; Cho, A.Y. *Appl. Phys. Lett.* 1992, 60, 2415–2417.
[24] Balkan, N.; Teke, A.; Straw, A.; Gupta, R.; Wolter, J.H.; Vleuten, W. *Appl. Phys. Lett.* 1995, 67, 935–937.
[25] Balkan, N.; Gupta, R.; Daniels, M.E.; Ridley, B.K.; Emeny, M. *Semicond. Sci. and Tech.* 1990, 5, 986–990.
[26] Sceat, R.; Dyson, A.; Hepburn, C.J.; Potter, R.; Boland-Thoms, A.; Hill, G.; Button, C.; Pinches, S. *Phyica E* 2003, 17, 607–609.
[27] Wah, J.Y.; Balkan, N. *IEE Proc. J. Optoelectron.* 2004, 151, 482–485.
[28] Wah, J.Y.; Yenidunya, R.; Dyson, A.; Boland-Thoms, A., Balkan, N.; Adams, M. *Proc. SPIE* 2001, 4283, 733–736.
[29] Mukai, T.; Yamamoto, Y.; Kimura, T. *Semiconductors and Semimetals,* 1985, 22-E, 265–319.
[30] Saitoh, T.; Suzuki, Y.; Tanaka, H. *IEEE J. Photon. Tech. Lett.* 1990, 2, 794–796.
[31] Piprek, J.; Bjorlin, S.; Bowers, E. *IEEE J. Quantum. Electron.* 2001, 37, 127–134.
[32] Bouche, N.; Corbett, B.; kuszelewcz, R.; Raj, R. *IEEE J. Photon. Tech. Lett.* 1996, 8, 1035–1037.
[33] Lewen, R.; Streubel, K.; Karlsson, A.; Rapp, S. *IEEE J. Phot. Tech. Lett.,* 1998, 8, 1067–1069.
[34] Jorlin, E.S.; Riou, B.; Keating, A.; Abraham, P.; Chiu, Y.J.; Piprek, J.; Bowers, J.E. *IEEE J. Phot. Tech. Lett.,* 2000, 12, 951–953.
[35] Tombling, C.; Saitoh, T.; Mukai, T. *IEEE J. Quantum Electron.* 1994, 30, 2491–2499.
[36] Wah, J.Y.; Loubet, N.; Potters, R.; Mazzucata, S.; Arnoult, A.; Carrer, H.; Marie, X.; Perena, E.B.; Balkan, N. *IEE Proc. J. Optoelectronics* 2003, 150, 72–74.
[37] Wah. J.Y.; Balkan, N.; Boland-Thoms, A. *Phys. Stat. Sol. (C)* 2005, 2, 3100–3103.
[38] Hepburn, C.J.; Wah, J.Y.; Boland-Thoms, A.; Balkan, N. *Phys. Stat. Sol. (C)* 2005, 2, 3096–3099.

J.Y. Wah received his B.Sc. in Science from the Department of Physics, National Normal Kaoshiung University (Taiwan) in 1996. He got M.Sc. in the Physics of Laser Communications with distinction and Ph.D. degrees from the Department of Physics, University of Essex, UK in 1999 and 2003, respectively. Soon after, he joined Multimedia University Malaysia as a Lecturer in the Faculty of Engineering. He is also active in doing online business using semiconductor laser physics. His current research interests are optoelectronic devices, such as vertical cavity emitters, HELLISH, THH, DBRs, VCSELs, VECSELs, VCSOA and SESAM.

Naci Balkan has been active in semiconductor research since 1975, when he started his Ph.D. at the University of Glasgow. At that time, his main interest was in the field of amorphous semiconductors, in particular, Ge and Si grown with thermal evaporation or RF sputtering. He continued with research in amorphous semiconductors until 1983 and published widely on low frequency AC loss in amorphous Si and Ge. Between 1980 and 1983 he worked as a Lecturer at Hacettepe, and Bogazici universities in Turkey. He joined the semiconductors group at Essex as a senior research officer in 1984, where his research interests moved to hot-electron transport, optical spectroscopy and instabilities in low dimensional III-V material systems such as GaAs/GaAlAs and GaInAs/GaAs quantum wells. In 1990, he became a Lecturer and established his own research group in semiconductor optoelectronics. The application of hot-electrons in optoelectronic devices became the focal point of his interest. This research led to the discovery of hot-electron vertical cavity surface emitting lasers (HELLISH-VCSEL) and top-hat HELLISH for wavelength conversion and amplification. He is also engaged in the study of novel material systems for applications in optoelectronic devices, including GaN, GaInAsN, and GaInAsP for displays, lasers and detectors. He has participated in over 25 national and international projects as either principal investigator or co-investigator, or consultant with funding from a wide range of sources including: EPSRC, NATO, TUBITAK-Turkey, the British Council, the European Commission, the Royal Society and the University of Essex. He has published over 140 research papers and book chapters, edited and co-edited special issue scientific journals, two reference books and co-authored a popular science book. He is a Fellow of the Institute of Physics and a member of EPSRC Review College.

In: Frontiers in Optical Technology: Materials and Devices ISBN: 1-60021-084-8
Editors: P.K. Choudhury and O.N. Singh, pp. 213-234

Chapter 8

TECHNIQUES AND ANALYSES OF COUPLING BETWEEN LASER DIODE AND FIBER-BASED MICROLENS

Faidz Abd. Rahman
Center for Photonics Research, Innovation and Application
Faculty of Engineering, Multimedia University,
63100 Cyberjaya, Selangor, Malaysia

Abstract

The use of LDs as signal source in optical communication systems requires that the emission be coupled into other optical components such as optical fiber, amplifier, waveguide or lens. This article reviews on the coupling efficiency between laser diodes (LDs) and single mode fibers (SMFs). An approach based on integrated lens or fiber lens is the focus of the article. It is found that the resulting coupling efficiency depends on the profile mismatch of LD and SMF. Thus, in this article, an analysis based on overlap integral and transmission matrix is discussed. Conical and hemispherical types of microlenses are considered here. Examples of applying the analysis are discussed, and results are compared to experimentally measured values.

1. Introduction

Since the laser diode (LD) was first demonstrated in the early 1960s, there has been rapid growth in the field of optical communications. LDs have been used as sources for long haul communication networks, cable TV backbones and short distance data communication networks. Recently, there has been a huge interest in increasing the available bandwidth to consumers by bringing fiber to the home or to the curb. This enables high-speed internet access, video on demand and other multimedia related applications. Due to this interest, the global photonic market is expected to grow up to \$12 billion by 2005, and over \$50 billion by 2010. The overall market for lasers in the telecommunications sector grew by 132% between 1999 and 2000, from \$2.8 billion to \$5.05 billion.

In order for this technology to be feasible, the cost of optoelectronic transceiver modules has to be lowered. The major factor that contributes to the cost of such modules comes from the packaging requirements between the optoelectronic and the passive waveguiding components [1,2]. The primary cost contribution is the active alignment due to the tight alignment tolerances between the source and the waveguide. Therefore, there has been a growing interest in developing passive coupling schemes using a silicon optical bench in conjunction with vision systems for a manufacturable process [3–5]. Passive coupling becomes more attractive for coupling the array of lasers to an array of single mode fibers (SMFs). This is because, active array alignment and fixing require substantially more advanced positioning systems and optimisation techniques compared to single device alignment. Generally, passive alignment systems use vision based pick and place systems or solder self-alignment methods for the alignment process. These methods are robust and reproducible if the alignment tolerances are large. Thus, the focus of recent researches is on producing low loss and high tolerance coupling schemes.

The primary obstacle in the packaging of semiconductor active devices, such as the LDs and optical amplifiers with SMFs, is the mode mismatch both in size and shape. LDs generally have tightly confined elliptic modes, while optical fibers have circular and more loosely confined modes. The mismatch in the shape and the size of the modes leads to poor butt-coupling efficiencies (≅10–15%) [6,7]. Generally, the direct end coupling configuration, i.e. when the LD and SMF or waveguide are positioned end-to-end, is the most suitable coupling configuration from the point of view of integrated optics. To overcome the mismatch in the modes, various mode conversion configurations or schemes have been developed. Basically, mode conversion configuration can be classified into three classes, i.e. mode conversion in the active device, using discrete optics and integrated lens.

For mode conversion in active device, expanded mode LDs have been proposed as a solution to improve the coupling efficiencies and misalignment tolerances. The challenge lies in modifying the modal shape of the LD to match the fiber mode better, without compromising on the laser performance. There have been two main approaches toward achieving mode matching, i.e. large optical mode approach and integration of an adiabatic mode converter, referred to as taper approach. In the large optical mode method [8–11], the optical mode is expanded by waveguide design. The size of the mode is constant throughout the entire laser cavity. It offers a rather simple and straightforward method of realising a narrow divergence laser beam. In the case of taper approach, a gradual mode expansion occurs by an integrated taper in the output facet [12–14]. Although these approaches improved the coupling efficiency between the laser and fiber, the working distance is still typically within 20 μm, and optical backreflection may impose an issue on laser performance, if an uncoated cleaved fiber is used.

The second class of coupling configuration utilises discreet or bulk lenses. This is the traditional way of coupling light from the LD into the optical fiber, i.e. by approximately matching the mode sizes by introducing a magnification factor. Several types of discreet lenses have been used, namely, ball lenses, plano-convex lenses, graded-index lenses and cylindrical lenses [15–24]. The problem associated with this approach is its requirement of a sophisticated alignment technique, because the size of the lenses is often not suitable for integrated optic applications.

To overcome this problem, an approach based on integrated lens has been introduced. Integrated lens has features that are suitable for optical coupling schemes, such as compactness, simplicity, stability, and freedom from bulky lenses. Among the techniques based on integrated lens are taper ended fiber, hemispherical ended fiber, spherical ended fiber, wedge ended fiber, conical end fiber, etc. [25–39]. However, these lensed fibers have disadvantages in their small working distance (less than 10 μm) and the tight transverse tolerance (about 0.3–0.6 μm) for 1 dB loss. The short working distance of these conventional lensed fibers is due to the limitations of the small core diameter of the optical fiber. Several schemes and configurations, based on the hemispherical ended fiber, have been found to improve on the working distance and tolerances [40–45].

2. Theory of Coupling

Basic theories of coupling between optical waveguides are presented in this section. The overlap integral method together with the ray transfer matrix has been used to calculate the coupling efficiency.

2.1. Coupling Efficiency Calculation

The coupling efficiency between a LD and a SMF is essentially given by the complex overlap integral of the laser and the fiber modes. In situations where the laser and fiber are not butt-joint, the mode of either the laser or the fiber has to be propagated using free-space Gaussian beam analysis to obtain the mode at the separation distance.

To derive the expressions for the coupling efficiency, it is assumed that the beam shape on the laser is Gaussian in both transverse and lateral directions. Thus, the coupling efficiency η for an elliptical LD is given by $(\eta_x\eta_y)^{1/2}$, where η_x and η_y are the components of coupling efficiencies in the x-z and y-z planes, respectively.

For the general coupling configuration of fig. 1, with the LD aperture in the plane A_1 and the SMF endface in the plane A_2, the 2D normalised power coupling efficiency is given by [46]:

$$\eta = \eta_x\eta_y = \left| \int \psi_{sz}\psi^*_{FL} dA_2 \right|^2 \tag{1}$$

where

$$\psi_{sz}(x_2, y_2, z_2 = z) = \psi^{(x)}_{sz} T \psi^{(y)}_{sz} \exp(-ikz) \tag{2a}$$

and

$$\psi_{FL}(x_2, y_2, z_2 = 0) = \psi_{Fx}(x_2, 0)\psi_{Fy}(y_2, 0) \tag{2b}$$

are the electric field profiles of the LD and SMF, respectively. The electric field profiles of the LD and SMF in the x-direction can be written as [47]

$$\psi_{sz}^{(x)} = \left(\frac{2}{\pi}\right)^{\frac{1}{4}} \frac{1}{(\omega_x(z))^{\frac{1}{2}}} \exp\left[-\frac{x_2^2}{\omega_x^2(z)} - ik\frac{x_2^2}{2R_x(z)}\right] \tag{3a}$$

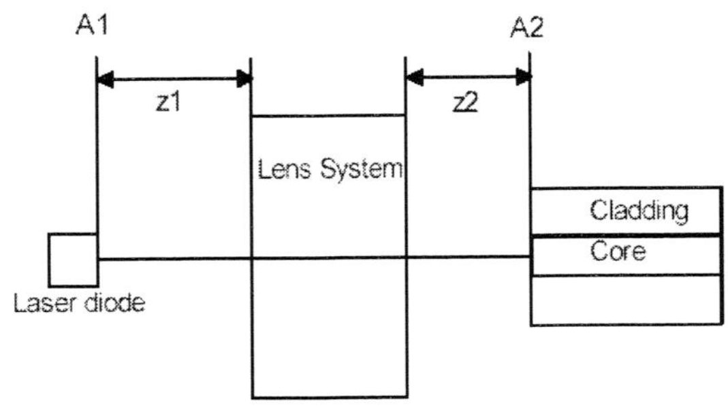

Figure 1. Coupling configuration.

$$\psi_{Fx}(x_2,0) = \left(\frac{2}{\pi}\right)^{\frac{1}{4}} \frac{1}{(\omega_{of})^{\frac{1}{2}}} \exp\left[-\frac{x_2^2}{\omega_{of}^2}\right] \tag{3b}$$

where $\omega_x(z)$ is the spot size and $R_x(z)$ is the radius of curvature of the LD at a distance z from the laser facet in the x-z plane, while ω_{of} is the spot size of the SMF at its endface. Both, the spot size $\omega_x(z)$ and radius of curvature $R_x(z)$, can be determined by the expressions [47]

$$\omega_x(z) = \omega_{ox}\left(1+\left(\frac{2z}{k\omega_{ox}^2}\right)^2\right)^{\frac{1}{2}} \tag{4a}$$

and

$$R_x(z) = z\left(1+\left(\frac{k\omega_{ox}^2}{2z}\right)^2\right) \tag{4b}$$

where ω_{ox} is the spot size of the LD at its endface. The expressions in terms of y are obtained by replacing x in above equations with y. The parameter T in eq. (2a) is the transmission coefficient for any lensing scheme present in between the LD and the SMF.

2.2. Ray Transfer Matrix

Matrix optics is a technique for tracing paraxial rays. The rays are assumed to travel only within a single plane, so that the formalism is applicable to systems with planar geometry, and to meridonal rays in circularly symmetric systems. A ray is described by its position and its angle with respect to the optic axis. These variables are altered as the ray travels through the system, as shown in fig. 2. In paraxial approximation, the position and the angle at the input and the output planes of an optical system are related by two linear algebraic equations. As a result, the optical system is described by a 2×2 matrix, called the ray transfer matrix, and it can be written as [47]

$$\begin{bmatrix} y_2 \\ \theta_2 \end{bmatrix} = \begin{bmatrix} A & B \\ C & D \end{bmatrix} \cdot \begin{bmatrix} y_1 \\ \theta_1 \end{bmatrix} \tag{5}$$

where the matrix M, whose elements are A, B, C and D characterizes the optical system.

2.2.1. The ABCD Law

As mentioned in the preceding section, an optical system can be completely characterised by the 2×2 matrix relating the position and inclination of the transmitted ray to those of the incident ray. The modification of a Gaussian beam by an arbitrary paraxial optical system is characterised by a matrix M of elements (A, B, C, D).

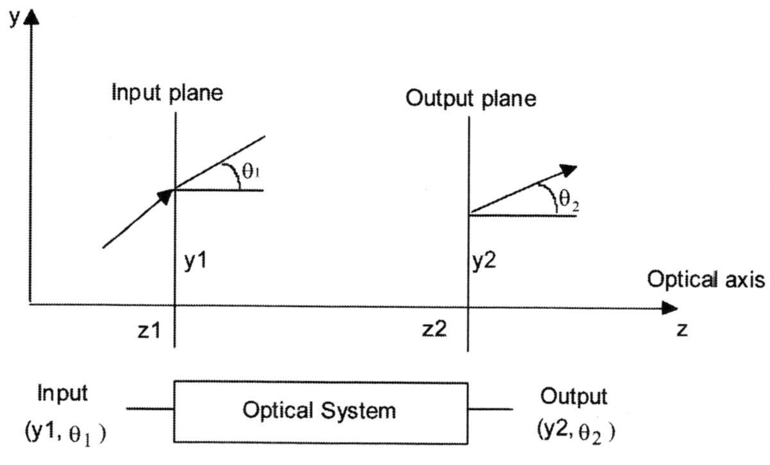

Figure 2. A ray enters an optical system at position y_1 and angle θ_1 and leaves at position y_2 and angle θ_2.

The q-parameters, q_1 and q_2, of the incident and transmitted Gaussian beams at the input and the output planes of a paraxial optical system, described by the (A, B, C, D) matrix, are related by [48]

$$q_2 = \frac{Aq_1 + B}{Cq_1 + D} \tag{6}$$

The parameters q identify the spot size $\omega_{x,y}(z)$ and the radius of curvature $R_{x,y}(z)$ of the Gaussian beam, and the relationship is given by [48]

$$\frac{1}{q_{x,y}(z)} = \frac{1}{R_{x,y}(z)} - i\frac{\lambda}{\pi\omega_{x,y}^{2} n} \tag{7}$$

where λ is the free-space wavelength and n is the refractive index (RI) of medium 1.

3. Conical Microlensed Fiber

One of the methods of forming a microlens is through chemical etching, which produces conical microlenses [49,52]. This technique is easily replicated. Moreover, it produces automatic alignment of the microlens with the fiber core. The lowest coupling losses reported, using this technique, are 1.38 dB at 1.55 μm [65] and 2 dB at 0.82 μm [50]. A coupling efficiency of 3 dB at 1.3 μm has also been reported [52] where this value is 4 dB less than that obtained with butt-coupling.

3.1. Fabrication

When a VAD or MCVD fabricated type of SMF is dipped into an etchant solution of hydrofluoric acid (HF) buffered in ammonium fluoride (NH_4F), the etching rates in the core and in the cladding are unequal, due to the different glass composition in each region [49,52]. This results in a self-centerd conical proturbance that is directly related to the doping profile, as shown in fig. 3. This proturbance acts as a phase shifter with lens-like properties. In the case of a step-index VAD fiber, the etching rate in the core is constant, since the doping concentration is uniform throughout the core section. On the other hand, the doping profile in the core of step-index MCVD fibers shows periodic variation of the doping concentration and a dip on the fiber axis.

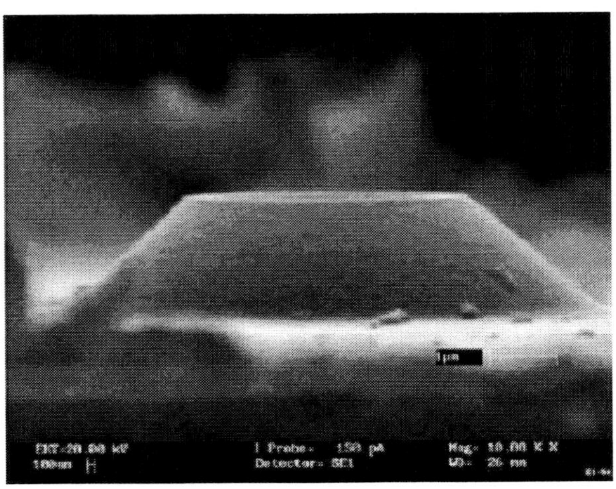

Figure 3. The side view of the conical microlens etched for 60 minutes.

3.2. Analysis

In the paper by Faidz *et al* [53], the authors have used the *ABCD* matrix to analyse the propagation of the Gaussian beam from the LD through the conical microlens. The parameters obtained from the matrix method, together with a method based on the overlap integral of LD and SMF electric field profiles, are then used to determine the coupling efficiency. To analyse the conically lensed fiber, the authors [53] have divided the microlens structure into four regions, (fig. 4). The ray matrix in the gap (M_o), conical microlens (M_1) surface and the microlens depth (M_2) are given by [53]

Region I

$$\begin{aligned} M^I{}_{10} &= M^I{}_2 M^I{}_1 M^I{}_0 \\ &= \begin{bmatrix} 1 & n_a l(r) \\ 0 & 1 \end{bmatrix} \cdot \begin{bmatrix} 1 & 0 \\ -\dfrac{1}{K^I{}_r(r)} \dfrac{n_a - 1}{n_a} & \dfrac{1}{n_a} \end{bmatrix} \cdot \begin{bmatrix} 1 & z \\ 0 & 1 \end{bmatrix} \end{aligned} \tag{8}$$

Region II

$$\begin{aligned} M^{II}{}_{10} &= M^{II}{}_1 M^{II}{}_0 \\ &= \begin{bmatrix} 1 & n_a H \\ 0 & 1 \end{bmatrix} \cdot \begin{bmatrix} 1 & z \\ 0 & 1 \end{bmatrix} \end{aligned} \tag{9}$$

Region III

$$\begin{aligned} M^{III}{}_{10} &= M^{III}{}_1 M^{III}{}_0 \\ &= \begin{bmatrix} 1 & n_a l(r) \\ 0 & 1 \end{bmatrix} \cdot \begin{bmatrix} 1 & 0 \\ \dfrac{1}{K^{III}{}_r(r)} \dfrac{n_a - 1}{n_a} & \dfrac{1}{n_a} \end{bmatrix} \cdot \begin{bmatrix} 1 & z \\ 0 & 1 \end{bmatrix} \end{aligned} \tag{10}$$

Region 0

$$\begin{aligned} M^0{}_{10} &= M^0{}_1 M^0{}_0 \\ &= \begin{bmatrix} 1 & H \\ 0 & 1 \end{bmatrix} \cdot \begin{bmatrix} 1 & z \\ 0 & 1 \end{bmatrix} \end{aligned} \tag{11}$$

where n_a is the RI of core region, z is the working distance between the LD and the fiber tip, and $K^I{}_r(r)$ and $K^{III}{}_r(r)$ are the lengths normal to the oblique segments in regions I and III to the central axis of the fiber, as given by $K^g{}_r(r)$ in fig. 4. The symbol g stands for regions 0, I, II or III, while r represents either x- or y- direction. These parameters are given as [53]

$$K^I{}_r(r) = \frac{r}{\sin\sigma} \tag{12a}$$

and

$$K^{III}{}_r(r) = \frac{r}{\sin\left(90^0 - \phi\right)} \tag{12b}$$

where σ is the cone base angle and $\phi = \tan^{-1}(D/c)$, with D as the depth and c as the radius of the central dip. Because the oblique segments are flat, the angles, which $K^I{}_r(r)$ and $K^{III}{}_r(r)$ make with the central axis, are constant. The refraction matrix at the oblique regions I and II are developed from Snell's law [54]. The parameter $l(r)$ in eqs. (8) and (10) can be written as [30,49,53,55,56]

$$l(r) = \begin{cases} H + D\left(\dfrac{r}{c} - 1\right) & 0 \le r \le c \\ \\ \dfrac{a - r}{a - b} H & b \le r \le a \end{cases} \tag{13}$$

where $r = x$ or y, and other symbols are defined in fig. 4.

According to the *ABCD* law discussed in Section 2, the spot size and the curvature of the Gaussian beam can be obtained from the matrix. These parameters are crucial for the analysis of coupling efficiency. In using this technique, the complex beam parameter q_j of the Gaussian beam is denoted by the curvature radius R_j of the wavefront, and the beam radius ω_j is given as in eq. (7). The relationship between q_1, the beam parameter at the LD facet, and q_2, the beam parameter at the SMF tip, is given in eq. (6). The parameters A, B, C and D can be obtained from

$$M^g = \begin{pmatrix} A^g & B^g \\ C^g & D^g \end{pmatrix} \tag{14}$$

where the subscript g denotes the regions 0, I, II or III, while M^g can be obtained utilising the matrix developed in eqs. (8)–(11). The transformed radii of curvature, after passing through the lens and at plane P_1, can be expressed as

$$R^g_{x2}(z) = \frac{A^g_2(z)^2 \omega^4_x(z)\pi^2 + \lambda^2 B^g}{A^g_2(z)^2 C^g_2(z)\omega^4_x(z)\pi^2 + \lambda^2 B^g D^g} \tag{15}$$

where

$$A^g_2(z) = A^g + \frac{B^g}{R_x(z)} \tag{16a}$$

$$C_2^g(z) = C^g + \frac{D^g}{R_x(z)} \tag{16b}$$

The y-component of transformed radii of curvature can be obtained by substituting x with y in eqs. (16a) and (16b). In the analysis, it is assumed that the spot sizes of the laser at the lens surface plane P_0 and at plane P_1 are the same. Only the radii of curvature of the laser would experience changes after passing through the lens.

Substitution of eqs. (2)–(4) and (8)–(16) into eq. (1) yields,

$$\eta_x = \frac{8}{\omega_{of}\pi}\left(I_x^2(x,0) + I_x^2(x,\frac{\pi}{2}) \right) \tag{17}$$

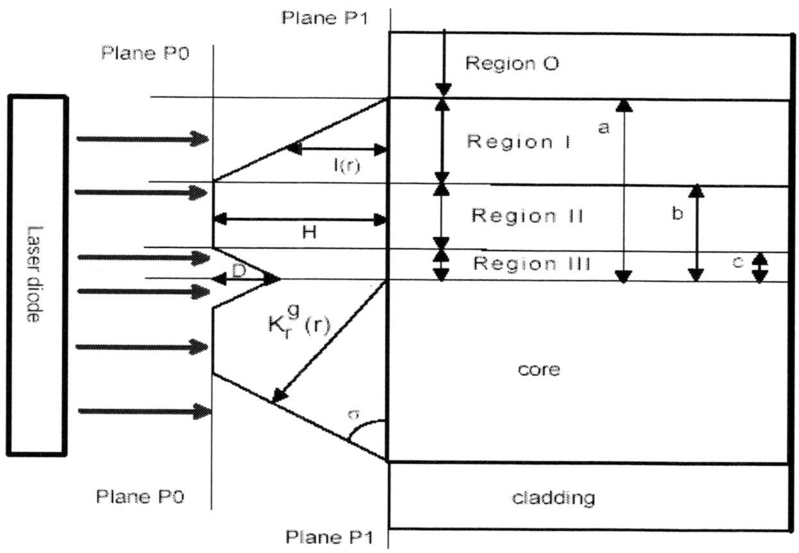

Figure 4. Geometrical illustration of the analytical model for coupling between a LD and a SMF with a truncated conical microlens.

where

$$I_x(x,\vartheta) = \int_0^\infty \frac{\exp\left(-\left[\frac{x^2}{\omega_x^2(z)} + \frac{x^2}{\omega_{of}^2} \right] \right)}{\omega_x^{1/2}(z)} \cos(q_{xi}(z)(x^2 + \vartheta) dx \tag{18a}$$

and

$$q_{xi}^g(z) = \frac{k}{2R_{x2}^g(z)} \tag{18b}$$

The corresponding expression for η_y is obtained by replacing x by y in the above equations, and the total efficiency then follows from eq. (1).

Figures 5 and 6 show results for the measurement of the coupling efficiency together with the analysis for microlens fabricated at $t = 20$ and 30 minutes, respectively [57]. It can be seen from the graph that the theoretical calculations (in solid lines) are accurate enough to predict the coupling efficiency and working distance of the conical microlens. Experimental results show that lensed fiber etched for $t = 20$ minutes has a coupling efficiency of about 41.4% and having a working distance of about 8.2 μm, while the etching time of 30 minutes gives a coupling efficiency of about 45% and a working distance of about 8.2 μm [57]. The experiment was conducted using a Fabry-Perot type LD operating at a wavelength of 1.31 μm, where the spot sizes of the LD were 0.921 μm and 0.632 μm in the parallel and the perpendicular directions, respectively, whereas the fiber spot size was 4.492 μm.

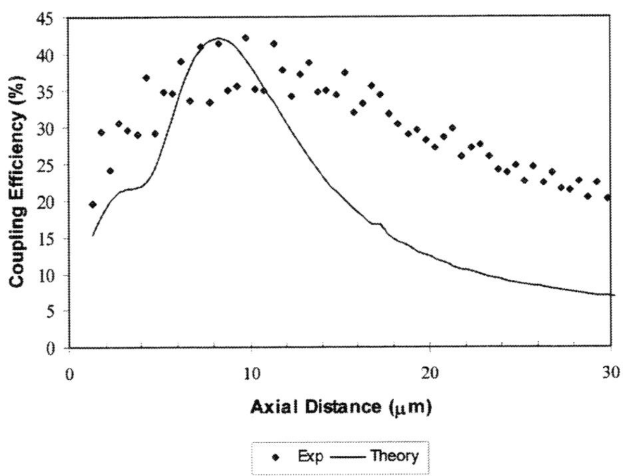

Figure 5. Theoretical and measured coupling efficiencies for etching time $t = 20$ minutes.

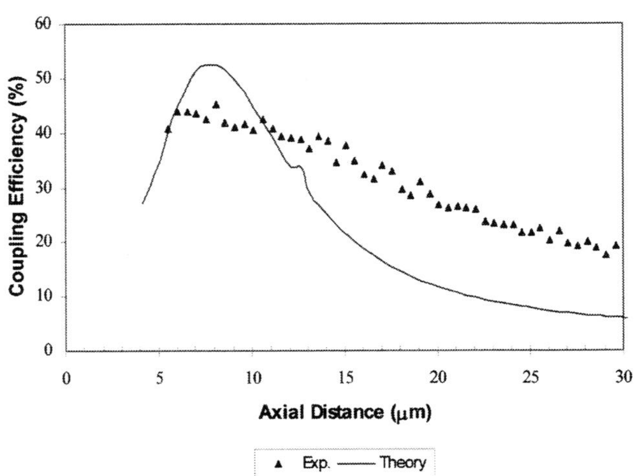

Figure 6. Theoretical and measured coupling efficiencies for etching time $t = 30$ minutes.

In the analysis, it is found that the coupling efficiency calculations are in good agreement with the measured values, and the discrepancies are probably due to losses associated with the texture of the microlens surface and deviation of laser beam from the Gaussian profile.

4. Hemispherically Lensed Tapered Fiber

In this section, a theoretical analysis will be discussed to determine the coupling efficiency between a LD and a hemispherically lensed tapered SMF, which takes into account the effect of spherical aberration on the lens surface. The results obtained clearly show that the proposed theoretical method gave a good estimation of the experimental data.

4.1. Mode Field Distribution in Tapered Fiber

For a weakly guiding step-index fiber, the spot size ω_f is defined as the radial distance at which the field amplitude is e^{-1} of its maximum, and can be approximated by [58]

$$\omega_f = a\left(0.65 + \frac{1.619}{V^{1.5}} + \frac{2.879}{V^6}\right) \tag{19}$$

where a is the core radius. The term V is defined by

$$V = \frac{2\pi a}{\lambda}\left(n_1^2 - n_2^2\right)^{1.5} \tag{20}$$

where λ is the wavelength and n_1 and n_2 are the core and the cladding RIs, respectively.

For a drawn taper, it is assumed that, in the tapered region, the cladding and the core radii decrease with the same ratio while the fiber is drawn. Initially the field is guided and confined by the core. As the core diameter decreases, the mode field first contracts slightly, then spreads out into the cladding, and gradually couples to the cladding mode. Finally, the point is reached when the mode field is no longer guided by the core, but is effectively guided by the waveguide consisting of the cladding and the surrounding medium. The mode field radius in the cladding-guiding section of the taper can then be estimated from eqs. (19) and (20) by substituting the cladding radius for a, the cladding RI for n_1 and RI of air for n_2. In the analysis, it was assumed that the wavefront in the taper region to be a plane wave.

4.2. Analysis

The coupling efficiency calculation is being done at the coupling plane (P_1), as shown in fig. 7. By using the *ABCD* matrix method, the ray matrix in the air gap (M_o) and hemispherical lens (M_1), is given by [59,60]

$$
\begin{aligned}
M_{10} &= M_1 M_0 \\
&= \begin{bmatrix} 1 & 0 \\ -\frac{1}{R}\frac{n_1 - 1}{n_1} - Abb & \frac{1}{n_1} \end{bmatrix} \cdot \begin{bmatrix} 1 & D \\ 0 & 1 \end{bmatrix}
\end{aligned} \tag{21}
$$

where n_1 is the cladding RI for the SMF, R_{eff} is the effective radius of the lens, and D is the working distance between the LD and the fiber tip. Finally, Abb is the spherical aberration caused by the hemispherical lens, and is given as [59,60]

$$
Abb = \frac{x^2 R_{eff}}{2 f n_1^{\,2}} \cdot \left(\frac{1}{D} + \frac{1}{R_{eff}} \right)^2 \cdot \left(\frac{1}{R_{eff}} + \frac{n_1 - 1}{D} \right) \tag{22}
$$

where x is the height of the ray at the fiber tip, and f is the focal length.

The complex beam parameter q_j of the Gaussian beam is denoted by the curvature radius R_j of the wavefront, and the beam radius ω_j is given in eq. (7). The relationship between q_1 (the beam parameter at the LD facet) and q_2 (the beam parameter at the SMF tip) is given in eq. (6). The parameters A, B, C and D can be obtained by solving eq. (21).

Again, by using eqs. (2)–(7) and (21) in eq. (1), the normalized power coupling efficiency can be obtained.

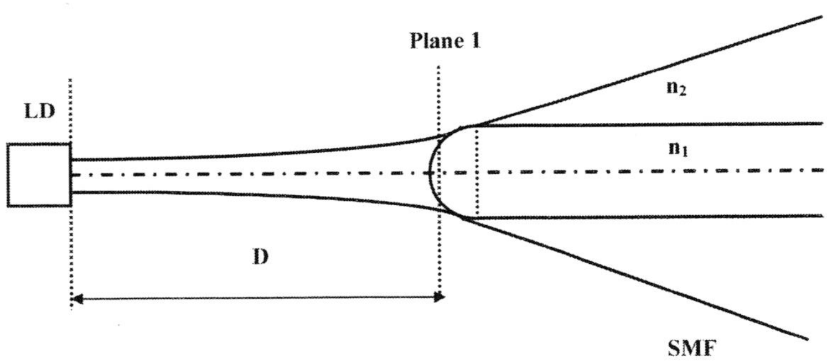

Figure 7. Schematic diagram of LD to tapered fiber coupling configuration.

A Fabry-Perot type LD operating at a wavelength of 1.55 μm is used in the experiment [60]. The beam waists of the LD are estimated to be 1.31 μm and 1.084 μm in the parallel and perpendicular directions, respectively. The hemispherically lensed tapered fiber is fabricated using the electric arc heating and pulling technique. Different lens radius (R), tapered angle (α) and tapered length (L) can be produced by varying the pulling parameter. Figures 8(a) and 8(b) show the plot of coupling efficiency measurement, together with the theoretical results obtained using the overlap integral and *ABCD* matrix techniques [60]. It can be seen from the graph that the theoretical calculations (in solid lines) are accurate enough to predict the coupling efficiency and working distance of the tapered fiber, as compared to the analysis without the spherical aberration (dashed line).

The highest coupling efficiency, that can be achieved in the present work, is 55% at a working distance of 20 μm, as shown in fig. 8(a). There is a compromise between coupling efficiency and working distance, whereby in fig. 8(b), the coupling efficiency is 43% at a working distance of 28 μm. A longer working distance, i.e. distance between the LD facet to the fiber, gives the advantage of positioning the coupling fiber farther away from the LD, thus decreasing the risk of striking the fiber against the LD during the assembly process, but with lower coupling efficiency. There is a slight discrepancy between the results of theoretical and experimental coupling efficiencies due to the assumption made in the analysis whereby the fiber core-guided mode is coupled to the fundamental cladding-guided mode while passing through the taper. However, the core mode can be coupled to higher order cladding guided modes, which will result in higher energy loss, and thus, the lower coupling efficiency. The working distance, however, could be predicted almost accurately, which is mainly influenced by the lens radius.

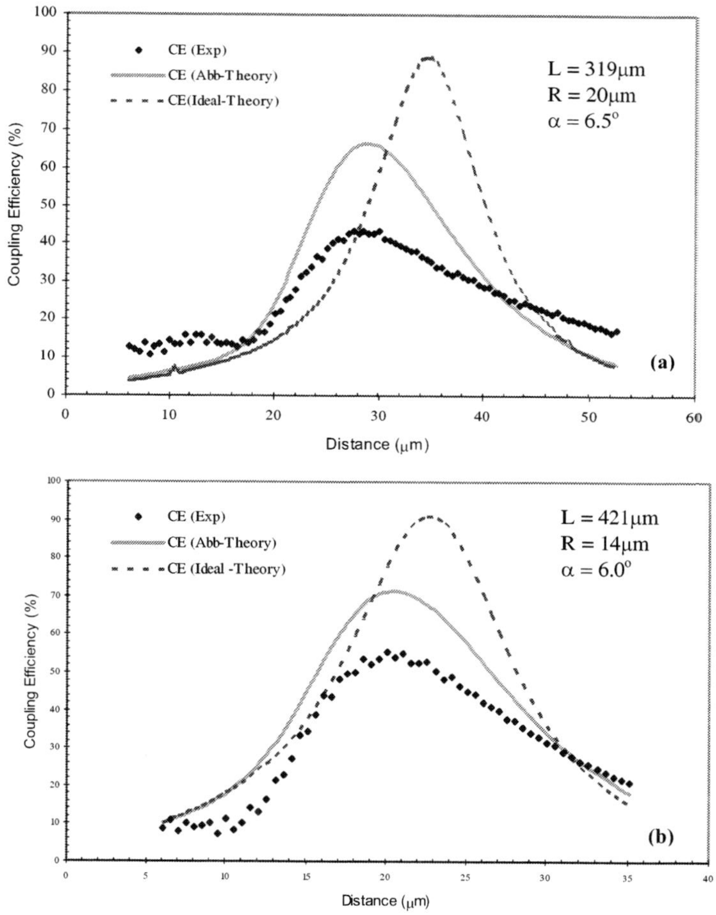

Figure 8. The measurement and calculation results for the tapered fiber. The solid lines are the theoretical calculation curve for samples (a) and (b).

The results obtained clearly show that the proposed analysis in [60] gave a good estimation to the experimental data, in terms of the working distance, for the hemispherically lensed tapered fiber.

5. Hemispherical Fiber Lens: A Scheme to Improve the Coupling Efficiency and Working Distance

In previous sections, we discussed the analysis of coupling efficiencies of conical microlens and hemispherically lensed tapered core fiber. However, the disadvantages of using these lensed fibers are due to their small working distance. The short working distance of these conventional lensed fibers is due to the limitation resulted from the small core diameter of the SMF. Several schemes and configurations, based on the hemispherical ended fiber, have been introduced to improve the working distance and transverse tolerance [41,43]. In this section, we will discuss a scheme that suggests the combination of multimode fibers with a hemispherical lensed fiber tip. The scheme, as shown schematically in fig. 9, would have a long working distance (i.e. the gap between the laser and the fiber) compared to the conventional lensed fiber [59].

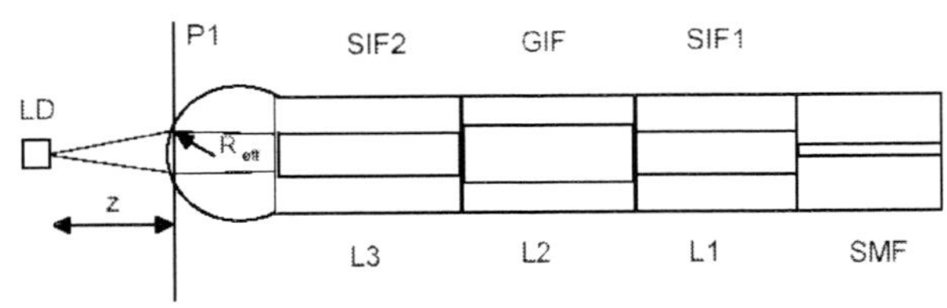

Figure 9. Configuration of the new hemispherical lensed coupling scheme that consists of step-index and graded-index fibers integrated with the SMF.

5.1. Principle of Coupling Scheme

The coupling scheme proposed consists of the hemispherical lens at the tip, step-index multimode fiber (SIF) and graded-index multimode fiber (GIF), which are integrated with the SMF [59]. In the proposed configuration, the hemispherically lensed fiber SIF_2 functions as a collimator to collimate the laser light into the GIF, which in turn, functions as a GRIN microlens to focus the laser light. The fiber SIF_1 has the function of optimising the numerical aperture (NA) to be as low as the SMF at the output end of the coupling configuration. SIF_1 also makes it possible to maintain and align the GIF at an appropriate distance from the SMF for focusing purposes.

The construction of the proposed SIF-GIF-SIF-SMF coupling scheme consists of splicing the SIF_2 and GIF together and cleaving the GIF to a length L_2. The end of SMF is also spliced to a SIF_1, in which, the SIF_1 is then cleaved to a length L_1. The next step involves the splicing of GIF and SIF_1 cleaved ends together. The configuration, which now consists of SMF-SIF_1-GIF-SIF_2, is then cleaved at a length L_3 at the SIF_2 part of the configuration. Finally, the hemispherical lens is fabricated at the tip of the cleaved SIF_2 end by using plasma-heating technique. The final configuration is shown in fig. 9.

The SIF used in the construction is a 50 μm core diameter multimode fiber with a 125 μm cladding diameter and a step-index profile with $\Delta n = 3\%$, while the SMF used has a core diameter of 8.2 μm, cladding diameter of 125 μm and a step-index profile of $\Delta n = 0.36\%$. The graded-index fiber used has a core diameter of 62.5 μm and cladding diameter of 125 μm. The inverse focusing parameter for this fiber is A_g = 128 μm.

Calculated ray trajectories in the fiber along the SIF_2-GIF-SIF_1 length as a function of the LD emission angle θ are shown in fig. 10. In the analysis, initial assumptions are made for the structural parameters, i.e. working distance z = 60 μm, SIF_2 length L_3 = 400 μm and an effective radius of the hemisphere R_{eff} = 45 μm. The initial assumptions are based on the work by Shiraishi *et al* [42], whereby with an A_g of about 200 μm, the minimum loss is obtained at lens radius of about 40 μm with the coreless fiber length in the range of 200–400 μm.

It can be seen from fig. 10 that there are specific axial positions L_2 = 50 μm or 450 μm where rays are almost parallel to the fiber axis. The GIF that is used in this work has very high focusing property (A_g = 128 μm). Because the numerical aperture (NA) of the output end of the configuration needs to be as low as the SMF, a GIF with length L_2 = 450 μm is chosen, and an SIF_1 with length L_1 = 600 μm should be placed between the GIF and SMF. Rays that are coupled to the SMF should satisfy the conditions that the input position h and the input angle ϕ at the output of the SIF_2-GIF-SIF_1 length are less than specific values, namely $|h| <$ 4.1 and $|\phi| < 7°$.

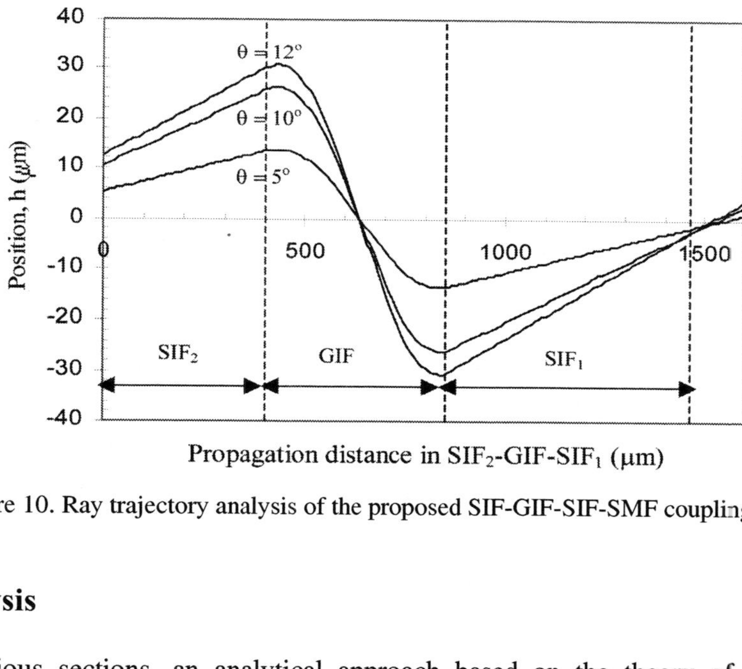

Figure 10. Ray trajectory analysis of the proposed SIF-GIF-SIF-SMF coupling scheme.

5.2. Analysis

As in previous sections, an analytical approach based on the theory of Gaussian beam propagation is being used to predict the coupling between the LD and SMF. The approach based on the *ABCD* matrix and overlap integral method produces a simple and general design for the scheme. The ray matrices for the beam are given by

$$M_{sgs(1)} = M_1 M_0$$

$$= \begin{bmatrix} 1 & 0 \\ -\frac{1}{R_{eff}}\frac{n-1}{n} - Abb & \frac{1}{n} \end{bmatrix} \cdot \begin{bmatrix} 1 & z \\ 0 & 1 \end{bmatrix} = \begin{bmatrix} A_{sgs1} & B_{sgs1} \\ C_{sgs1} & D_{sgs1} \end{bmatrix} \qquad (23)$$

$$M_{sgs(2)} = M_4 M_r M_3 M_r M_2$$

$$= \begin{bmatrix} 1 & L3 \\ 0 & 1 \end{bmatrix} \cdot \begin{bmatrix} 1 & 0 \\ 0 & \frac{n_0}{n} \end{bmatrix} \cdot \begin{bmatrix} \cos\gamma_2 & A_g \sin\gamma_2 \\ -\frac{1}{A_g}\sin\gamma_2 & \cos\gamma_2 \end{bmatrix} \cdot \begin{bmatrix} 1 & 0 \\ 0 & \frac{n}{n_0} \end{bmatrix} \cdot \begin{bmatrix} 1 & L1 \\ 0 & 1 \end{bmatrix}$$

$$= \begin{bmatrix} A_{sgs2} & B_{sgs2} \\ C_{sgs2} & D_{sgs2} \end{bmatrix} \tag{24}$$

where $M_{sgs(1)}$ and $M_{sgs(2)}$ are the ray matrices of the LD and fiber fields, respectively, at the coupling plane (P_1), shown in fig. 9. The ray matrices $M_{sgs(1)}$ and $M_{sgs(2)}$ consist of matrices that describe the transformation of the ray in the air gap (M_o), hemispherical lens (M_1), the SIF$_1$ (M_2), the GIF (M_3), the SIF$_2$ (M_4) and refraction at the boundary (M_r). In the above equations, n is the RI for SIF$_1$ and SIF$_2$, n_o is the central RI for the GIF, $\gamma_2 = L_2/A_g$ with A_g as the inverse focusing parameter for the GIF, R_{eff} is the effective radius of the lens, z is the working distance between the LD and fiber tip, L_1 is the length of SIF$_1$, L_2 is the length of GIF, L_3 length of SIF$_2$, and finally, Abb is the spherical aberration caused by the hemispherical lens, and is given in eq. (22).

The spot size and the curvature of the Gaussian beam can be obtained from the $M_{sgs(1)}$ and $M_{sgs(2)}$ matrices. These parameters are crucial for the analysis of coupling efficiency. In the analysis of coupling efficiency, the normalized overlap integral method was used [refer to eq. (1)]. The LD and SMF profiles are approximated using the Gaussian approximation, and are given by

$$\psi_{sz}^{(x)} = \left(\frac{2}{\pi}\right)^{\frac{1}{4}} \frac{1}{(\omega_x(z))^{\frac{1}{2}}} \exp\left[-\frac{x_2^2}{\omega_x^2(z)} - ik\frac{x_2^2}{2R_x(z)}\right] \tag{25a}$$

$$\psi_{Fx} = \left(\frac{2}{\pi}\right)^{\frac{1}{4}} \frac{1}{(\omega_f(L))^{\frac{1}{2}}} \exp\left[-\frac{x_2^2}{\omega_f^2(L)} - ik\frac{x_2^2}{2R_f(L)}\right] \tag{25b}$$

where the spot sizes $\omega_x(z)$ and $\omega_f(L)$ and radii of curvatures $R_x(z)$ and $R_f(L)$ of the laser and fiber, respectively, are at the lens surface plane P_1. The parameter L, which consists of L_1, L_2 and L_3, is the length of the SIF-GIF-SIF configuration under consideration. The expressions in terms of y are obtained by replacing x in equations above with y. The complex beam parameter q_j of the Gaussian beam is represented by the curvature radius R_j of the wavefront and the beam radius ω_j, and is given in eq (7).

The transformed spot size and radii of curvatures of the laser and fiber fields at plane P_1 for both the configurations can be expressed as

$$R_{x(sgs(1))}(z) = \frac{A_{sgs(1)}{}^2 \omega_{ox}^4 \pi^2 + \lambda^2 B_{sgs(1)}(z)}{A_{sgs(1)}{}^2 C_{sgs(1)}(z) \omega_{ox}^4 \pi^2 + \lambda^2 B_{sgs(1)}(z) D_{sgs(1)}(z)} \tag{26}$$

and

$$R_{f(sgs(2))}(L1, L2, L3) = \frac{A_{sgs(2)}(L2, L3)^2 \omega_{of}^4 \pi^2 n^2 + \lambda^2 B_{sgs(2)}(L1, L2, L3)}{A_{sgs(2)}(L2, L3)^2 C_{sgs(2)}(L2) \omega_{of}^4 \pi^2 n^2 + \lambda^2 B_{sgs(2)}(L1, L2, L3) D_{sgs(2)}(L1, L2)} \tag{27}$$

The spot size of the fiber field after passing through the lengths L_1, L_2 and L_3 can be written as:

$$\omega_{f(sgs2)}(L1, L2, L3) = \sqrt{\frac{\left(A_{sgs2}(L2, L3)\pi\omega_{of}^2 n\right)^2 + \left(B_{sgs2}(L1, L2, L3)\lambda\right)^2}{(A_{sgs2}(L2, L3) D_{sgs2}(L1, L2) - B_{sgs2}(L1, L2, L3) C_{sgs2}(L2))\pi^2 \omega_{of}^2 n^2}} \tag{28}$$

In the analysis, it is assumed that the spot sizes of the laser at the lens surface (i.e. plane P_1) do not change after passing through the lens.

The y-component of transformed radii of curvature can be obtained by substituting x with y in the above equations. The equations given and the respective equations in y-component can then be used to determine the coupling efficiency.

A Fabry-Perot type LD operating at a wavelength of 1.31 μm is used in the experiment. The beam waists, defined as $1/e^2$ of the maximum intensity, are estimated to be 0.921 μm and 0.632 μm in the parallel and perpendicular directions, respectively. The SIF-GIF-SIF-SMF coupling configuration is constructed with various non-optimum structural parameters, as listed in Table 1. For the measurement, the SMF with the coupling scheme constructed at one of its ends, and the LDs placed on the holders of an alignment system equipped with a microscopy device, are used and aligned to each other.

Table 1. Structural parameters of the fabricated lensed fiber scheme. The SIF diameter used is 50 μm, while the diameter for the GIF is 62.5 μm

Sample	***R*** **(μm)**	***L*₁** **(μm)**	***L*₂** **(μm)**	***L*₃** **(μm)**	***A*g** **(μm)**
A		650	440	390	
B	~ 45	520	410	140	128
C		590	430	350	

Figure 11 shows the results of the coupling efficiency measurement together with the analytical results obtained using the overlap integral and *ABCD* matrix techniques. It can be seen from the graph that the theoretical calculations, in solid lines, are accurate enough to predict the coupling efficiency and working distance of the proposed scheme. Results show that the Sample A has a coupling efficiency of about 55% or coupling loss of 2.6 dB, and having a working distance of 65 μm. The working distance between the laser and fiber can be improved by changing the structural parameters of the configuration, as shown by Samples B and C, however, their coupling efficiencies dropped to 31% (at 90 μm) and 39% (at 70 μm), respectively. The transverse tolerances are about ±0.5 μm to ±1.0 μm in the parallel, and ±0.5 μm to ±0.7 μm in the perpendicular directions, while the axial tolerances are about 12–20 μm.

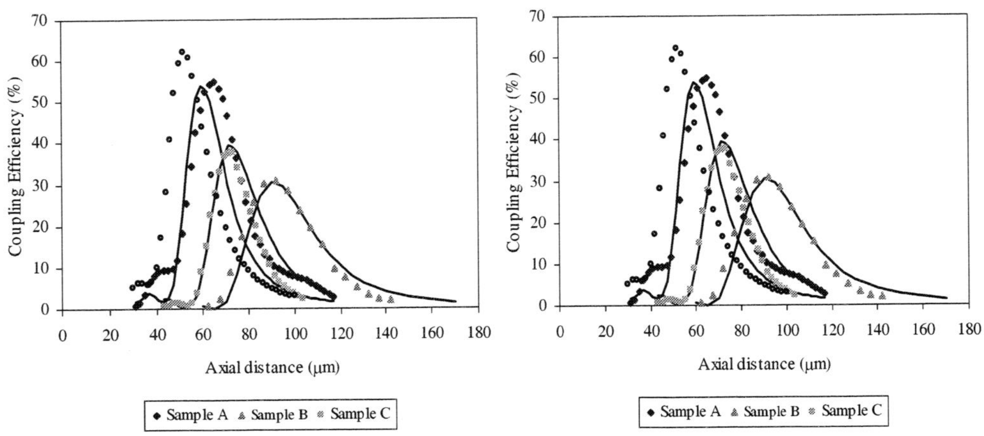

Figure .11. The measurement and calculation results for the coupling scheme. The solid lines are the theoretical curve for each of the samples. The hollow circles represent the theoretical curve for the optimum lengths of L_1, L_2 and L_3 obtained from the ray trajectory analysis.

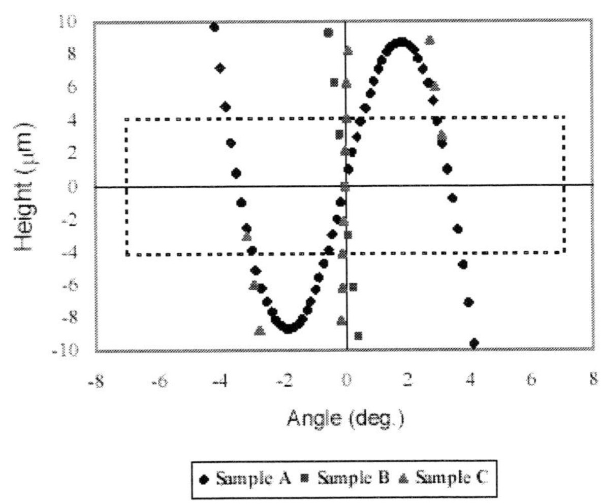

Figure 12. Focusing properties of the coupling scheme illustrated in fig. 9. The rectangular region indicated by the dashed line represents the domain for the specific values of h and ϕ. Analyses were conducted at $z = 65$ μm at a wavelength of 1.31 μm.

Figure 11 also shows the coupling efficiency, given by the hollow circles, of the coupling configuration calculated for optimum lengths of L_1 = 600 μm, L_2 = 450 μm and L_3 = 400 μm, that were determined from the ray trajectory analysis. The coupling efficiency is about 62% at a working distance of 52 μm. There is a discrepancy in the working distance, whereby initial assumption made was z = 60 μm. This is probably due to the aberration effect that causes the rays not to focus at a unique point, thus influencing the working distance.

To further strengthen the result, a qualitative analysis of the coupling scheme is conducted. Rays that are coupled to the SMF should satisfy the conditions that the input height h and the input angle ϕ are less than specific values, i.e. $|h| < 4.1$ μm and $|\phi| < 7°$, indicated by the rectangular dashed lines in fig. 12. The result of the focusing parameter observed from fig. 12 indicates that Sample A has the most rays coupled into the SMF. This is based on the number of points that falls into the rectangular region, as compared to Samples B and C.

6. Conclusion

The objective of this article is two-fold, i.e. to introduce analytical models to analyse the coupling between LDs and SMFs, and to experimentally determine the coupling efficiency between LDs and SMFs. Conical and hemispherical types of mirolenses are considered in this article, and results obtained are compared to experimentally measured values.

In Section 3, the analytical model to calculate the coupling efficiency of conical microlens is discussed. The transfer matrix is used to analyse the propagation of the Gaussian beam from the LD through the conical microlens. It is found that the calculations are accurate in terms of the peak position, but the efficiency is slightly higher compared to the experimental results. This is probably, as mentioned before, due to losses associated with the texture of the microlens surface, deviation of laser beam from the Gaussian profile and misalignments during measurement. For fiber distance above 15 μm, the experimental values are higher compared to the predicted results. This is probably due to the assumption made in the analysis that the laser beam is parallel to the optical axis. However, in fact, the laser beam diverges from the optical axis farther away from the laser facet, which might cause the discrepancy at longer distances. Generally, the analytical technique discussed is more reliable, at wider range of single mode fiber core dimension, compared to method used in previous work [49,52].

In Section 4, the discussion is focused on hemispherically lensed tapered core fiber. Again, the transfer matrix is used to analyse the propagation of the Gaussian beam, and the effect of spherical aberration on the lens surface is taken into account. The results obtained clearly show that the theoretical method gave a good estimation to the experimental data as compared to the method used in [58]. There is a slight discrepancy between the theoretical and the experimental results for coupling efficiency. This is due to the assumption made in the analysis whereby the fiber core-guided mode is coupled to the fundamental cladding-guided mode when passing through the taper. However, the core mode can be coupled to higher order cladding-guided modes, which will result in higher energy loss, and thus, the lower coupling efficiency. The working distance, however, could be predicted almost accurately, which is mainly influenced by the lens radius.

In Section 5, analysis and measurement of the coupling scheme shows that there is an improvement in the working distance compared to a conventional lensed fiber scheme. The distance between the LD and the fiber is improved, i.e. to about 65 μm to 90 μm, depending on the structural parameter used, which is important in terms of the effect of multiple reflection, and also, desirable for the automatic packaging process of an optical subassembly. Multiple reflections between the facets would create power variation or fluctuation due to the effect of interference. It is reported that the estimated reflections are about –30 dB and –60 dB at working distances of 10 μm and 100 μm, respectively, and multiple reflection effect is reduced at longer distances. Besides the power variation, the sub-micron alignment requirement in the conventional lensed fiber (between ±0.3 to ±0.5 μm) would make the packaging process more complicated. However, this problem is reduced if the proposed coupling scheme is being used.

References

[1] Boudrean, R.A. *J. of Metals* 1994, 46, 41–45.

[2] Dagenais, M.; Vusirkala, V.; Merritt, S.A.; Bartolo, R.E.; Saini, S.S. *Proc. SPIE* 1997, 3005, 38-47.

[3] Yamada, Y.; Suzuki, S.; Moriwaki, K.; Hibino, Y.; Tohmori, Y.; Akatsu, Y.; Nakasuga, Y.; Hashimoto, T.; Terni, H.; Yanagisawa, M.; Inone, Y.; Akahori, Y.; Nagase, R. *Electron. Lett.* 1995, 31, 1366–1367.

[4] Armentio, C.A.; Negri, A.J.; Tabasky, M.J.; Boudreau, R.A.; Rothman, M.A.; Fitzgarald, T.W.; Haugsjaa, P.O. *IEEE Trans. Components, Hybrids and Manufac. Tech.* 1992, 15, 1072–2079.

[5] Wong, Y.-M. *J. Light. Tech.* 1995, 13, 995–1013.

[6] Wang, L.A.; Su, C.D. *J. Light. Tech.* 1996, 14, 2757–2762.

[7] Sidorin, Y.; Howe, D. *Appl. Opt.* 1997, 36, 4273–4277.

[8] Verdiell, J.M.; Ziari, M.; Welch, D.F. *Electron. Lett.* 1996, 32, 1817–1818.

[9] Shih, M.H.; Choa, F.S.; Kapre, R.M.; Tsang, W.T.; Logan, R.A.; Chu, S.N.G. *Electron. Lett.* 1995, 31, 1058–1060.

[10] Vusirikala, V.; Merritt, S.A; Kareenahalli, S.; Gopalan, B.; Stone, D.; Wood, C.E.C; Dagenais, M. *IEEE J. Phot. Tech. Lett.* 1996, 9, 1130–1133.

[11] Vakhshoori, D.; Hobson, W.S.; Han, H.; Lopata, J.; Henien, G.E.; Wynn, J.D.; deJong, J.; Schnoes, M.L.; Zydik, G.J. *Electron. Lett.* 1996, 32, 1007–1008.

[12] Lealman, I..F.; Rivers, L.J.; Harlow, M.J.; Perrin, S.D. *Electron. Lett.* 1994, 30, 1685–1687.

[13] Mersali, B.; Bruckner, H.J.; Feulillade, M.; Sainson, S.; Ougazzaden, A.; Carenco, A. *J. Light. Tech.* 1995, 13, 1865–1872.

[14] Zengerle, R.; Leminger, O.; Weiershausen, W.; Faltin, K.; Hubner, B. *IEEE J. Phot. Tech. Lett.* 1995, 7, 532–534.

[15] Wilson, R.G. *Appl. Opt.* 1998, 37, 3201–3205.

[16] Karstensen, H. *J. Opt. Commun.* 1988, 9, 42–49.

[17] Leslie, L.A.; Mann, J.W.; Lalk, G.R.; Krchnavek, R.R.; Andreadakis, N.C.; Zah, C. *J. Light. Tech.* 1991, 9, 477–484.

[18] Ratowsky, R.P.; Yang, L.; Deri, R.J.; Chang, K.W.; Kallman, J.S.; Trott, G. *Appl. Opt.* 1997, 36, 3435–3438.

[19] Kato, K.; Nishi, I.; Yoshino, K.; Hanafusa, H. *IEEE J. Phot. Tech. Lett.* 1991, 3, 469–470.

[20] Huang, S.-Y.; Gaebe, C.E.; Miller, K.A.; Wiand, G.T.; Stakelon, T.S. *IEEE Trans. Adv. Packg.* 2000, 23, 165–169.

[21] Karstensen, H.; Drogemuller, K. *J. Light. Tech.* 1990, 8, 739–747.

[22] Saruwatari, M.; Nawata, K. *Appl. Opt.* 1979, 18, 1847–1856.

[23] Kawano, K. *Appl. Opt.* 1986, 25, 2600–2605.

[24] Saruwatari, M.; Sugie, T. *IEEE J. Quantum. Electron.* 1981, QE-17, 1021–1027.

[25] Edwards, C.A.; Presby, H.M.; Dragone, C. *J. Light. Tech.* 1993, 11, 252–257.

[26] Shah, V.; Curtis, L.; Vodhanel, R.S.; Bour, D.P.; Young, W.C. *J. Light. Tech.* 1990, 8, 1313–1318.

[27] Presby, H.M.; Edwards, C.A. *Electron. Lett.* 1992, 28, 582–584.

[28] Presby, H.M.; Giles, C.R. *IEEE J. Phot. Tech. Lett.* 1993, 5, 184–186.

[29] Hunziker, W.; Bolz, E.; Melchior, H. *Electron. Lett.* 1992, 28, 1654–1656.

[30] Eisenstein, G.; Vitello, D. *Appl. Opt.* 1982, 21, 3470–3474.

[31] Gangopadhyay, S.; Sarkar, S.N. *Appl. Opt.* 1997, 36, 8582–8586.

[32] Gangopadhyay, S.; Sarkar, S.N. *Opt. Commun.* 1996, 132, 55–60.

[33] Modavis, R.A.; Webb, T.W. *IEEE J. Phot. Tech. Lett.* 1995, 7, 798–800.

[34] Sakai, J.I.; Kimura, T. *IEEE J. Quantum Electron.* 1980, QE-16, 1059–1067.

[35] Yamada, J.I.; Murakami, Y.; Sakai, J.I.; Kimura, T. *IEEE J. Quantum. Electron.* 1980, QE-16, 1067–1072.

[36] Hanmou, H.; Itoh, M. *Electron. Lett.* 1995, 31, 793–794.

[37] Ning, L.T.; Freude, W. *Electron. Lett.* 1991, 27, 1202–1203.

[38] Wang, Z.; Mikkelsen, B.; Pederson, B.; Stubjaer, K.E.; Olesen, D.S. *J. Light. Tech.* 1991, 9, 49–55.

[39] Alder, T.; Stohr, A.; Heinzelmann, R.; Jager, D. *IEEE J. Phot. Tech. Lett.* 2000, 12, 1016–1018.

[40] Chen, W.T.; Wang, L.A. *IEEE J. Phot. Tech. Lett.* 2000, 12, 501–503.

[41] Shiraishi, K.; Oyama, N.; Matsumura, K.; Ohishi, I.; Suga, S. *J. Light. Tech.* 1995, 13, 1736–1744.

[42] Shiraishi, K.; Ohnuki, H.; Matsumura, K.; Ohishi, I.; Morichi, H.; Kazami, H. *J. Light. Tech.* 1997, 15, 356–363.

[43] Shiraishi, K.; Kuroo, S.I. *J. Light. Tech.* 2000, 18, 787–794.

[44] Kweon, G.; Park, I.; Shim, J. *Appl. Opt.* 1998, 37, 4789–4796.

[45] Kalonji, N.; Semo, J. *Electron. Lett.* 1994, 30, 892–894.

[46] Kogelnik, H.; *Microwave Research Institute Symposia Series* (Ed. Fox, J.) 1964, 333–347.

[47] Saleh, B.E.A.; Teich, M.C. Fundamental of Photonics; Wiley-Interscience: NY, 1991.

[48] Emkey, W.L.; Jack, C.A. *J. Light. Tech.* 1987, 5, 1156–1164.

[49] John, J.; Maclean, T.S.M.; Ghafouri-Shiraz, H.; Niblett, J. *IEE Proc. J. Optoelectron.* 1994, 141, 178–184.

[50] Ghafouri-Shiraz, H.; Kotsas, A.; Maclean, T.S.M. *Microw. Opt. Tech. Lett.* 1989, 2, 170–172.
[51] Kotsas, H.; Ghafouri-Shiraz, H.; Maclean, T.S.M. *Opt. Quantum Electron.* 1991, 23, 367–378.
[52] Ghafouri-Shiraz, H.; Asano, T. *Opt. Lett.* 1986, 11, 537–539.
[53] Rahman, F.A.; Takahashi, K.; Chuah, H.T. *Opt. Commun.* 2003, 215, 61–68.
[54] Jenkins, F.A.; White, H.E. Fundamentals of Optics; McGraw-Hill: NY, 1976.
[55] Ghafouri-Shiraz, H.; Cameron, K.; Fatah, R.M.A. *Microw. Opt. Tech. Lett.* 1990, 3, 214–217.
[56] Ghafouri-Shiraz, H.; Aruga, T. *Opt. and Laser Tech.* 1996, 28, 367–371.
[57] Faidz, A.R.; Mishra, V.N. *Optik* 2004, 114, 457– 460.
[58] An, H.L. *Opt. Commun.* 2000, 181, 89–95.
[59] Rahman, F.A.; Takahashi, K.; Chuah, H.T. *Opt. Commun.* 2002, 208, 103–110.
[60] Kumaran, S; Faidz, A.R. Microw. *Opt. Tech. Lett.* 2004, 43, 341–343.

Faidz A. Rahman received the B.Sc. (Hons.) and M.Sc. degrees from the Department of Physics, University of Malaya (Malaysia) in 1994 and 1996, respectively, and the Ph.D. degree from the Faculty of Engineering, Multimedia University (Cyberjaya, Malaysia) in 2002. In 1998, he joined Multimedia University (Malaysia) as a Lecturer, and is presently an Associate Professor at the Faculty of Engineering. He is also the Chairman of the Center for Photonics Research, Innovation and Applications, a research center in the Faculty of Engineering of Multimedia University. His current research interests are in optical fiber devices, components, fiber sensors and optoelectronic packaging technology. His work has been published in various international journals and conference proceedings.

In: Frontiers in Optical Technology: Materials and Devices ISBN: 1-60021-084-8
Editors: P.K. Choudhury and O.N. Singh, pp. 235-271

Chapter 9

AN OVERVIEW OF OPTICAL SENSORS AND THEIR APPLICATIONS

P. K. Choudhury[1] and O. N. Singh[2]
[1] Faculty of Engineering, Multimedia University, Cyberjaya 63100, Malaysia
[2] Department of Applied Physics, Institute of Technology
Banaras Hindu University, Varanasi 221005, India

Abstract

In the developed nations of the world, optical sensor industry is a multimillion-dollar business. This chapter is devoted to the review of the fundamentals of fiber optic sensing techniques, and describes the different types of optical sensors and their applications in different industries. Descriptions are made of intensity based sensors, distributed sensors, and interferometric sensors, their different types and specific applications. Examples of some more forms of FOSs, and their use in the context of environment are also reviewed. In addition, discussions are made of the role of integrated optics technology in sensing applications. It is noteworthy that photonic crystals and STFs have been the emerging areas of research in the present day; the chapter describes the use of such materials and/or mediums in the area of optical sensing.

1. Introduction

Fiber optic sensors (FOSs) have been the focus of research and development for over twenty five years. In this period, varieties of sensors have been devised by the investigators for multifarious range of applications. Initial devices incorporating such sensors have been designed for very specific applications. However, because of peak research activity, situation has now changed very rapidly, and a variety of such sensors now appeared in the market. The basic conceptual building blocks of FOSs are now well-known, and presently, the principal areas of application of these sensors appear to be in medical systems, aerospace and hydrospace sciences, electrical power supply industries, and petrochemical, oil and gas industries [1–10]. In the developed nations of the world, optical sensor industry is a multimillion-dollar business and, by the end of this century, it is projected to grow over 600 million-dollar industry [11].

Optical technologies have long played a vital role in instrumentation and sensors, especially for non-contact measurements. Although various forms of chemical and medical sensors were invented in the past, such sensors implementing the optical fiber technology offer significant advantages over conventional systems. The basic reason is that the energy source and the information processing involve the transmission of lightwave signals through optical fibers which are made of chemically and electrically inert mediums. Todate highly sophisticated instruments for sensing measurements have been investigated which include remote measurement capabilities, multiple sensing, miniaturized spectrometers, miniaturized optical probes to adapt into hypodermic models for clinical and biomedical applications etc. Although, so far as the sensors for chemical and medical applications are concerned, there are still some problems related to poor long-term stability, slow response time, mechanical design of the probe-to-fiber coupling etc., yet the optical fiber sensing technique has been recognized to have enormous potentials. Apart from this, the application of physical sensors to the biomedical area also appears to be extremely promising.

The key factors in the development of FOSs include the fabrication of sensors that are small and trendy, lightweight, immune to electromagnetic (EM) interference and require no electrical power at the point of sensing. Fiber optic sensing technology always faced the problem of competing with the established low cost and highly reliable conventional sensor technologies that provide adequate performance characteristics. In fact, the performance characteristics of FOSs are equally well as compared to their conventional sensor counterparts, but the overall technology becomes more expensive owing to their small market volume. As a result, FOSs are realized only when there are some specialized application needs, e.g. trendy and non-consumption of electrical power etc. Such fiber optic sensing technology is also implemented when there are requirements of multiplexing the sensors (optical fibers can be multiplexed efficiently).

A large number of components depending on the physical principles have been developed over the decade. Today some of the commercially available components are fused and polished couplers, absorption filters, amplifiers and sources made of rare earth doped fibers etc., which are greatly used in all-fiber sensing devices. Efforts are also put for the development of active fiber components such as sandwitchable couplers or fast light modulators. Several attempts have been made in this regard, but they have been inherently limited by the passive nature of fiber material. Factors related to materials are of high importance for both the sensing means and the practical implementation of FOS systems. A great many varieties of optical fiber materials and structures have been introduced leading to various forms of applications in sensing technology. Polished fiber substrate technology has emerged as a powerful tool to perform a large number of basic functions within the fiber itself by providing controlled access to its core, and thereby guiding the wave.

It is noteworthy that majority of the FOS products are based on multimode fiber systems, which have large core diameter and high numerical aperture, resulting thereby high light-collecting efficiency of fibers, and therefore, low cost light sources such as LEDs can be implemented. The output monitored by such FOS systems is typically related to the measurement of the intensity of light either as a direct intensity measurement or via the polarization properties of light. Monomode fiber systems are also developed which are normally based on interferometric techniques. But, for such FOS systems, laser diodes are used as the light source, which are more expansive than LEDs. Also, such systems incorporate costlier components such as fiber couplers, integrated optic (IO) modulators etc., and the measurements are based on the phase delay of the light in the sensor arrangement. A new class of sensors has been investigated by the introduction of monomode fiber components into optical devices. In its simplest form, the fiber itself serves basically as a flexible link for the transmission of optical signals to and from the optical arrangement that constitutes the sensor. In more complex arrangements, optical fiber itself serves not only as the medium of transmission, but also as the sensor. FOSs can be categorized on the basis of these two forms of sensing devices – i.e. intensity or interferometric techniques. Major advancements in this new exciting field are expected because novel fiber optic components are developed specifically for sensor applications.

FOSs are classified into the categories of optrodes, fiber coupled sensors and distributed FOSs. In all these configurations, optical fibers themselves serve as a passive pipe of light in spectroscopic analyses, or can be coated with a substance to produce an electrochemical response.

Optrodes are optical fiber based analogues of electrodes made of an optical fiber, and the tip of the fiber is sensitised via chemical indicators that respond to pH or a reflective diaphragm to change the amount of light reflected at the tip due to changes in chemical properties or pressure. In general, in such sensors, the sensitised tip reacts with the measurand. Several kinds of sensor configurations based on the single-ended transducer type, as shown in fig. 1, have been realized. Such kinds of sensors are generally applied to biomedical and chemical sensing applications, e.g. for measurands including temperature, pressure and chemical parameters.

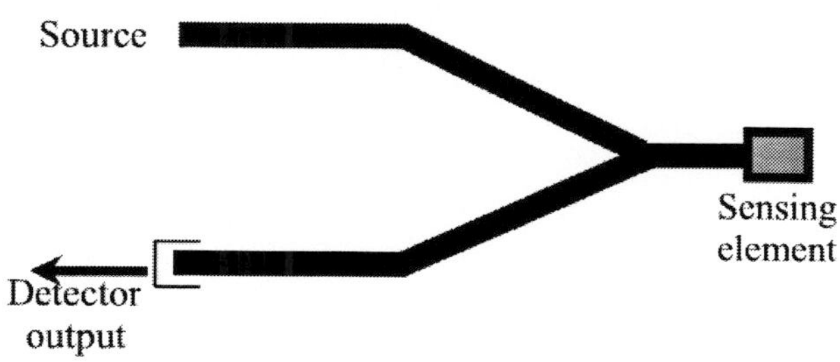

Figure 1. Single-ended sensor transducer.

In fiber-coupled sensors, fibers themselves act as passive waveguides. These waveguides take light to and from the points of measurement, and in this way, offer the best ability to perform quasi-remote measurements. In such sensors, the parameters of interest are measured at a specific location. In the case when a longer length of optical fiber is used for sensing measurements, the sensor can be configured as an extended path-integrating device (fig. 2). Alternatively, long sensing fiber can be wrapped on suitable transducers, and such techniques are used in interferometric sensors to provide increased sensitivity to weak physical fields, e.g. in acoustic sensing.

In distributed FOSs, all or a part of the fiber is sensitised that reacts with the measurand, and subsequently offer a special distributed measuring capability. That is, in such FOS systems, the measurand parameter of interest is monitored at many locations along a fiber, and this technique opens up new potential applications where a spatial mapping of the parameters are required.

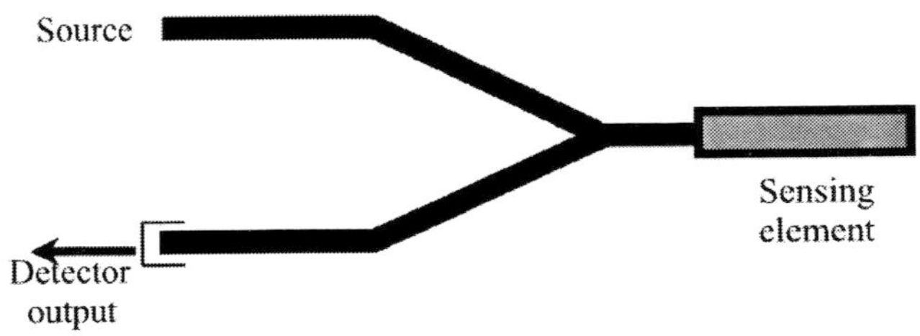

Figure 2. Extended path integrating device.

2. Intensity Based Sensors

A variety of intensity based FOSs implementing numerous sensing techniques viz. absorption, emission, luminescence, fluorescence, etc., have been devised so far in order to monitor various physical parameters such as temperature, pressure, displacement etc. The techniques involved may be, in general, point sensing or distributed sensing leading thereby to different degrees of sensing capabilities. The most basic configuration of FOS systems is shown in fig. 1 where the sensing element is designed to some change in some physical characteristics of light, e.g. wavelength, intensity, polarization etc., which are then monitored at the optical detector. There are several varieties of probes or transducers used in FOS systems, some of which are coupled fibers, birefringent probes and microbend probes [12–25].

In sensors employing birefringent probes, light from an input fiber is passed through a birefringent element which acts simply as a phase plate that is sensitive to lateral displacements. This birefringent element is placed between two polarizers, and the light signal coming out of the second polarizer is coupled to a detector (fig. 3). The birefringent material changes the state of polarization of the light passing through it, causing thereby a change in the optical power passing through the second polarizer. In the case of microbend sensors, the transmitted light in the fiber modulates periodic bends in the fiber which are produced by placing it between two corrugated plates (fig. 4). This results in a radiative out-coupling of light from the fiber, and the degree of coupling depends on the degree of bending. Thus, a kind of attenuation is introduced at a given location in the fiber, and the degree of attenuation depends on the pressure applied to the probe. Such FOSs can be classified into two types, viz., intrinsic and extrinsic. Intrinsic sensors are those where there occurs an interaction between light and measurand, and the light remains guided within the fiber, whereas in extrinsic sensors, the light is coupled from the fiber through a sensing region and back into fiber. The mechanisms of a few forms of intensity based sensors are discussed below.

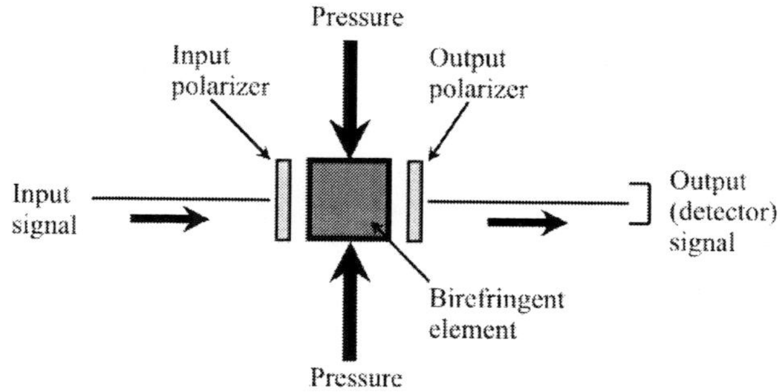

Figure 3. Arrangement implementing birefringent probes.

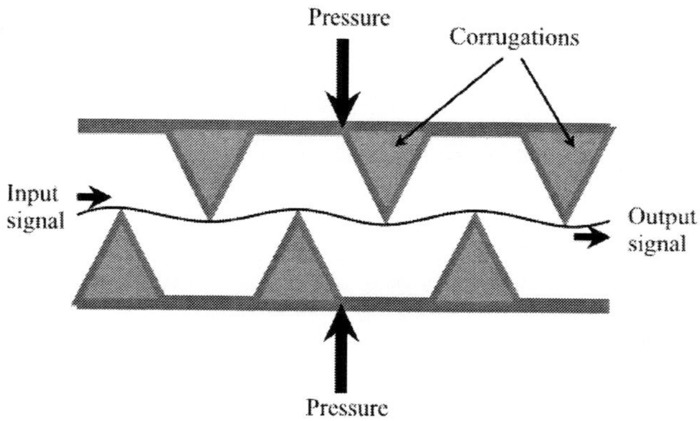

Figure 4. Microbend arrangement used in sensors.

2.1. Fabry-Perot Type Sensors

Several forms of Fabry-Perot sensors are described in the literature [26–29]. A rough sketch of the kind of sensor probe, that can be used with multimode fibers with a simple intensity based signal processing scheme, is demonstrated in fig. 5, where the light from an optical source is coupled to a micro dimensional Fabry-Perot element (consisting of two partially reflecting surfaces separated by a suitable gap of either air or some transparent material) located at the tip of the fiber. In the present configuration, light reflecting off the two surfaces is coupled back into the fiber, and therefore, the net reflectivity of light is essentially a function of the phase difference between the two paths. The interference may be either constructive or destructive depending on whether the relative phase difference between the reflected optical beams is in phase or out of phase. The reflected power P_R is given as

$$P_R = \frac{1}{2}\xi P_0\left[1+\cos\left(\frac{4\pi nd}{\lambda}\right)\right]$$

where ξ, P_0, n and d are coupling coefficient, input power, refractive index (RI) of the material between the two reflecting surfaces and the thickness of the gap, respectively. Thus, the effective reflection coefficient $P_R/(\xi P_0)$ is a function of wavelength λ, and therefore, on illumination, a spectral response (depending on d, n and λ) is obtained which changes with the applied pressure owing to the change in the thickness of the gap. Shift in the spectral response is a measure of the applied pressure.

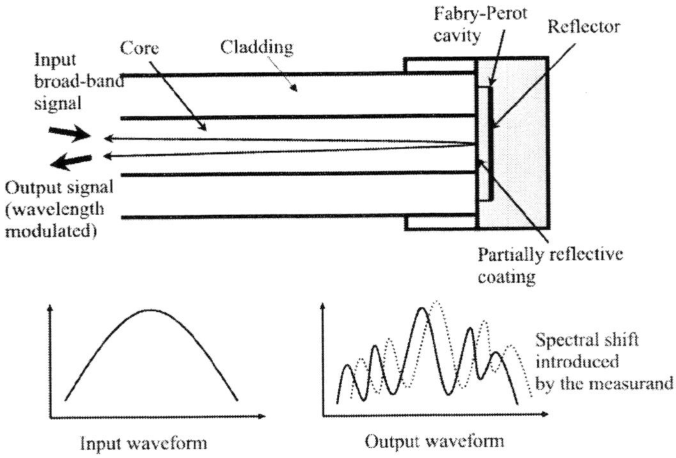

Figure 5. Fabry-Perot type sensor.

2.2. Emission Based Sensors

Functioning of emission based sensors primarily depends on the spectral shift in the blackbody radiation emitted by a body at a certain temperature. Such sensors use a kind of metallic coating at the tip of the fiber [30–32]. The blackbody radiation follows Planck's law (i.e. the total radiation W emitted by a body is proportional to the fourth power of the temperature T at which the radiation is emitted; $W \propto T^4$), and the temperature of the emitting surface can be obtained by the ratio of radiance detected in two band-limited sections of the spectrum. This kind of sensor system works particularly well at higher tepmeratures, i.e. more than 100 ^{0}C.

2.3. Differential Absorption Based Sensors

A simple fiber optic temperature sensor, based on differential absorption phenomenon, is shown in fig. 6, where the light emitting from two optical sources (wavelengths λ_1 and λ_2) is injected into the fiber. The injected light falls on the tip of the fiber where it passes through an absorptive thermochroic material, reflected by a reflective surface, and then returns into the fiber. The amount of light returned to the input end of the fiber is detected by a fiber coupler. The thermochroic material introduces a change in the spectral absorption characteristics with temperature producing thereby a spectral shift. In this kind of sensor system, in general, a solution is encapsulated at the tip of the fiber. However, semiconductor materials such as GaAs or heavily doped glass, that exhibit a shift in the absorption band, can also be used.

Varieties of systems employing optical absorption and differential optical absorption methods are reported in the literature [33–40]. These generally require an open cell configuration, and the light propagating through it acts as the detection system. But these systems meet the disadvantage that a very precise alignment of light is required and, in addition, the light in the near infrared region is transmitted causing thereby to monitor the weaker overtone bands only. However, the low-loss optical fibers are advantageous in the sense that the light can be transmitted to a longer distance, and the sensor head can be remotely located from the monitoring station.

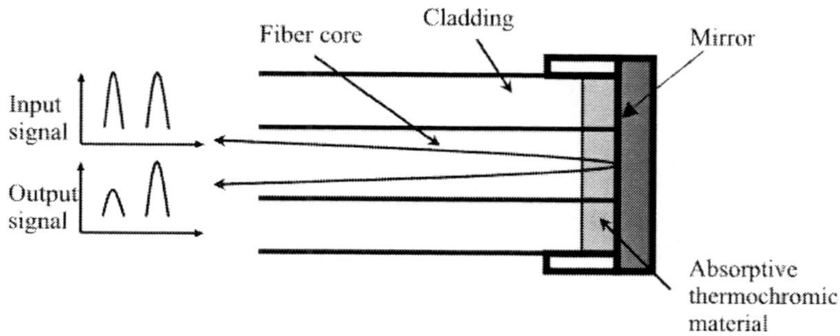

Figure 6. Sensor arrangement based on differential absorption.

2.4. Flourescence Luminsecence Type Sensors

The fluorescence and luminescence phenomena may also be implemented for sensing [41–48]. Such optical signals of many materials exhibit spectral and temporal changes with temperature. Principle of spectral based sensors depends on the detection of shifts in the fluorescence spectrum introduced by the measurand. The temporal based sensors involve the measurement of the lifetime of a fluorescence signal with the strength of the measurand. Figure 7 represents a typical example of temporal based such kind of FOS where a fiber-tip is sensitized by a fluorescent material (to be sensed), which is excited by a laser light source. In this process, a portion of the emitted light is collected by the fiber, which is then passed through a detector. The emitted fluorescence signal decays with a characteristic time constant that depends on the material and the host substrate. As a result, for a given material, the decay of the fluorescence signal can be measured with time, which ultimately yields the measurement of the change in temperature.

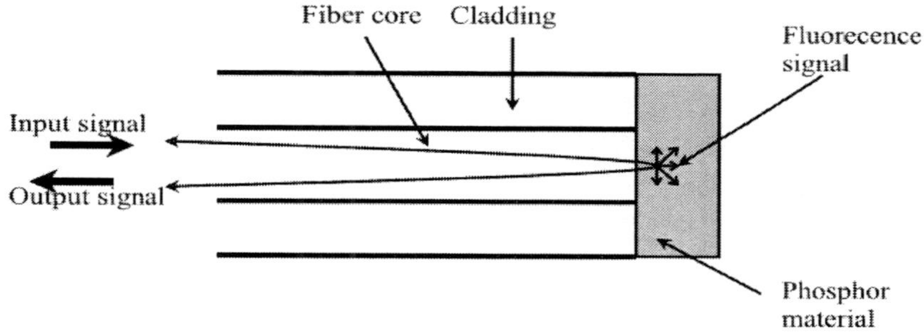

Figure 7. Temporal based fluorescence/luminescence sensor.

2.5. Evanescent Field Sensors

Sensors employing modified cladding elements are primarily based on the study of the evanescent wave spectroscopy, which play a very important role in the area of sensing technology. The technique involved in this kind of sensor incorporates the study of evanescent wave spectroscopy where the evanescent waves, penetrating the cladding region, are absorbed by the fluid (under testing) which is placed on the uncladded region of an optical fiber. Thus, a part of the fiber is sensitised by the fluid, and the study of the spectral composition of these evanescent waves, after getting absorbed by the fluid, give the estimation of their concentration. Normally He-Ne laser is used as the optical source, and the uncladed portion of the fiber is sealed in a glass tube, which has facilities for continuous filling and drawing with the fluids of various concentrations. However, the selection of light source essentially depends on the peak absorption wavelength of the measurand. The output end of the fiber is connected to a power meter, and the determination of the transmitted power gives the concentration of the absorbing fluid. In this way a continuous monitoring of the absorbing fluid is done. The sensitivity of this kind of sensor depends on the type of the source used, numerical aperture (NA) of the fiber and the RI of the fluid. This also depends on the launching condition of the beam, and on the geometry of the uncladed portion of the fiber. However, if the uncladed region is tapered and only the selected rays are allowed to launch into the fiber, there occurs a many fold increase in sensitivity. Some particular forms of evanescent field sensors are discussed in the later sections of this review article.

2.6. Photometric Sensors

Types of sensors described so far involves the use of probe by which the measuring parameter is sensed. Photometric sensors are those where no any probe or transducer is used at the fiber end, and are useful for the measurement of several biological parameters in human body. As for example, such sensors can be used to measure the relative velocity of the blood flow in vessels which is desirable for a wide range of medical purposes [49,50]. The velocity of blood flow in vessels can be determined by using a fiber coupled laser-Doppler velocimeter. The method is based on the Doppler frequency shift of laser light due to scattering from moving red cells. In this method, a graded-index optical fiber is inserted into a blood vessel through a catheter at a fixed angle to the axis of the vessel, and a laser light (of low intensity) is propagated through the fiber, which gets scattered by the blood cells at the outer end. Now the blood cells, moving in the artery, will introduce a change in the frequency of the scattered light from that of the original laser light, and this change in frequency will be higher for the cells moving with higher velocity towards the fiber tip. Some of the light is back-scattered, and is collected by the same fiber. The Doppler shift in frequency is determined by the relationship

$$\Delta \nu = \frac{2nv\cos\theta}{\lambda}$$

where n (= 1.33) is the RI of the blood, λ is the free-space wavelength of laser light, v is the velocity of the blood flow and θ is the angle between the direction of the flow of blood and the fiber axis. Now, using this equation, velocity of the flow of blood can be evaluated by measuring the shift in frequency.

Photometric sensors are also used to determine the oxygen content of the blood by the measurement of the oxyhaemoglobin concentration [51–53], which is of high importance for human life. Blood contains two kinds of haemoglobins among which, one carries oxygen and the other does not. Both these kinds of haemoglobins reflect IR light equally, but the former kind reflects red light more than the latter one. Therefore, two light sources, one emitting IR (805 nm) and the other emitting red (660 nm), are used [54]. A fiber coupler is used where the 660 nm wavelength light is coupled to one fiber while 805 nm wavelength light is coupled to the other, and the lights exiting from the other end of the coupler are reflected by the haemoglobins present in the blood (fig. 8). The reflected light is collected by another fiber, the other end of which is connected to a detector. Now the intensity of the reflected red light gives the amount of oxyhaemoglobin while the intensity of the reflected IR light gives the total quantity of haemoglobin present in the blood. The concentration of oxyhaemoglobin, $HbNO_2$, is given by

$$\%\ HbNO_2 = A - B \times (R_1/R_2)$$

where A and B are constants that depend on the geometry of the probe, and R_1 and R_2 are, respectively, the reflectance at the first and the second wavelengths [54].

The similar technique can also be employed for the determination of the cardiac output, which is of high importance for the treatment of cardiac surgery patients because, by this method, one can estimate the ability of the heart and lungs to supply oxygen.

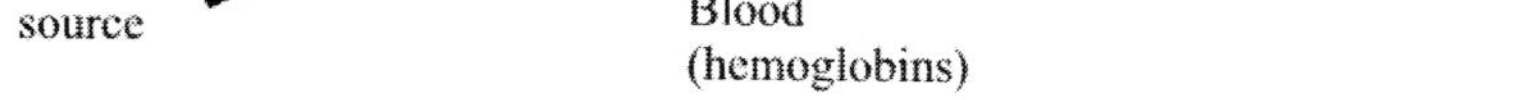

Figure 8. A fiber coupler arrangement.

3. Distributed Sensors

As stated earlier, distributed FOSs use all or a part of the sensitised fiber that reacts with the measurand, i.e. the measurand under investigation is monitored at many locations along the length of the fiber. Intrinsic distributed sensing and quasi-distributed sensing are the two techniques that fall under this category. In the former method, the measurand can be determined at any point along the fiber length, i.e. the fiber is sensitized along its entire length, whereas in the latter system, the measurand is determined at a number of predefined locations along the fiber length, i.e. the fiber is sensitive to the measurand at certain numbers of locations. Nowadays a rapid increase is seen in the number of techniques to achieve distributed sensing along optical fibers [55–62]. However, it is rather difficult to generalize on preferred approaches, but the optical time domain reflectometry (OTDR) method, that includes Raman and fluorescent variants, appear to present major advantages for distributed measurements. The achievement of more sensing range-cells per monitoring station, and greater ease of making compressions of the measurand value at each position, are the factors that increase the practical suitability of optical sensor systems. Some of the sensors based on this principle are described below.

3.1. Rayleigh Scattering Based Sensors

In optical fibers, Rayleigh scattering is the process by which light is scattered form the microscopic density fluctuations in the fiber material [63]. The loss due to this kind of scattering represents the fundamental limitation to loss in fibers at around 1.3 μm and 1.5 μm in silica glass fibers. As the loss in fiber at these wavelengths is very low, the amount of (back-) scattered light is also very low, which can be detected back at the input end of the fiber [64–66]. If light pulses are injected into an optical fiber having uniform loss and scattering characteristics, the power of the back-scattered light falls down exponentially. However, this decay appears as a straight line when viewed on a logarithmic scale; the slope of the line being proportional to the product of the loss and scattering coefficient of the glass. Thus, the spatial variation in these parameters gives rise to a change in the slope of this curve. This technique is termed as OTDR, and is widely implemented for the analysis of loss in optical fiber communications [67–71]. Intrinsic distributed fiber sensors typically utilize either basic loss or the loss due to scattering in the sensing part of the fiber. The change in the scattering coefficient has been used in distributed temperature sensing. However, this kind of dependence is very weak in solid core fibers; liquid core fibers are more suitable to perform this kind of sensing measurement. There are several other mechanisms related to loss which can also be implemented in distributed temperature sensing, e.g. the temperature dependence of the radiative loss due to external RI changes, bending loss in plastic clad fibers, absorption in appropriately rare-earth elements doped fibers.

3.2. Raman Scattering Based Sensors

Apart from Rayleigh scattering, Raman scattering has also been used for the demonstration of distributed fiber optic temperature sensors [72–77]. In this process, components consisting of Stokes and anti-Stokes lines (corresponding to the lower and the higher photon energies, respectively) are produced in a broad band about the exciting wavelength, and the amount of the backscattered light is analyzed by a modified OTDR method in which the ratio of the Stokes to anti-Stokes lines are measured. It is to be noted down that the magnitude of the coefficient of Raman scattering [78] is far less than that of Rayleigh scattering, and that is why, significantly high input power and signal with longer times are required; this is the major drawback of this kind of measurement technique.

3.3. Brillouin Scattering Based Sensors

Brillouin scattering is another phenomenon implemented for sensing applications [79–81]. As the light propagates through an optical fiber, there occurs an interaction between photons and phonons (generated by the photorestrictive processes in the fiber), and this causes a kind of scattering called Brillouin scattering of light. Light scattered by phonons presents a 'shift' by an amount determined by the velocity of the acoustic waves, which in turn, depends on the density of the glass, and hence the temperature of the material. The Brillouin shift in a fiber is given by the expression

$$\nu_B = \frac{2n_{eff}\nu_m}{\lambda}$$

where n_{eff} is the effective RI of the fiber, v_m is the velocity of the acoustic waves in the medium, and λ is the operating wavelength. The Brillouin shift in the frequency depends on two parameters, i.e. temperature and strain in the fiber, which is because the effective RI of the fiber and the velocity of phonons (in the fiber) both depend on these two parameters. It has been observed that the efficiency of spontaneous Brillouin scattering is approximately 20 dB weaker than that of Rayleigh scattering. Apart from these, several other forms of distributed fiber optic sensors have been demonstrated by the investigators that include phenomena like non-linear effects, mode coupling etc., e.g. Kerr effect, mode coupling in birefringent fibers etc.

3.4. Fiber Bragg Grating Sensors

The phenomenon of RI change in the core (of an optical fiber) due to the photosensitivity of optical fiber was first discovered in 1978 [82]. Fiber Bragg grating (FBG) is a longitudinal periodic variation of RI in the core of a monomode optical fiber, and fiber Bragg reflectors are written onto the core using a high intensity UV laser source. Optical fibers can be made photosensitive by doping the core of the fiber with materials like germanium (Ge), potassium (K) or boron (B) during the fabrication of preform. It has been found that conventional Ge-doped fibers, when illuminated with UV light in the region of 248 nm, exhibit a relatively strong photosensitive response [83]. There are several techniques to fabricate FBGs of which holographic and phase mask methods are generally implemented [84–88]. In both the cases, a quasi-permanent kind of grating structure is developed in the fiber core with the Bragg wavelength given as $\lambda_B = 2n_{eff}D$, n_{eff} being the effective RI of the fiber core and D being the grating pitch. Such a fiber with grating, when illuminated with a broadband source of light, reflects a narrowband component at the Bragg wavelength, and a perturbation of grating results in a shift in the Bragg wavelength (fig. 9).

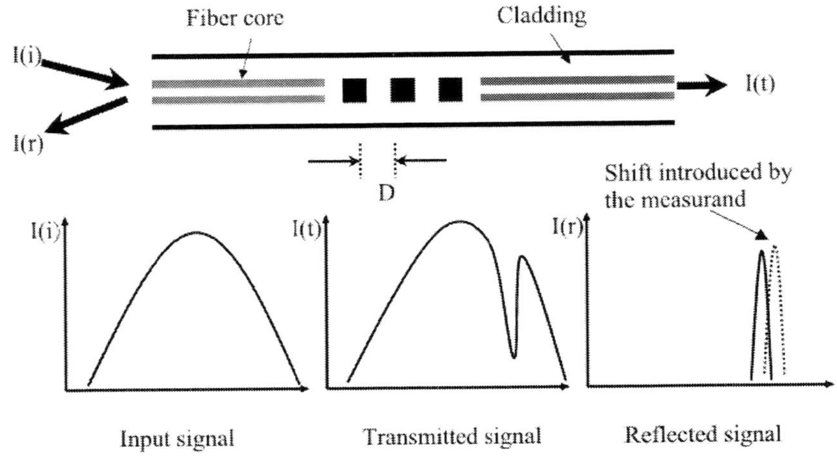

Figure 9. Fiber Bragg grating sensor.

The use of FBG as sensing elements has been very well established, and by exploiting their inherent properties, these are used for a variety of sensing applications [89–96]. FBG sensors are the class of sensing devices based on the photosensitivity of optical fibers, and are widely used for the measurement of strain [97–100] because it (the strain) causes both the physical elongation of the sensors as well as change in the effective RI of the fiber [92]. The physical elongation results in fractional change in the grating pitch, and the change in RI is caused due to the photoelastic effect. The shift in Bragg wavelength is determined by the expression

$$\delta\lambda_B = (1 - p_e)\lambda_B \varepsilon$$

where p_e is effective photoelastic coefficient ($\cong 0.22$ for fused silica) and ε is the applied strain; p_e is given by

$$p_e = \frac{n^2}{2}\{P_{12} - \eta(P_{11} + P_{12})\}$$

where η is Poisson's ratio and P_{ij} are Pockel's coefficients of strain optic sensor.

FBG sensors are also used for the measurement of temperature. Bragg wavelength depends on temperature as follows:

$$\delta\lambda_B = \left\{\left(\frac{dD}{dT}\right)D + \left(\frac{dn}{dT}\right)n\right\}\lambda_B \delta T$$

where $(dD/dT)/D$ represents the thermal expansion whereas the term $(dn/dT)/n$ relates to the thermal dependence of RI. It has been observed that in silica fibers that the latter part has more dominant effects on the observed shift.

FBG sensors have several advantages over the other sensing techniques, e.g. the output doesn't depend directly on the total light, input power of the source, attenuation, etc. In addition, wavelength division multiplexing (WDM) can also be achieved by assigning each sensor to a different part of the available source spectrum. This is illustrated in fig. 10 where two (or many) grating structures are shown that can be used for a large number of devices; the maximum number of gratings that can be addressed in this way, however, depends on the width of the source profile [90,95,96].

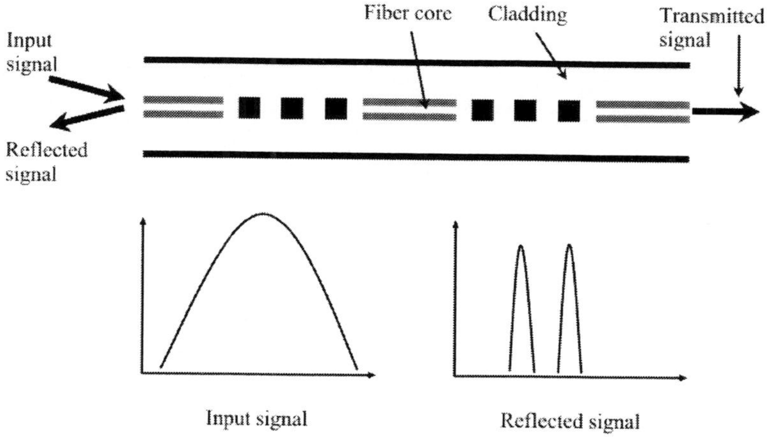

Figure 10. Multiple Bragg-grating fiber sensor.

Sensing devices implementing FBGs are based on passive broadband illumination, where light with a broadband spectrum is input to the system, and the narrowband component of the light, reflected by the FBG, undergoes detection. There are several techniques to measure such Bragg reflected wavelength, e.g. interferometric detection, passive optical filtering etc. [99,100] The general principle of the interferometric method is shown in fig. 11. An imbalanced interferometer has always a term related to the phase that depends on the wavelength of the input light signal. Light reflected by the grating is directed through the interferometer having unequal paths, and since the phase of the unbalanced interferometer depends on the input wavelength, shifts in Bragg wavelength are naturally converted into phase shifts. The phase shift of the output signal is related to the Bragg wavelength shift as

$$\delta\varphi = \left(\frac{2\pi nD}{\lambda^2}\right)\frac{1}{\delta\lambda}$$

where nD is the optical path difference. Proper choice of the factor nD can make the system sensitive even to smaller shifts in Bragg wavelength.

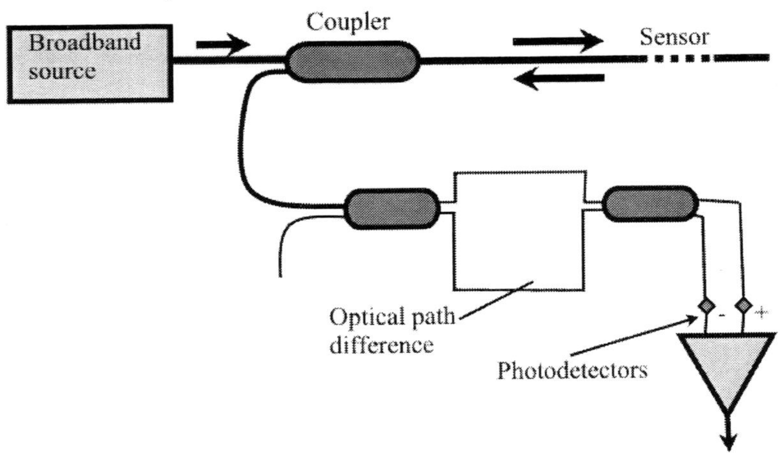

Figure 11. Fiber based interferometric sensor.

Another useful area of FBGs has been the fiber lasers where FBGs are used as spectrally narrowband reflectors for certain in-fiber cavities. Such fiber lasers find useful applications for tunable single frequency devices for WDM networks. These are also used for strain and temperature monitoring [95,96,101–104]. FBG lasers basically utilize two gratings of properly matched Bragg wavelength, which creates an in-fiber cavity. The doped section of the fiber between the gratings allows the optical pumping of the system which ultimately results in the cavity, and therefore, the lasing action starts. Figure 12 represents a typical WDM array of FBG laser sensors. However, several other configurations employing FBG laser sensors have been presented by the investigators.

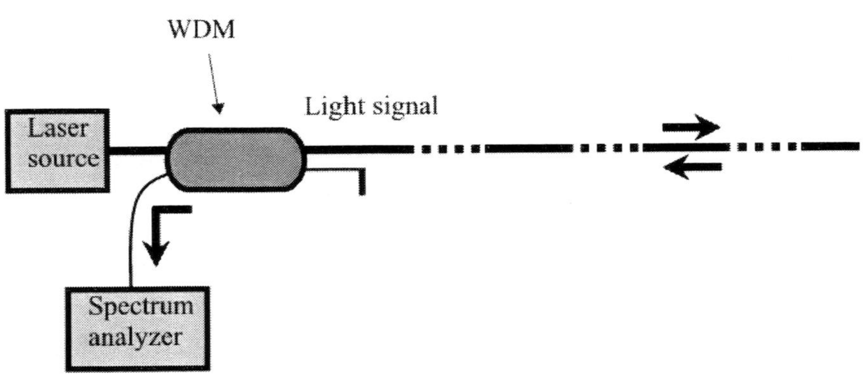

Figure 12. WDM based FBG laser sensor.

4. Interferometric Sensors

Interferometric sensors are of great importance because these are able to provide the sensitivity to a wider range of weak fields. These sensors find significant applications in detecting acoustic perturbations. Much work in the fields of interferometric demodulation, multiplexing etc., has appeared in the literature which motivated the investigators to implement this technology for sensing [105–114].

Several forms of interferometer configurations have been proposed by the investigators that include Mach-Zehender, Michelson, Fabry-Perot, and ring resonator types, among which the Mach-Zehender type configuration is widely implemented for sensing measurements. It is to be noted that Mach-Zehender and Michelson type configurations are two-beam systems, whereas Fabry-Perot- and ring resonator-type configurations are multiple-beam systems. Such interferometric sensors yield a non-linear output described by a cosine interference function, and the phase shift introduced by the interferometer (proportional to the strength of the measurand) is encoded non-linearly into an intensity change at the detector stage. Techniques based on the modulation of laser frequency were in use to introduce carrier phase shifts in an unbalanced interferometer, and such methods were found much more useful in the technological point of view. Earlier investigators used He-Ne lasers or 0.8 μm semiconductor lasers for interferometric sensing applications [115]. Subsequent invention of a number of lasers with narrower linewidth made the interferometric systems much more sensitive to the phase detection [116]. Typically, for example, diode-pumped ring cavity Nd-YAG laser operates at 1.319 μm, and exhibits linewidths of the order less than 10 KHz. Such a laser system allows the use of interferometers with long path differences between the arms without affecting the phase sensitivity. Fiber lasers are of another class which are also proved to be suitable optical sources for certain sensing applications implementing interferometric configurations, e.g. short-cavity Bragg grating based erbium-doped fiber lasers exhibit linewidths of the order less than 10 KHz [117, 118].

In connection with the techniques to introduce internal mirrors in a fiber by fusion splicing, end-coated fibers were reported [119,120]. Earlier work in the area of multiple beam interferometer used coatings of monolayer dielectric materials which, after splicing, showed reflectance less than 10%. Fabry-Perot interferometers fabricated with such reflectors generated interference fringes. This is shown in fig. 13. Later multilayer deposited films on the ends of fibers were also tested, and it was found that the fusion splicing of multilayer coated fibers to uncoated fibers produced internal mirrors with much improved reflectance. One of the most widely recognized multiple beam interferometric configurations is the extrinsic Fabry-Perot interferometer (EFPI) in which two fibers, one carrying light from optical source and the other acting simply as a reflector, are inserted together into an alignment tube (fig. 14) [121]. In this system, Fabry-Perot cavity is formed by the air gap between the two uncoated fiber end faces; the interference signal is detected in the reflected light back at the source end of the input fiber. These sensors are used to monitor strain. In such systems, in general, both the fibers are epoxied into the alignment tube. Application of strain results in a change in the air gap between the fiber end facets which are then detected in terms of the shifts in the signal. An almost similar kind of sensor configuration can also be formed by allowing one of the fibers to move freely within the alignment tube.

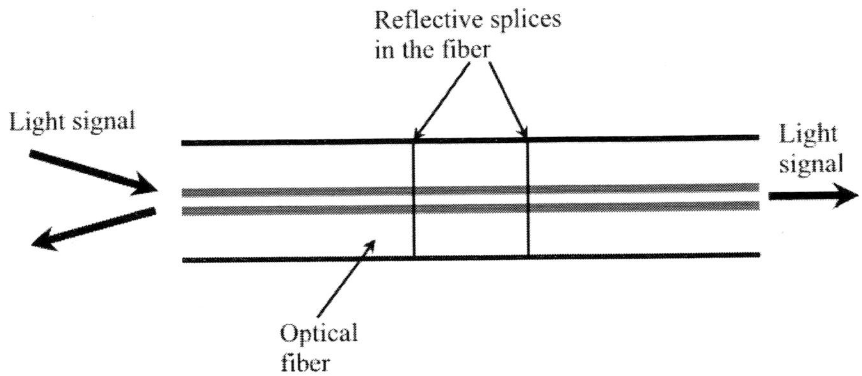

Figure 13. Fiber Fabry-Perot sensor.

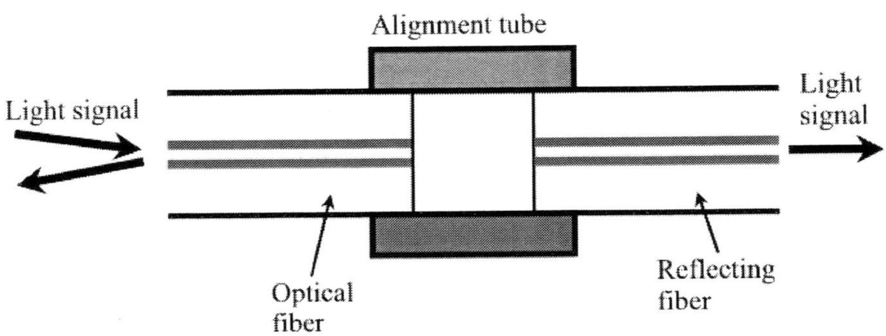

Figure 14. Extrinsic fiber Fabry-Perot sensor.

5. Some other Forms of FOSs and their Applications

The design of a simple and low cost liquid level sensor has been presented that can be fabricated by using an optical source emitting a narrow beam and a photodetector, and a tube containing some liquid. The light beam from the source may travel either through the tube (containing the liquid) or reflected by the surface of the liquid [122]. In this kind of sensor, the intensity of light is changed due to its reflection/refraction from the liquid surface and/or absorption in the liquid, and the intensity (detected by the photodetector) changes as a function of the liquid level. Investigators claim that the sensor can be used in automobile and aeronautical industries, and in the measurement of the level of oil in electrical transformers.

The operation of Bragg displacement sensor is based on the principle of distributed feedback dye lasers. Using Nd:YAG laser as the optical source, experimental demonstration of microdisplacement sensors based on the periodicity of Bragg grating and interference overlap length modulation has been presented by K. Nasrullah [123]. It was found that the displacement length varies linearly with the change of wavelength as well as the optical pulse length. This kind of ultrasmall displacement sensor can be used to determine the frequency of vibration or simple displacements in ultraprecise applications.

Faraday effect current sensors and Sagnac effect fiber optic gyroscopes are also of practical applications [124]. Both the Faraday and Sagnac effects can be detected in terms of phase changes between two counter-propagating lights in ring interferometers [125,126]. Investigators have presented fiber ring lasers for current or rotation sensing in which the non reciprocal phase shift due to the magnetically induced circular birefringence or the Sagnac effect is detected in terms of the beat frequency between the counter-travelling laser beams [127–132]. For such devices, fiber Brillouin lasers or erbium-doped fibers lasers were used.

Low birefringence optical fibers are very useful for Faraday effect current sensors because stable sensitivity can be achieved. Lead glass monomode fibers present six-times larger Verdet constant than a silica fiber with very low photoelastic constant as well as low intrinsic birefringence. As such, these fibers are able to make the polarization of light (propagating in itself) stable against external mechanical perturbations and the effects of temperature, and therefore, such fibers are used for polarization modulation current sensor [133,134].

Fiber ring lasers using laser diode amplifiers and flint glass fibers are also studied where the use of a laser diode amplifier (instead of erbium-doped fiber amplifier) for a gain medium makes the set up economic as well as provides the opportunity to use the device to sense various wavelengths and the integration with other semiconductor based devices [135–137]. The application of the device was demonstrated as frequency-domain fiber current sensor and fiber gyroscope. It was found that, because of the involvement of a Faraday bias element in the cavity, the laser was substantially free from frequency locking, and was able to detect the polarity of current as well as the direction of rotation.

It is rather important to detect the liquid-phase water in many applications. It has strong absorption bands in the near IR region of the EM spectrum, and therefore, near IR spectroscopy is certainly of great importance in the context of water sensing [138–141]. Demonstrations have been done for the application of FTIR absorption spectroscopy in sensing water concentration in milk [142]. Plastic fibers with fluorescent-dye-doped cladding have been reported to monitor breathing conditions and humidity [143]. Although several kinds of optical sources have been proposed for such purposes [144–147], a Tm^{3+} doped YAG crystal seems to be a more suitable one as its small crystal can even yield high pumping efficiency. Investigation of optical fiber water-droplet sensor has been reported which is based on the principle of absorption spectroscopy, and uses the fluorescence from a laser diode pumped Tm^{3+}:YAG crystal as a compact and high power fluorescent light source [148,149]. In the experiment, the fluorescent light was launched into an optical fiber, which acts for both sending the fluorescent light and picking up the perturbed optical signal. It is found that such sensors can be used for the estimation of both the size and the position of water droplets. Also, it can be used to distinguish clearly water from other perturbations such as oil and surface irregularities.

The suitability of a fiber optic portable Raman spectrometer for in-situ measurement of the percentage of water by volume in methanol has been demonstrated by Asundi *et al* [150]. The basic measurements involved are the measurements of intensities of scattered radiation, from the material under investigation, as a function of wavelength or frequency. Also, it is possible to represent the measured intensity as a function of Raman shift. By exciting the test sample with a suitable laser beam, the scattered radiation was detected with an array of detectors. The intensity of the scattered radiation was represented as a function of Raman shift, which gives the signature of each material (that is unique to that particular material). Raman spectra diagnostic tests show the obtained spectra on par with those obtained by conventional spectrometers. It has been found that the measured intensities at the characteristic Raman shift corresponding to methanol are inversely proportional to the volume of water present in methanol. The experimental procedure can even be applied to measure and quantify the methanol content in gasoline, and can be directly used in the quality control where there are the possibilities of adulteration of gasoline with methanol.

Utilizing the thermal and optical properties of Ta_2O_5 (tantalum pentoxide), a robust, compact and reliable fiber optic low-coherence wavelength modulated thermometric sensor has been presented by Inci *et al* [151]. The principle of operation of the sensor is based on the temperature dependence characteristics of wavelength, and the sensor is free from vibrational effects and intensity fluctuations, and also, has the potential for multiplexing. Their design was based on the spectral shifts introduced by the thin films of Ta_2O_5. A monolayer film of Ta_2O_5 was deposited on the cleaved end face of a monomode optical fiber, and the film was illuminated with a broad-band super luminescent diode source through a directional coupler. Interference fringes were obtained on reflection from the film, which show shifts with the rise in temperature. This kind of sensor can be used to measure higher temperature ranges up to 650 °C. For the measurement of such a high temperature, Ta_2O_5 is found to be a good candidate as it has a very high melting point, strong adhesion to glass and most other oxides, high RI (typically 2.15 at 1.55 μm), high thermal optical coefficient (about ten times greater than that of fused silica) and low absorption in the near UV (350 nm) to IR (8 μm) region of the EM spectrum.

A fiber microbend sensor (MBS) is an inexpensive intensity based sensor used to detect deformation and several other related quantities such as strain, temperature, pressure etc. [152–156]. However, its sensing performances are rather complicated, and are the function of fiber parameters and deformation. This feature limits their applications up to simple monitoring or on-off sensors. Investigators have reported MBSs based on non-uniform deformations of fiber caused by some means like corrugated plates etc. [157,158]. Yoshino *et al* [159,160] reported a spiral microbend-loss sensor, and it was shown that, by means of monomode fiber embedded metal springs, strain and deformation can be measured with good linearity and reproducibility, and also, with large dynamic range in strain/deformation measurements. Their spiral-type MBS was based on the uniform change in the bend radius of the fiber so that the characteristics of the sensor can be specified simply from the circular bending loss of the used monomode fiber. Analyses were also made of the effect of spring parameters (e.g. radius, pitch and length) on the sensing performances.

Optical gas sensing methods have proved several advantages (over conventional chemical methods) such as the possibility of fast, remote, selective, nondestructive and intrinsically safe detection [161–168]. An optical method to sense toxic gases (like methane etc.) utilizing evanescent waves of optical fibers has been demonstrated by Tai *et al* [169]. In their experiment, a single multimode step-index silica optical fiber having 50 μm and 125 μm as core and cladding diameters, respectively, was used as both a sensor and an optical transmission medium, and therefore, the detection of gas could be performed in a simple, flexible, and possibly distributed way. They presented an experimental demonstration for the detection of methane gas by its strong optical absorption at 3.392 μm of a He-Ne laser light. The sensing part of the fiber was made by heating and expanding some short section of the fiber, and the length and the diameter of the sensing part ranged from 5–10 mm and 1.8–7 μm, respectively. Optical transmittance of the fiber for the 3.392 μm-line was measured for various lengths of the fiber, and for various concentrations of methane gas. It was observed that an evanescent wave of 5–40% of the total propagating power was generated outside the fiber, and the use of 1.8 μm diameter and 10 mm length of the sensor fiber showed that the minimum detectable concentration of methane gas was less than the lowest explosive limit.

A variety of FOSs based on immuno-fluorescence spectroscopy has been developed which can detect very low concentration of specific chemicals in complex samples [170]. The chemical selectivity of these sensors depends on the antibody-antigen reaction. A fiber optic microprobe is located at the end of a fiber; the surface of the microprobe contains the antibody. If the microprobe is inserted into the sample containing antigens, these will be captured by the antibodies. When the original fiber is excited by a laser source, antigens fluoresce; the intensity of fluorescence is used for the detection of several viral infections, e.g. screening of cancers etc.

pH sensor is also a kind of FOS which is based on the principle that the absorbance of the phenol red dye depends on its pH value. In such sensors, an optical fiber retroscatter monitor interrogates the indicator at two different wavelengths; one measures the change in reflectance as the dye responds to pH changes, whereas the other provides a normalization signal at a wavelength that is insensitive to pH. In general, the probe contains polyacrylamide microspheres which are covalently bound to the red dye in order to prevent these from diffusing out of the probe [171]. Typically, the probe diameter is kept about 5–10 μm. These polyacrylamide spheres are sealed in a cellulosic tube that is attached at the ends of two fibers, and permeable to hydrogen ions. Now two lights having their wavelengths 560 nm and 600 nm are passed through one fiber. The 560 nm wavelength light is absorbed by the dye, and the amount of absorption is a function of the pH value of the solution. The 600 nm wavelength light acts as a reference beam, and its absorption does not depend on the pH of the solution. Both the light beams, after multiple scattering and absorption in the cellulosic tube, ultimately return to the other fiber connected to the detector. The ratio of the intensities of two lights reveals the pH of the solution.

The response of fiber optic evanescent field sensors to pH variations has also been presented by Choudhury and Yoshino [172]. In the experiment, a bent probe was used because the sensitivity is remarkably improved in this case. The bend was particularly chosen to be of the U-shape. At this point, it is of worth to mention that Takeo and Hattori [173] reported sensors with a U-shaped probe for the measurement of RIs of liquids. In order to study the sensitivity, an experimental investigation has also been carried out with fiber probes of different values of the radius of bend [172]. Figure 15 shows the evanescent absorption measured for straight probe as well as bent probes. Aqueous solutions of patent blue VF dye with different concentrations serve the purpose of measurand. The RI of these samples is found to be 1.332 (as measured by KEM RA-130 refractometer). A typical absorption spectrum of the aqueous solution of patent blue VF dye is shown in fig. 16, which is obtained by a Hitachi U-3000 spectrophotometer. We observe the peak absorption wavelength near 635 nm, which is very close to the He-Ne laser wavelength. As such, a He-Ne laser can be used as the light source.

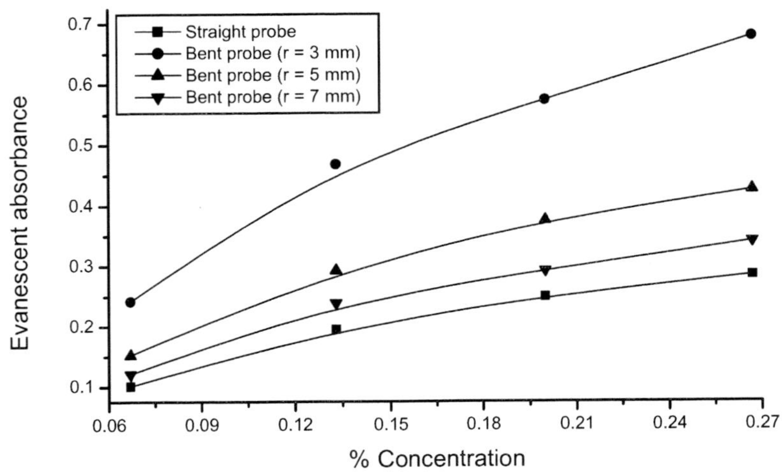

Figure 15. Variation of the evanescent absorption of straight and bent probes.

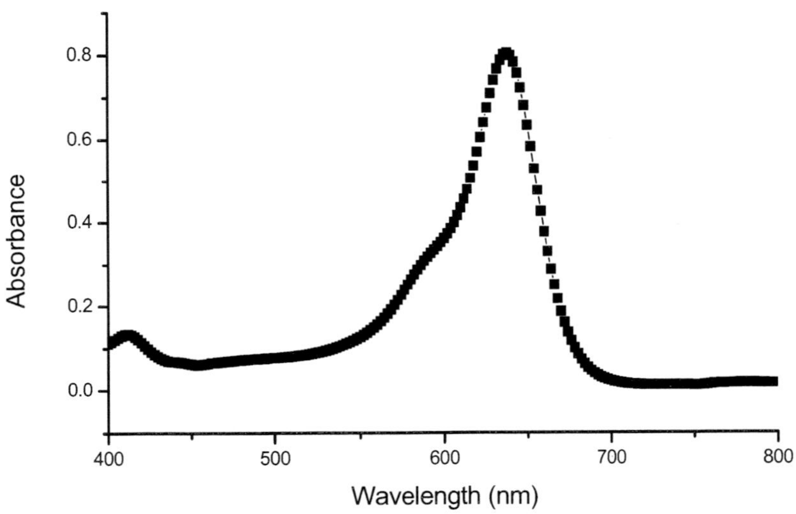

Figure 16. Absorption spectrum of aqueous patent blue VF dye.

Figure 15 shows that the absorption increases with the decrease in bending radius. Compared to the straight probe, we observe that the absorption is increased roughly by 1.2 times for a bent probe of radius of curvature $r = 7$ mm. For $r = 5$ mm, the absorption increases by 1.6 times approximately, and for $r = 3$ mm, the absorption is increased roughly by 2.4 times. Therefore, with the decrease in radius of curvature, the absorption, and hence the sensitivity of the probe increases. However, the value of optimal bending (in order to get the maximum sensitivity) will be a function of several other factors, viz. the measurand RI, the penetration depth and loss. Thus, to maximize the sensitivity, a trade-off among these factors will have to be considered. Also, the maximum sensitivity will be limited to the value of bend (of the probe) beyond which the process of scattering starts taking place.

Study of the variation of evanescent absorbance for a fixed concentration of patent blue VF dye is shown in fig. 17 for different values of the uncladded fiber length. A linear relationship between the absorbance and the interaction length is observed, which predicts that for a fluid, obeying the Beer-Lambert law of absorption, the evanescent absorbance is linearly related to both the exposed fiber length and the fluid concentration. Paul and Kychakoff [174], however, presented earlier the study of the absorption behavior of rhodamine 6G laser dye, and reported a linear kind of dependence of the absorption coefficient on the measurand concentration. In their experiment, they used organic solvents.

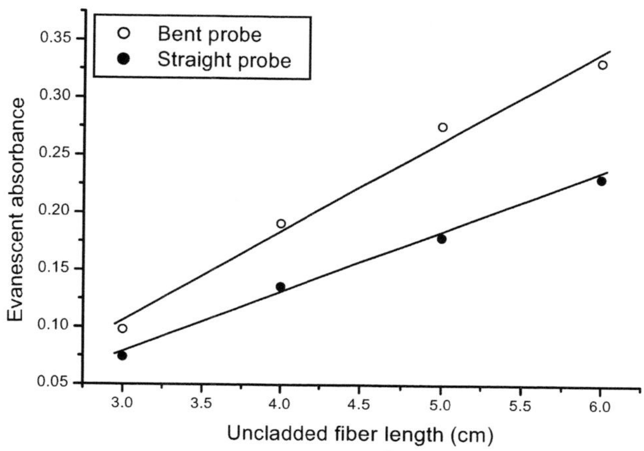

Figure 17. Variation of the evanescent absorption with different probe length.

Figure 18 represents the plot of the absorption coefficient with the percentage concentration of aqueous dye solution corresponding to two individual cases, viz. free-beam as well as the method of fiber optics; the latter one implements silica fiber with fixed uncladded length (5 cm). It is seen that, in the case of free-beam method, the absorption coefficient depends nonlinearly on the measurand concentration. This kind of behaviour cannot be easily seen for the case of fiber optics method, which is because of compact vertical scale in fig. 18. However, the same curve in fig. 15, corresponding to the case of $r = 5$ mm, shows the nonlinear dependence of the evanescent absorbance on the measurand concentration. A similar kind of nonlinear dependence of the evanescent absorbance on the measurand concentration was reported earlier by Ruddy *et al* [175], which is in contrast with the Beer-Lambert law.

Choudhury and Yoshino [176] reported the nonlinear kind of dependence of absorbance on the measurand concentration which is in contrast with the Beer-Lambert type linear dependence. This is attributed to the accumulation of high density of dye molecules on the bare fiber surface, which is due to the electrostatic attraction between polar silica surface and the ionic measurand solution, i.e. the positively charged chromophores. This is justified through the experiments performed under the free-beam condition too [176]. It is reported that the pH has a significant role in determining the nonlinearty of the absorbance response. The degree of nonlinearity decreases with the decrease in pH value, which is owing to the less accumulation of dye molecules on the fiber surface as the value of measurand pH decreases.

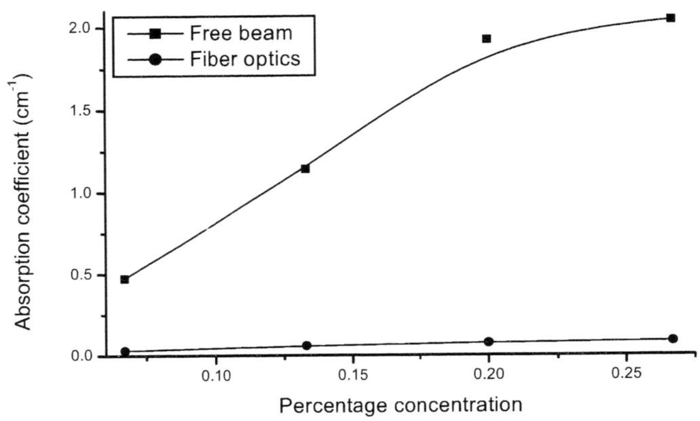

Figure 18. Plot of absorption coefficient with percentage dye concentration.

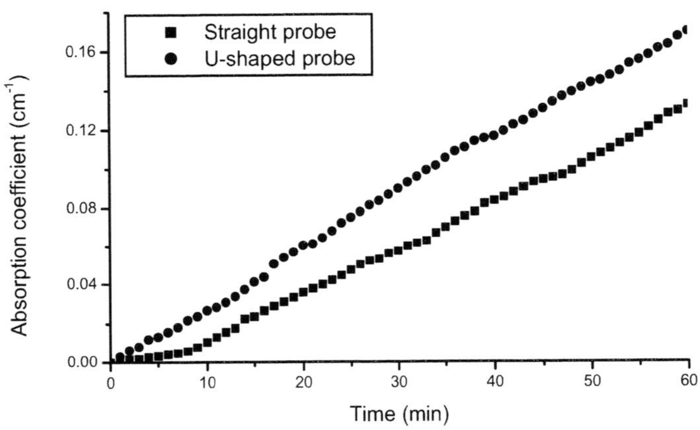

Figure 19. Variation of the evanescent absorption with time for straight and bent probes.

Evanescent field sensors can also be used for detecting rate coefficients [176,177] of time dependent chemical reactions. Industrial applications of chemical kinetics require the knowledge of chemical reaction rates, which is essential to establish the optimum physical conditions to perform a reaction. The conventional methods for finding rate coefficients are volumetric analysis [178,179] and the optical transmission [180]. The method of volumetric analysis is much complicated whereas the method of optical transmission is quite expensive. The reported method of evanescent fields would be potentially useful for the study of different chemical kinetics owing to the low cost and simple design. As an illustrative case, typically the reaction between potassium iodide (KI) and potassium peroxdisulfate ($K_2S_2O_8$), which is a time dependent reaction, is reported [176,177]. Equimolar aqueous solutions (0.052 M) of both of these reagents are used. The reaction results in the evolution of iodine, and the quantity of evolution is time dependent. A violet colour solution is formed, and the colour deepens with the increase in time, indicating the increase in the concentration of iodine with time. The absorption spectrum of iodine is presented in ref. [181] and [182]. Both straight as well as bent probes are used to compare the sensitivity (fig. 19). It is found that the result obtained with the U-shaped probe (rate coefficient = 0.0145 min^{-1}) is better than that obtained with a straight probe (rate coefficient = 0.0114 min^{-1}), as confirmed by using a Hitachi U-3200 spectrophotometer (fig. 20; rate coefficient = 0.015 min^{-1}), which is attributed to the increased sensitivity of the U-shaped probe.

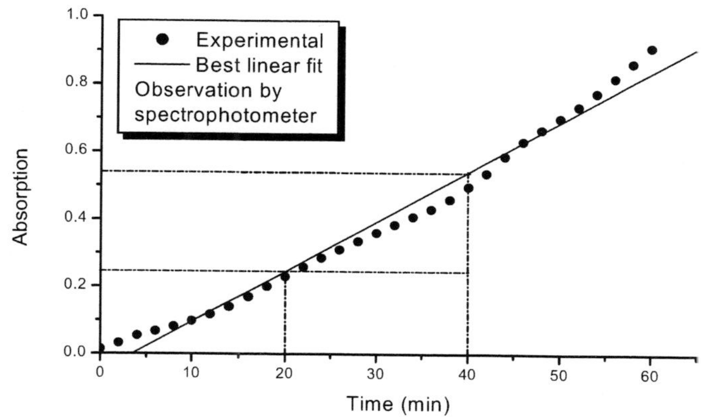

Figure 20. Variation of absorption as observed by spectrophotometer.

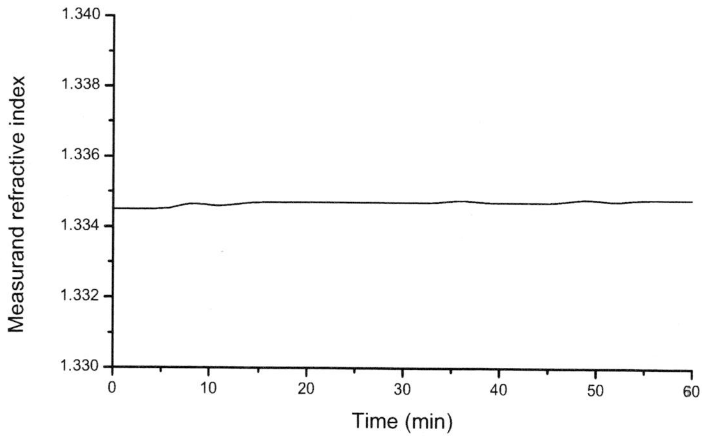

Figure 21. Variation of measurand RI with time as recorded by refractometer.

In this context, it is also of worth to observe if there is any significant variation in the measurand RI during the chemical reaction, because the studies related to evanescent wave spectroscopy are based on constant RI of the measurand. During the chemical reaction, the measurand RI is monitored by a refractometer (KEM RA-130), and it is seen that the RI remains almost unchanged (fig. 21). This shows the validity of the results obtained by the absorption of evanescent waves. The concept of such an arrangement can be implemented for detecting a variety of other time dependent chemical reactions, and in general, for the study of chemical kinetics. Further, it is only an illustrative case of chemical reaction presented in the foregoing discussion. However, the process is very general, and can be applicable to monitor any time dependent reaction. The simplicity of the system and the low cost are the additional features of this method for the determination of rate coefficients.

Another important potential market for fiber optic sensing technology is in the areas of aerospace and marine applications [183,184], which is principally because optical fibers are immune to electromagnetic interference. Current technologies are advanced to implement the transfer between the measurand and the modulation function applied to the fiber, and to retrieve the information from all the transducer elements in an array. The main potential applications of optical fibers in the area of aerospace sensing are the displacement transducers to monitor the position of control surfaces, embedded FOSs for monitoring the performance and fabrication of carbon fiber-reinforced composite materials, and in the advanced testing of gas turbine engines. FOSs find similar applications in respect of marine technology, and high sensitivity interferometric fiber sensors.

Rotationally nonhomogeneous thin films comprised of parallel columns with nano-engineered shapes are a new class of materials addressed as sculptured thin films (STFs) [185–187]. Many new applications of STFs in the area of optical technology are coming up in the horizon [188–190]. The porous characteristic of STFs can be exploited for the purpose of optical sensing. STFs can be morphologically classified into a type called thin-film helicoidal bianisotropic mediums (TFHBMs), which are STFs having cholesteric morphology [191,192]. The other type of STF morphology is nematic [193]. Dielectric TFHBMs can be used for the determination of gas concentrations, and the relevant theory has been developed by A. Lakhtakia [194]. It was reported that bilayers comprised of two matched layers of dielectric TFHBMs with opposite structural handedness have the property to reflect arbitrarily polarized, normally incident, plane waves within a specific frequency range. Porous characteristic of TFHBMs can be exploited to devise a model such that the infiltration of a TFHBM bilayer by a gas would shift as well as reduce the frequency range. Thus, a properly designed TFHBM bilayer can serve as the active element of an optical device for sensing gas concentrations. The mechanism may be understood in a way that the tunability of the notch-filter response of the TFHBM bilayers [195] can be used for the detection of gas concentration. If the microcolumns of the TFHBM bilayers, which are supposed to be made of chemically inert isotropic materials, the bilayer itself, on the immersion in the gaseous environment, would cause gas molecules (to be sensed) to diffuse in the intermicrocolumnar void regions. This would ultimately result in a change in the permittivity dyadic, resulting thereby a shift in the notch in the transmission spectrum. The amount of shift may be used to estimate the concentration of gas. Thus, in this kind of sensor, nanoscopic-to-continuum models are devised to predict the electromagnetic response of TFHBM bilayers. For sensing some specific gases, such sensors may not be necessarily superior to the existing ones. However, these may be implemented for gases whose sensing is currently infeasible or problematic. TFHBM bilayers may provide the capabilities of real-time monitoring for sensing several toxic as well as non-toxic gases. Patches of such bilayers may be compactly mounted on the same platform in order to sense different gases simultaneously. Further, it is expected that TFHBM bilayer sensors are reusable, which is essentially a key feature toward the logical development of such sensors.

STF Šolc filter has also been proposed for sensing gas concentrations [196]. The gas, to be sensed, is allowed to enter the filter (which is porous); the electromagnetic response of the filter is modified by the gas itself. Thus, in such sensors again, the transmissivity is found to be strongly dependent on the concentration of the gas present in the inter microcolumnar void region of the thin film. Ultra-narrow band-pass features in the transmission spectrum of the Šolc filter shows a shift with the change in gas concentration. As such, by using STF Šolc filters, concentration of gas can be determined.

It has been shown by Lakhtakia *et al* [197] theoretically as well as experimentally that, by cascading two identical chiral STF sections with one section twisted with respect to the other by 90° about their common helicoidal axis, a special reflection hole may be realized. Such an arrangement can be used for optical sensing of fluids [198]. In such a device, a chiral STF is treated as a two-phase composit material with locally biaxial dielectric properties, and it has been predicted that the presence of a fluid in the porous film results in a red-shift of the spectral holes.

Chiral STFs have been proposed for biosensing applications, which are essentially based on their emission characteristics [199–202]. The ref. [202] particularly presents the theory of second-harmonic-generated radiation from a chiral STF, which can be applied for bio-sensing. In such a system, a linear emission is supposedly due to luminescence-producing bio reactions occurring in the void region of the chiral STF platform. In conclusion, theoretical and experimental investigations indicate that chiral STFs have established a strong platform in the area of bio-sensing, and their commercialization will be undoubtedly realized in the near future.

6. FOSs in the Environmental Context

The current technology related to FOSs involves their use in process control to monitor parameters like temperature, liquid level, gas density and image. Advantages of these sensors have been confirmed through their practical use in industrial plants. Such sensors are found to be expensive, and the efforts are underway to improve the fabrication techniques, reliability, and cost reduction. The development of sensor networks plays very important role in cost reduction. Multiplexed and distributed sensing systems enable simultaneous measurements of several points, and these are the key technology to expand the use of such sensors in process control technology.

Nowadays environmental assessment and control market has come in the limelite because it has become a global issue how technology can solve the rapidly growing problems in the environmental context, in order to maintain and improve the quality of life of the nation. Scientists have found that the hottest application of FOSs could be in the emerging area of environmental safety monitoring where a variety of chemical reactions are essentially required. Fiber optic chemical sensors are extremely used in monitoring various chemical processes taking place in the environment. The environmental sensor industry is aware of this major issue, and in the present world of technological improvements, it is rather difficult to classify various environmental sensors by one single technology because the recent trend is now to fabricate hybrid devices, i.e. the entire system involves a combination of various technologies [170,203].

Scientists have invented a variety of gas sensors for the health and safety monitoring purposes. These sensors are used for the detection of various toxic gases and air pollutants like methane and other hydrocarbons, which are deemed to be environmentally threatening

[161–169,204]. These toxic gases exist in the air at ppm level, and are hazardous for human beings. For such sensing applications, different methods like evanescent wave spectroscopy, optical absorption, differential optical absorption, infrared fiber optics etc., are usually employed. Several other methods have been reported for the detection of toxic gases like methane which, with the mixture of air, is highly explosive. One of the major advantages of the optical gas detection systems is that these are intrinsically safe. Different methods have been employed for the detection of methane gas using the absorbance of optical power at various wavelengths viz. 1.33 μm, 1.66 μm, 2.25 μm, 3.33 μm, 3.39 μm and 7.7 μm. Systems like fiber optic multipoint sensors are also investigated [205].

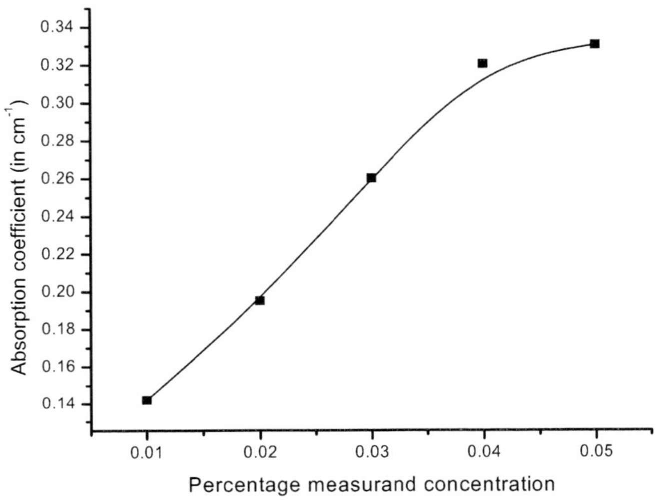

Figure 22. Variation of absorption coefficient against percentage concentrations of chlorine water.

The presence of chlorine in drinking water is highly hazardous for health. Using a U-shaped probe, Choudhury and Yoshino [206,207] reported the use of evanescent field FOSs for the detection of chlorine content in drinking water. For this, a green LED with its peak emission wavelength around 530 nm is used as the light source. Different concentrations of chlorine water are used as measurands, in which a little amount of diethyl phenylene diamine (DPD) organic compound is added. DPD reacts with chlorine forming an azo compound. For illustration, the measurements are made for the concentration range from 0.01% – 0.05%, and it is found that, during the experiment, the RI values of the measurands remain almost constant at 1.335 (as measured by KEM RA-130 refractometer). Figure 22 shows the variation of the evanescent absorption coefficient against the different percentage concentrations of chlorine water. It is found that, corresponding to the lower concentration values, the absorption coefficient varies approximately linearly with the percentage concentration. However, for higher concentration values, the dependence is nearly nonlinear, which is attributed to the higher deposition of azo dye over the silica surface. At lower concentrations, less azo dye molecules are deposited on the bare fiber core. However, for higher concentration values, more azo dye molecules are formed which get deposited on the silica surface of the bare fiber core, reducing thereby the absorption of light. Thus, the characteristic of the sensor is greatly affected due to the loading of azo dye molecules on the fiber surface. Figure 23 shows the reproducibility of the results, which shows almost similar evanescent response curves, though there exists a little deviation owing to the possible presence of azo molecules over the bare fiber surface even after rinse. Thus, the chlorine content in drinking water can be detected by colour formation.

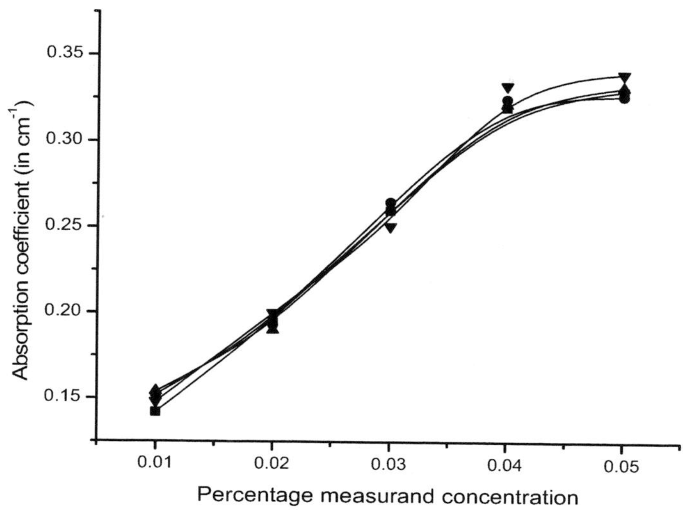

Figure 23. Reproducibility of the results of fig. 22.

Sensor systems employing evanescent wave spectroscopy has the advantage that it can be coupled (optically and mechanically) directly to an optical fiber. D-shaped fibers are used as evanescent wave methane gas sensor, and for this, attenuation measurements on guided waves are done at a wavelength corresponding to 1.33 μm and 1.66 μm. Such systems are more rugged and less expansive arrangements [208–210]. Parallel and sequential measurements are also made by employing a number of pairs of optical fibers connected to the individual sample cell for monitoring methane and carbon mono-oxide in respect of their concentrations.

Electrochemical sensors are those that produce electrical signals induced by chemical reactions [180,211]. Bio-sensors are also a kind of electrochemical sensors used in the context of environmental monitoring [212,213]. The detection of gases is achieved by measuring changes in electrical parameters (e.g. resistance or capacitance) that happens due to the absorption of gas molecules by organic polymers like polyanilene mixed with polypyrrole. Distributed FOSs are used for the detection of methane, which exploit the interaction of the evanescent light wave in a distributed fiber with a methane sensitive coating. Thallium-doped fiber laser operating at a wavelength of 1.648 μm is also a kind of methane sensor, and these can also be used for several gases lying in the wavelength range from 1.65 μm to 2.05 μm. Infrared methanometer is another methane sensor that uses a Febry-Perot etalon filter. One of the most commercially available environmental sensors involves the use of smart skin technology, and is used for leak detection of underground storage tanks [203]. Varieties of sensors, using novel membrane technology, have also been invented for environmental measurements and analyses. However, the use of novel chiral STFs in the area of biosensing is described in Sec.6 above [200–202].

7. The Role of Integrated Optic (IO) Technology in Sensing Applications

There has been a promising role of IO too in the FOS research [214–220]. However, still many problems like fiber coupling, production of suitable masks, reduction in loss, packaging etc., are to be resolved before the IO technology can find commercial implementation to its fullest extent. Silicon IO appears to be highly promising in context of sensors because it offers fairly simple and inexpensive means for precise fabrication of waveguides with much more flexibility than that achievable by glass waveguides formed by the ion exchange method. Apart from this, silicon technology offers the evolution of high-speed modulators, sources as well as detectors too. However, direct combination of silicon microcircuits with silicon optics is of high significance giving substantial rewards. Truly speaking, silicon has to play several significant roles in sensing technology because its mechanical properties are also the best among other semiconductor materials. Consequently, it can be implemented wherever both the mechanical integrity as well as the ability to guide light are required; e.g. optical modulators. Just for an example, as a resonator, it can measure slowly varying quantities through the mechanical resonant frequency [216]. It can also be used in accurately defined alignment structures for assembling FOSs and other related precision assemblies [217]. Undoubtedly silicon offers its remarkable scope as optical, mechanical and electronic material, and the RandD community has yet to fully appreciate its true significance.

IO technology based sensors have also their applications in the environmental sensor context. Integrated optical acoustically tunable filters and optochemical microsensors also find applications in environmental monitoring, and are used for the detection of ammonium nitrate and heavy metals. Planar gas and silicon sensors are the other classes that also find extensive use. Semiconductor gas sensors, that use semiconducting oxides of zinc and tin with a small percentage of promoter like platinum, antimony, aluminium etc., find wide applications in the detection of reducing gases like H_2, CO etc. IR gas sensors, that involve specific IR absorption by the rotational-vibrational transition of gas molecules, are used for meteorological measurements, ambient air monitoring etc. The sensitivity of such sensors depends upon the choice of the spectral range and the optical path in closed or open configuration that enhances the limit of detection up to the ppm range. The open configuration system is a kind of compact arrangement of multiple reflections, the length of which may vary from a few centimetres to several meters. Modern technology introduced systems having programmable controls to maintain various parameters like temperature, pollutant concentrations, relative humidity etc., constant even for a period of one year or more, and can be used for simultaneous testing of four gases at a time.

8. Conclusion

Optical fibers are immune to electromagnetic interference, chemically inert and mechanically flexible which enable FOSs to have the unique capability to monitor several parameters as functions of position along the length of fibers even in the worst environmental conditions. These important features of optical fibers combine to provide an attractive baseline technology for several kinds of structural monitoring, and as a result, FOS area has grown to become very broad over the past two decades. Research in the basic sensing techniques and their applications reflects that this technology can be exploited in active structures as the input to a localized control system. These sensors are useful for a diverse range of applications from biomedical science, underwater acoustics to structural health monitoring, safety assessment, and wide range of other related operations.

In this review article, attempt has been made to describe some of the most common methods implemented in FOS technology, and some kinds of FOSs are also presented, which find applications in environmental and biomedical sciences. However, the future of these sensing systems in medical area appears to lie in the integration of several kinds of sensors into a single catheter that should fit through the smallest vessels in a human body. The field of FOS technology is experiencing rapid developments in the area of smart structures with several demonstrations of embedded sensor systems.

Research and investigations in this area also reflect that such sensor systems can be undoubtedly of great importance so far as the environmental sector is concerned. To date most of the commercially available sensor systems are fiber-coupled instruments in which fibers themselves play the important passive role by taking light to and from remote measuring locations. During the recent years various organizations manufacturing fiber optic technology based sensor systems have shown their big impact, which is certainly evident from their rapid growth. Undoubtedly, it is needless to say that the current and emerging safety and environmental legislation will continue to drive the need for increasingly versatile monitoring systems, and FOSs will play an important part in the evolution of such structures and prove their major impact on future generations of various environmental products.

Although it is not possible to cover all types of FOSs, it is hoped that a flavour of the fascinating subject has been presented in this review article.

References

[1] Peterson, J.I.; Vurek, C.G. *Science* 1984, 224, 123–127.

[2] Brenci, M.; Conforti, G.; Falciai, R; Mignani, A.G.; Scheggi, A.M. *Opt. Sensors* 1986, 1, 163–169.

[3] Conforti, G.; Bacci, M.; Brenci, M.; Falciai, R., Mignani, A.G.; Scheggi, A.M. *Proc. SPIE* 1985, 576, 66.

[4] Nuttul, J.D. *Proc IEE J. Optoelectron.* 1985, 132, 250.

[5] Gardnier, P.T.; Edwards, R.A. In *Fibre Optic Sensors* 1987, IoP Short Meeting Series, Bristol.

[6] Rashleigh, S.C.; Ulrich R. *Appl. Phys. Lett.* 1979, 34, 768–770.

[7] Hale, K.F. *Phys. Technol.*1984, 15, 129–135.

[8] Kuribara, M. *Electron. Lett.* 1983, 19, 133–135.

[9] Berthold, J.W. *Proc. SPIE* 1985, 566, 37–44.

[10] Nagai, M.; Shimizu, M.; Oghi N. Proc. of the 2nd Int. Conf. on Optical Fiber Sensors, OFS '84, Stuttgart 1984, 207–210.

[11] Dakin, J.; Culshaw, B. Optical fiber sensors: applications, analysis and future trends; Artech House: Boston, 1997, Vol. 4.

[12] Bucaro, J.A.; Darty, H.D.; Caron, E.F. *Appl. Opt.* 1977, 16, 1761–1762.

[13] Budiansky, B.; Drucker, D.C.; Kino, G.S.; Rice, J.*R. Appl. Opt.* 1979, 18, 4085–4088.
[14] Hocker, G.B. *Appl. Opt.* 1979, 18, 3674–3683.
[15] Hocker, G.B. *Appl. Opt.* 1979, 18, 1445–1448.
[16] Cielo, P.G. *Appl. Opt.* 1979, 18, 2933–2937.
[17] Tateda, M.; Tanaka, S.; Sugiyama, Y. *Appl. Opt.* 1980, 19, 770–773.
[18] Lagekos, N.; Bucaro, J.A.; Jarzynski, J. *Appl. Opt.* 1981, 20, 2305–2308.
[19] Lyle, J.H.; Pitt, C.W. *Electron. Lett.* 1981, 17, 244–245.
[20] Dakin, J.P.; Liddicoat, T.J. J. *Meas. Control.* 1982, 15, 176–177.
[21] Dakin, J.P. *Proc SPIE* 1984, 468, 219–226.
[22] Marvin, D.C.; Ives, N.A. *Appl. Opt.* 1984, 23, 4212–4217.
[23] Grattan, K.T.V.; Manwell, J.D.; Sim, S.M.L.; Wilson, C.A. *IEE Proc. J. Optoelectron.* 1987, 134, 291–294.
[24] Grattan, K.T.V.; Palmer, A.W.; Saini, D.P.S. *IEEE J Light. Tech.* 1987, LT-5, 972–979.
[25] Spooncer, R.C.; Jones, B.E.; Ohba, R. P*roc. SPIE* 1987, 798, 137–141.
[26] Petuchowski, S.J.; Giallorenji, T.G.; Sheem, S.K. *IEEE J. Quantum Electron.* 1981, QE-17, 2168–2170.
[27] Franzen, D.L.; Kim, E.M. *Appl. Opt.* 1981, 20, 3991–3992.
[28] Yoshino, T.; Kurosawa, K.; Itoh, K.; Ose, T. *IEEE J. Quantum Electron.* 1982, QE-18, 1624–1633.
[29] Beheim, G. *Electron. Lett.* 1986, 5, 238–239.
[30] Diles, R.A. *J. Appl. Phys.* 1983, 54, 1198–1021.
[31] Dixon, J. *Meas. and Control* 1987, 20, 11–16.
[32] Grattan, K.T.V. *Meas. and Control* 1987, 20, 32–39.
[33] Inaba, H.; Kobayashi, T.; Hirama, M.; Hamza, M. *Electron. Lett.* 1979, 15, 749–751.
[34] Kobayashi, T.; Hirama, M.; Inaba, H.; *Appl. Opt.* 1981, 20, 3279–3280.
[35] Snell, D.; Pitt, G.D. Proc. BHRA Int. Conf. on Opt. Tech. in Process Control 1983, 27–41.
[36] Chan, K.; Ito, H.; Inaba, H. *Appl. Phys. Lett.*1983, 43, 634–636.
[37] Chan, K.; Ito, H.; Inaba, H. *Appl. Opt.* 1983, 22, 3802–3804.
[38] Chan, K.; Ito, H.; Inaba, H. Tech. Dig. 9th Nat. Laser Radar Symp. Japan 1983, 74–75 (in Japanese).
[39] Chan, K.; Ito, H.; Inaba, H. *IEEE J. Light. Tech.* 1984, LT-2, 234–237.
[40] Chan, K.; Ito, H.; Inaba, H. *Appl. Opt.* 1984, 23, 3415–3420.
[41] Sholes, R.R.; Small, J.G. *Rev. Sci. Instruments* 1980, 51, 882–884.
[42] Bosselmann, Th.; Reule, A.; Schröder, J. *Proc. SPIE* 1984, 514, 151–154.
[43] Grattan, K.T.V.; Palmer, A.W. *Rev. Sci. Instruments* 1985, 56, 1784–1787.
[44] Augousti, A.T.; Grattan, K.T.V.; Palmer, A.W. *IEEE J. Light. Tech.* 1987, LT-5, 759–762.
[45] Grattan, K.T.V.; Selli, R.K.; Palmer, A.W. *Rev. Sci. Instruments* 1988, 59, 1328–1335.
[46] Grattan, K.T.V.; Palmer, A.W.; Zhang, Z.Y. *Rev. Sci. Instruments* 1991, 62, 1210–1213.

[47] Zhang, Z.Y.; Grattan, K.T.V.; Palmer, A.W. *Rev. Sci. Instruments* 1991, 62, 1735–1742.

[48] Zhang, Z.Y.; Grattan, K.T.V.; Palmer, A.W. *Rev. Sci. Instruments* 1992, 63, 3177–3181.

[49] Tanaka, T.; Benedeck, G.B. *Appl. Opt.* 1975, 14, 189–196.

[50] Kilpatrick, D.; Tyberg, J.V.; Parmley, W.W. *IEEE Trans. Biomed. Eng.* 1982, BME-29, 142–145.

[51] Johnson, C.C.; Palm, R.D.; Stewart, D.C.; Martin, W.E. *J. Assoc. Advance Instrum.* 1971, 5, 77–83.

[52] Woodroff E.A.; Koorajian, S. *Med. Instrum.* 1973, 7, 287–292.

[53] Landsmann, M.H.J. Fiber optic reflection photometry; Verenidge Reproductive Bedrijve, Groningen: Netherlands, 1975.

[54] Gamble, B.G.; Hugenholtz, P.G.; Monroe, R.G.; Polanyi, M.; Nadas, A.S. *J Intracardiac Oximetry* 1965, 31, 328–343.

[55] Yoshino, T.; Nara, M. Proc. of the 13th Cong. of Int. Comm. for Opt., Sapporo (1984), 324–325.

[56] Bjarklev, A. *IEEE J. Light. Tech.* 1986, LT-4, 342–346.

[57] Dakin, J.P. *Proc. SPIE* 1987, 798, 149–156.

[58] Cotter, D. *Opt. and Quantum Electron.* 1987, 19, 1–6.

[59] Appleyard, A.P.; Scrivener, P.L.; Maton, P.D. *Rev. Sci. Instruments* 1990, 61, 2650–2654.

[60] Hartog, A.H. in Optical fiber sensor technology; Grattan, K.T.V.; Meggitt, B.T. (eds.); Chapman and Hall: UK, 1995.

[61] Fei, L.; Aidong, M.; Deying, Z. *Proc. SPIE* 1996, 2828, 296–300.

[62] Guangping, X.; Keey, S.L.; Asundi, A. *Proc. SPIE* 1999, 3897, 497–504.

[63] Marcuse, D. *Principles of optical fiber measurements;* Academic Press: NY, 1981.

[64] Vita, P.D.; Rossi, U. *Opt. Quantum Electron.* 1980, 11, 17–22.

[65] Kim, B.Y.; Choi, S.S. *Electron. Lett.* 1981, 17, 193–195.

[66] Conduit, A.J.; Payne, D.N.; Hartog, A.H.; Gold, M.P. *Electron. Lett.* 1981, 17, 308–310.

[67] Everard, A.K.J. *Proc. SPIE* 1987, 798, 42–46.

[68] Hartog, A.H.; Payne, D.N.; Conduit, A.J. Proc. 6th ECOC (post-deadline paper) 1980, York: UK.

[69] Rogers, A.J. *Electron. Lett.* 1980, 16, 489–490.

[70] Danielson, B.L. *Appl. Opt.* 1985, 24, 2313–2322.

[71] Healey, P. J. *Phys. E: Sci. Inst.* 1986, 19, 334–341.

[72] Ippen, E.P. *Appl. Phys. Lett.* 1970, 16, 303–306.

[73] Stollen, R.H. *Appl. Phys. Lett.* 1972, 20, 62–64.

[74] Stollen, R.H. in Optical fiber telecommunications; Miller, S.E.; Chymoweth, A.G. (eds.); Academic Press: London, 1979.

[75] Farries, M.C.; Rogers, A.J. *Proc. OFS* 1984, 2, 121–132.

[76] Dakin, J.P.; Pratt, D.J.; Bibby, G.W.; Ross, J.N. *Electron. Lett.* 1985, 21, 569–570.

[77] Dakin, J.P.; Pratt, D.J.; Bibby, G.W.; Ross, J.N. *Proc. OFS* (post-deadline paper) 1985, San Diego.

[78] Ferraro, J.R.; Nakamoto, K. Introductory Raman Spectroscopy; Academic Press: London, 1994.
[79] Ippen, E.P.; Stollen, R.H. *Appl. Phys. Lett.* 1972 21, 539–542.
[80] Cotter, D. *Electron. Lett.* 1982, 18, 495–496.
[81] Rogers, *Proc. SPIE* 1986, 718, 65.
[82] Hill, K.O.; Fuji, Y.; Johnson, D.C.; Kawasaki, B.S. *Appl. Phys. Lett.* 1978 32, 647–649.
[83] Hand, D.P.; Russell, R.S.J. *Opt. Lett.* 1990, 15, 102–105.
[84] Melz, G.; Morey, W.W.; Glenn, W.H. *Opt. Lett.* 1989, 14, 823–825.
[85] Anderson, D.Z.; Mizrahi, V.; Erogan, T.; White, A.E. *Electron. Lett.* 1990, 26, 730–732.
[86] Kashyap, R.; Armitage, J.R.; Campbell, R.J.; Williams, D.L.; Maxwell, G.D.; Ainslie, B.J.; Miller, *Bell Syst. Tech. J.* 1993, 11, 150–159.
[87] Askins, C.G.; Tsai, T.; Williams, G.M.; Friebele, E.J. *Opt. Lett.* 1994, 19, 147–149.
[88] Putnam, M.A.; Askins, C.G.; Williams, G.M.; Friebele, E.J.; Baskansky, M.; Reintjes, J. *Proc. SPIE* 1995, 2444, 403–409.
[89] Morey, W.W.; Melz, G.; Glenn, W.H. *Proc. SPIE* 1989, 1169, 98–107.
[90] Morey, W.W.; Dunphy, J.R.; Melz, G. *Proc. SPIE* 1991, 1586, 216–224.
[91] Measures, R.M.; Melle, S.M.; Lui, K. *Smart. Mater. Struct.* 1992, 1, 36–44.
[92] Crossby, P.A.; Powell, G.R.; Fernado, G.F.; Waters, D.N.; France, C.M.; Spooncer, R.C. *Proc. SPIE* 1997, 3042, 141–153.
[93] Ramaswamy, N.S.; Murukeshan, V.M.; Asundi, A.K. *Proc. SPIE* 1999, 3897, 388–396.
[94] Chan, P.K.C.; Jin, W.; Demokan, M.S. *Proc. SPIE* 1999, 3897, 488–496.
[95] Srinivas, T.; Das, I.S.; Selvarajan, A. *Proc. SPIE* 1999, 3897, 480–487.
[96] Jin, W. *Proc SPIE* 1999, 3897, 468–479.
[97] Melle, S.M.; Liu, K.; Measures, R.M. *Appl. Opt.* 1993, 32, 3601–3609.
[98] Tang, L.; Tao, X.; Choy, C.L. *Smart Mater. Struct.* 1999, 8, 154–160.
[99] Kersey, A.D.; Berkoff, T.A.; Morey, W.W. *Electron. Lett.* 1992, 28, 236–238.
[100] Kersey, A.D.; Berkoff, T.A.; Morey, W.W. *Opt. Lett.* 1993, 18, 72–74.
[101] Jackson, D.A.; Riberio, A.B.L.; Reekie, L.; Archambult, J.L. *Opt. Lett.* 1993, 15, 1192–1194.
[102] Kersey, A.D.; Berkoff, T.A.; Morey, W.W. *Opt. Lett.* 1993 18, 1370–1372.
[103] Davis M.A.; Kersey, A.D. *Electron. Lett.* 1994 30, 75–77.
[104] Koo, K.P.; Tveten, A.B.; Vohra, S.T. *Electron. Lett.* 1999 35, 165–167.
[105] Shupe, D.M. *Appl. Opt.* 1980, 19, 654–655.
[106] Bosselman, Th.; Ulrich, R. Proc. of the 2nd Int. Conf. on Opt. Fibre Sensors, OFS '84 (Stuttgart) 1984, 361–365.
[107] Boheim, G. *Appl. Opt.* 1985, 16, 2335–2340.
[108] Dabikewicz, Ph.; Ulrich, R *Opt. Lett.* 1986, 1986, 11, 5435–5445.
[109] Kersey, A.D.; Dandridge, A. *Electron. Lett.* 1986, 22, 616–617.
[110] Velluet, M.T.; Graingorge, Ph.; Arditty, H.J.; *Proc. SPIE* 1987, 838, 78–83.
[111] Harl, J.C.; Saaski, E.W.; Mitchell, G.L. *Proc. SPIE* 1987, 838, 257–261.

[112] Boheim, G. *Rev. Sci. Instrum.* 1987, 58, 1655–1659.
[113] Mariller, C.; Lequime, M. *Proc. SPIE* 1987, 798, 121–130.
[114] Kock, A.; Ulrich, R.; *Proc. SPIE* 1990, 1267, 128–133.
[115] Dandridge, A.; Tveten, A.B. *Appl. Phys. Lett.* 1981, 39, 530–532.
[116] Dandridge, A.; Tveten, A.B. *Opt. Lett.* 1982, 7, 279.
[117] Ball, G.A.; Morey, W.W. *Opt. Lett.* 1992, 17, 1992, 420.
[118] Kersey, A.D.; Berkoff, T.A.; Morey, W.W. *Electron. Lett.* 1992, 28, 236–238.
[119] Lee, C.E.; Taylor, H.F. *Electron. Lett.* 1988, 24, 193–194.
[120] Lee, C.E.; Taylor, H.F. *IEEE J. Light. Tech.* 1991 9, 129–134.
[121] Murphy, K.A. *Opt. Lett.* 1991, 16, 273.
[122] Ghosh, A.K.; Bedi, N.S.; Paul, P. *Proc. SPIE* 1999, 3897, 522–533.
[123] Nasrullah, K. *Proc. SPIE* 1999, 3897, 578–581.
[124] Rochford, K.B.; Rose, A.H.; Deeter, M.N.; Day, G.W. *Opt. Lett.* 1994, 19, 1903–1905.
[125] Vali V.; Shorthill, R.W. *Appl. Opt.* 1976, 15, 1099–1100.
[126] Arditty, H.J.; Bourbin, Y.; Mapouchon, M.; Puech, C. Proc. of the Tech. Dig., 3rd Int. Conf. on Integrated Optics and Opt. Fiber Commun. (San Francisco) 1981, 128–130.
[127] Kadiwar, R.K.; Giles, I.P. *Electron. Lett.* 1989, 25, 1729–1731.
[128] Zarinetchi, F.; Smith, S.P.; Ezekiel, S. *Opt. Lett.* 1991, 16, 229–231.
[129] Raab, M.; Quasat, T. *Opt. Lett.* 1994, 19, 1492–1494.
[130] Kim, S.K.; Kim, H.K.; Kim, B.Y. *Opt. Lett.* 1994, 19, 1810–1812.
[131] Tanaka, Y.; Yamasaki, S.; Hotate, K. *IEEE Phot. Tech Lett.* 1996, 8, 1376–1369.
[132] Kung, A.; Nicati, P.-A.; Robert, P.A. *IEEE Phot. Tech. Lett.* 1996, 8, 1680–1682.
[133] Böhm, K.; Peterman, K.; Weidel, W. *Opt. Lett.* 1982, 7, 180–182.
[134] Kurusawa, K.; Yoshida, S.; Sakamoto, K.; Masuda, I.; Yamashita, T. Trans. *IEE Japan* 1996, 116-B, 93–103.
[135] Takahashi, Y.; Yoshino, T.; Horie, K. *Opt. Rev.* 1997, 4, 417–422.
[136] Takahashi, Y.; Yoshino, T. *Appl. Opt.* 1997, 36, 6770–6773.
[137] Takahashi Y.; Yoshino, T. *IEEE J. Light. Tech.* 1999, 17, 591–597.
[138] Curcio, J.A.; Petty, C.C. J. *Opt. Soc. Am.* 1951, 41, 302–304.
[139] Libnau, F.O.; Kvalheim, O.M.; Chirsty, A.A.; Toft, J. *Vibrational Spectroscopy* 1994, 7, 243–254.
[140] Lin, J.; Brown, C.W. *Vibrational Spectroscopy* 1994, 7, 117–123.
[141] Sanghera, J.S.; Kung, F.H.; Pureza, P.C.; Nguyen, V.Q.; Miklos, R.E.; Aggarwal, I.D. *Appl. Opt.* 1994, 33, 6315–6322.
[142] Hop, E.; Juinge, H.J.; Hemert, H.V. *Appl Spectroscopy* 1993, 47, 1180–1182.
[143] Muto, S.; Sato, H.; Hosaka, T. *Jpn. J. Appl. Phys.* 1993, 33, 6060–6064.
[144] Fan, T.Y.; Huber, G.; Byer, R.L.; Mitzscherlich, P. *IEEE J. Quantum Electron.* 1988, 24, 924–933.
[145] Stonemann, R.C.; Esterowitz, L. *Opt. Lett.* 1990, 15, 486–488.
[146] McAleavey, F.J.; MacCraith, B.D. *Electron. Lett.* 1995, 31, 1379–1380.
[147] Zhu, X.; Cassidy, D.T. *Appl. Opt.* 1996, 35, 4689–4693.
[148] Yokota, M.; Yoshino, T. *Appl. Opt.* 1998, 37, 2526–2533.
[149] Yokota, M.; Yoshino, T. *Mea.s Sci. Technol.* 2000, 11, 152–156.

[150] Asundi, A.K.; Veeredhi, V.R. *Proc. SPIE* 1999, 3897, 187–192.
[151] Inci M.N.; Yoshino, T. *Opt. Rev.* 2000, 7, 205–208.
[152] Fields, J.N.; Cole, J.H. *Appl. Opt.* 1980, 19, 3236–3267.
[153] Asawa, C.K.; Yao, S.K.; Streans, R.C.; Mota, N.L.; Downs, J.W. *Electron. Lett.* 1982, 18, 362–364.
[154] Yoshino, T.; Nara, M.; Kurosawa, K. Proc. of the 13th Cong. of the Int. Commission for Opt. (Sapporo) 1984, 324–325.
[155] Tomita, S.; Tachino, H.; Kasahara, N. *IEEE J. Light. Tech.*1990, 8, 1829–1832.
[156] Berthold, J.W. *IEEE J. Light. Tech.* 1995, 13, 1193–1199.
[157] Oscroft, G. *Proc. SPIE* 1987, 734, 207–213.
[158] Lagakos, N.; Cole, J.H.; Bucaro, J.A. *Appl. Opt.* 1987, 26, 2171–2180.
[159] Yoshino, T.; Inoue, K.; Kobayashi, Y.; Takahashi, Y. Proc. of the 11th Opt. Fiber Sensors Conf. (Sapporo) 1996, 264–267.
[160] Yoshino, T.; Inoue, K.; Kobayashi, Y. *IEE Proc. J. Optoelectron.* 1997, 144, 145–150.
[161] Zhu, D.G.; Petty, M.C. Harris, M. *Sens. and Actuat.* B 1990, 2, 265–269.
[162] Qin, S. J. *Sens. and Actuat. B* 1991 3, 255–260.
[163] Vukusic P.S.; Sambles, J.R. *Thin Solid Films* 1992, 221, 311–317.
[164] Kim, S.R.; Choi, S.A.; Kim, J.D.; Choi, K.H.; Purk, S.K.; Chang, Y.H. *Synthetic Metals* 1995, 71, 2293–2294.
[165] Patel, N.G.; Makhija, K.K.; Panchal, C.J.; Dave, D.B.; Vaishnav, V.S. *Sens. and Actuators B* 1995, 23, 49–53.
[166] Campbell, D.; Collins, R.A. *Thin Solid Films* 1997, 295, 277–282.
[167] Azim-Araghi M.E.; Krier, A. *Appl. Surf. Sci.* 1997, 119, 260–266.
[168] Kim, S.R.; Kim, J.D.; Choi, K.H.; Chang, Y.H. *Sens. and Actuators* B 1997, 40, 39–45.
[169] Hideo Tai, H.; Tanaka, H.; Yoshino, T. *Opt. Lett.* 1987, 12, 437–439.
[170] Choudhury, P.K. *Curr. Sci.* 1998 74, 723–725.
[171] Peterson, J.I.; Goldstein, S.R.; Fitzgerald, R.V.; Buckhold, D.K. *Anal. Chem.* 1980, 52, 864.
[172] Choudhury P.K.; Yoshino, T. *Optik* 2003, 114, 13–18.
[173] Takeo, T.; Hattori, H. *Jpn. J. Appl. Phys.* 1992, 21, 1509–1512.
[174] Paul, P.H.; Kychakoff, G. *Appl. Phys. Lett.* 1987, 51, 12–14.
[175] Ruddy, V.; MacCraith, B.D.; Murphy, J.A. *J. Appl. Phys.* 1990, 67, 6070–6074.
[176] Choudhury, P.K.; Yoshino, T. 49th Spring Meeting of the *Jpn. Soc. of Appl. Phys.* (Tokyo) 2002, Extended Abstracts, No. 3, 1206.
[177] Choudhury, P.K.; Yoshino, T. *Meas. Sci. Technol.* 2002, 13, 1793–1797.
[178] Shoemaker, D.P.; Garland, C.W. *Experiments in Physical Chemistry;* McGraw-Hill Kogakusha Ltd.: Tokyo, 1967.
[179] Cotton, F.A.; Darlington, C.L.; Lynch, L.D. *Chemistry – An Investigative Approach*, Houghton Mifflin Co.: Boston, 1980.
[180] Walker, D.; Prudy, K.; Tarczynski, F. *Proc. SPIE* 1998, 3537, 26–33.
[181] Internet http://www.cudenver.edu//chemistry/classes/chem4538/I2DISS.htm.
[182] Internet http://www.mines.edu/fs_home/dwu/classes/pchem/lab/Wet%20Lab%201/WetLab1.htm.
[183] Henning, M.L.; Lamb, C. *Proc. OFS '88* (New Orleans) 1988, 84.

[184] Culshaw, B. Application of fiber optic sensors in the aerospace and marine industries. In *Optical Fiber Sensors – Systems and Applications,* Vol. 2 (eds. B. Culshaw and J.P. Dakin), Artech House, Inc.: 1989.
[185] Lakhtakia, A. *Optik* 1997, 107, 57–61.
[186] Lakhtakia, A. Messier, R. *Mater. Res. Innov.* 1997, 1, 145–148.
[187] Lakhtakia, A. Messier, R. Sculptured thin films: nanoengineered morphology and optics, *SPIE,* 2005, Vol. PM143.
[188] Lakhtakia, A. *Optik* 1997, 110, 289–293.
[189] McCall, M.W.; Lakhtakia, A. *J. Mod. Opt.* 2000, 47, 743–755.
[190] Venugopal, V.C.; Lakhtakia, A. in Singh, O.N.; Lakhtakia, A. (Eds.), *Electromagnetic fields in unconventional materials and structures,* Wiley: New York, 2000, Chapter 5.
[191] Young, N.O.; Kowal, J. *Nature* 1959, 183, 104–105.
[192] Robbie, K.; Brett, M.J.; Lakhtakia, A. *Nature* 1996, 384, 616.
[193] Messier, R.; Gehrke, T.; Frankel, C.; Venugopal, V.C.; Otaño, W.; Lakhtakia, A. *J. Vac. Sci. Technol.* 1997, 15, 2148–2152.
[194] Lakhtakia, A. *Sensors and Actuators* B 1998, 52, 243–250.
[195] Lakhtakia, A.; Venugopal, V.C. *Microw. Opt. Tech. Lett.* 1998, 17, 135–140.
[196] Ertekin, E.; Lakhtakia, A. *Eur. Phys. J.: Appl. Phys.* 1999, 5, 45–50.
[197] Hodgkinson, I.J.; Wu, Q.H.; Thorn, K.E.; Lakhtakia, A.; McCall, M.W. *Opt. Commun.* 2000, 184, 57.
[198] Lakhtakia, A.; McCall, M.W., Sherwin, J.A.; Wu, Q.H.; Hodgkinson, I.J. *Opt. Commun.* 2001, 194, 33–46.
[199] Lakhtakia, A. *Opt. Commun.* 2001, 188, 313–320.
[200] Lakhtakia, A. Opt. Commun. 2002, 202, 103–111. *Corrections in* Lakhtakia, A. *Opt. Commun.* 2002, 203, 447.
[201] Lakhtakia, A. *Microw. and Opt. Tech. Lett.* 2003, 37, 37–40.
[202] Stelz, E.E.; Lakhtakia, A. *Opt. Commun.* 2003, 216, 139–150.
[203] Pinchbeck, D.; Kitchen, D. *Proc. Conf. Electronics in Oil and Gas Industries* (London) 1985, 496–501.
[204] Chan, K.; Furuya, T.; Ito, H.; Inaba, H. *Opt. Quantum Electron.* 1985, 17, 153–155.
[205] Arakawa, Y.; Fukunaga, H.; Inaba, H. Proc. of the 4th Int. Conf. on *Opt. Fiber Sensors, OFS '86* (Tokyo) 1986, 135–138.
[206] Choudhury, P.K.; Yoshino, T. *Optik* 2004, 115, 329–333.
[207] Choudhury, P.K.; Yoshino, T. *Proc. SPIE* 2005, 5634, 136–139.
[208] Muhammed, F.A.; Stewart, G. *Int. J. Optoelectron.* 1992, 7, 705.
[209] Muhammed, F.A.; Stewart, G. *Electron. Lett.* 1992, 28, 1205–1026.
[210] Muhammed, F.A.; Stewart, G.; Jin, W. *IEE Proc. J. Optoelectron.* 1993, 140, 115.
[211] Bacci, M.; Balidini, F.; Brenci, M.; Conforti, G.; Falciai, R.; Mignani, A.G. *Proc. of the 4th Int. Conf. on Opt. Fiber Sensors (*Tokyo) 1986, 323–326.
[212] Butler, M.A. *Appl. Phys. Lett.* 1984, 45, 1007–1009.
[213] Andrade, J.D.; Vanwagen, R.A.; Gregonis, D.E.; Newby, K.; Lin, J.N. *IEEE Trans. Electron. Dev.* 1985 ED-32, 1175–1179.
[214] Petersen, K.E. *Proc. IEEE* 1982, 70, 420–457.
[215] Jones, B.E.; Philip, G.S. *Proc. of the Conf. on Sens and their Appl,* Manchester (London, Institute of Physics) 1983, 86–88.

[216] Langden, R.M. *J. Phys. E* 1985, 18, 103–115.

[217] Venkatesh, S.; Culshaw, B. *Electron. Lett.* 1985, 21, 315–317.

[218] Stearns, R.G.; Kino, G.S. *Appl. Phys. Lett.* 1985, 47, 1048–1050.

[219] Soref, R.A.; Lorenzo, J.P. *IEEE J. Quantum Electron.* 1986, QE-22, 873–879.

[220] Canham, L.T., Barraclough, K.G.; Robbins, D.J. *Appl. Phys. Lett.* 1987, 51, 1509–1511.

Pankaj Kr. Choudhury started his research career in 1989 from the Department of Applied Physics, Institute of Technology, Banaras Hindu University (BHU), Varanasi (India), and received Ph.D. degree in Physics in 1992. From 1992 to 1997, he worked as a Research Associate in the Department of Electronics Engineering at BHU. Also, during 1995–1996, he was a Guest Lecturer in Physics at the Faculty of Science and Technology, Mahatma Gandhi Kashi Vidyapith, Varanasi (India). In 1997 he joined the Department of Physics at Goa University, Goa (India) as a Lecturer, where he was until 1999. In 1999 he joined the position of Researcher at the Center of Optics, Photonics and Lasers (COPL), Laval University, Quebec (Canada), where he worked on the fabrication and characterization of ferroelectric optical waveguides. During 2000–2003 he was a Researcher at the Faculty of Engineering, Gunma University, Kiryu (Japan). Presently he is a Professor at the Faculty of Engineering, Multimedia University (Malaysia). His current research interests lie in the theory of optical waveguides and EM wave propagation, which include fiber and integrated optics, fiber optic devices and sensors. He has authored and co-authored over eighty five research publications (including three book chapters) in these areas. He is a Senior Member of IEEE and a Regular Member of the Optical Society of America.

Onkar N. Singh received his M.Sc. and Ph.D. degrees, respectively, in 1963 and 1967 in Physics (Spectroscopy) from the Banaras Hindu University (BHU), Varanasi (India). Joining the BHU faculty in 1963, he became a Professor of Applied Physics in 1986, and served as the Head of the Department of Applied Physics, Institute of Technology (BHU) from 1991 to 1993. He has worked as a visiting/research scientist at the International Center for Theoretical Physics (Trieste, Italy), FOM Institute for Atomic and Molecular Physics (Amsterdam, The Netherlands), Max-Planck Institute for Aeronomy (Lindau, Germany), Ludwig Maxmilian University (Munich, Germany), and the Technical University of Munich (Germany). He has published over 100 research papers and book chapters on molecular spectroscopy, electromagnetic fields, lower stratospheric pollution and climate change, and was co-editor of Vol. 4/E of the *Handbook of Environmental Chemistry* (Springer, Heidelberg, 1999). He was the representative of India to the Atmospheric Chemistry of Commission of IUPAC (1989–1995) and was also a member of the National Committee of World Climate Research Programme (1997–2000). He also co-edited the book *Electromagnetic fields in unconventional structures and materials* (Wiley, USA, 2000). He is Member/Fellow of several learned bodies.

In: Frontiers in Optical Technology: Materials and Devices ISBN: 1-60021-084-8
Editors: P.K. Choudhury and O.N. Singh, pp. 273-287

Chapter 10

OPTICAL CURRENT SENSORS

Toshihiko Yoshino[1] ***and Masayuki Yokota***[2]

[1] The Kaisei Academy, Nishi-Nippori 4-2-4, Arakawa-ku, Tokyo 116-0013, JAPAN

[2] Department of Electronic Engineering, Faculty of Engineering, Gunma University, 1-5-1 Tenjin-cho, Kiryu, Gunma 376-8515, Japan

Abstract

This review article describes the present state-of-the-art of fiber optic current sensing technology based on Faraday effect. The importance of the fulfillment of HICOC (homogeneous isotropic closed optical circuit) condition is addressed. The optical current sensors are categorized into three types, viz. all-fiber, bulk-optic and hybrid, and their available HICOC techniques, advantageous features, performances and applications are mentioned.

1. Introduction

In comparison with electric current transformers, an optical current sensor or transformer (OCT) has the invaluable features of non-contact, highly insulating and high-speed measurements [1,2]. Because the OCT technology is closely related to the optical magnetic-field sensing technology, a lot of optical magnetic-field sensors are potentially applicable to OCT as well. The most reliable OCT, however, should be based on Ampere's circuital law, according to which the line integration of current-induced magnetic fields along a closed loop is equal to the current which intersects the closed loop. In order to perform this integration optically, we use Faraday effect, which states that the plane of linear polarization of a light beam propagating in an isotropic material subject to magnetization is rotated in proportion to the magnetization. So far many works on OCT have been made from engineering as well as scientific interests, and some OCTs developed have reached the stage of practical use and are commercially available. This chapter is devoted to the description of the present state-of-the-art of the OCT technology. The basic principles, types, performances and applications of OCTs are addressed.

2. The Basis

2.1. Optical Ampere's Circuital Law

When a steady electric current I flows in 3D space, according to the Ampere's circuital law,

$$I = \oint \vec{H} \cdot d\vec{s}\,, \tag{1}$$

where $\vec{H}$ is the current-induced magnetic field vector and $\vec{s}$ is the line vector along which the loop integration is carried out. In terms of Faraday effect, eq. (1) becomes

$$n\oint V\vec{H} \cdot d\vec{s} = nVI = \phi\,, \tag{2}$$

where V is Verdet constant of the loop material (assumed to be constant along the entire optical path), ϕ is Faraday rotation angle due to one round trip of light along the loop, and n is the number of turns of the optical loop. Thus, in order that the OCT based on eq. (2) may function properly, the optical loop has to be made of a homogeneous, isotropic and closed optical circuit (HICOC).

2.2. The HICOC Condition

Unless the HICOC condition is fulfilled, the OCT undergoes the measurement uncertainty such that the output signal of the OCT is affected by external currents of the measuring current, and also, by the relative position of the measuring current within the optical loop. In order to evaluate the measurement error due to imperfect fulfillment of the HICOC condition, we assume that the optical loop of an OCT involves some part which has a different Verdet constant V' from the other part, as shown in fig. 1(a). Faraday output signal then undergoes a measurement error due to the cross-talk with (or isolation from) a surrounding current as:

$$e = \left|(V' - V)/V\right|(\theta/2\pi)\,, \tag{3}$$

where θ stands for the subtended angle from the current position to the path AB as shown in fig.1(a) [3,4]. Calculation of eq. (3) as a parameter of e is shown in fig. 1(b). We consider the case that the optical path is not closed but terminated at separated points A and B. This corresponds to $V' = 0$ on the AB line so that thereby-caused error due to cross-talk is [from eq. (3)] $e = \theta/2\pi$. In a particular case, in which the external current is located at a distance d from the AB line (AB = L), θ is then $\theta_1 = 2\tan^{-1}(L/2d)$. Therefore, referring to fig. 1(b), for $x = \theta_1/2\pi$ and $y = 1$, it follows that when d is, e.g. 10 mm, in order to achieve $e < 0.01$, L must be smaller than 0.63 mm. It is thus recognized that even the presence of such a slightly opened loop can cause a significant cross-talk error for close currents. On the other hand, when the traveling light has an elliptical polarization on the path AB, it corresponds to $V' = V\cos\Delta$ (Δ being the polarization retardation) so that $e < 0.01$ requires $\Delta \leq 12°$.

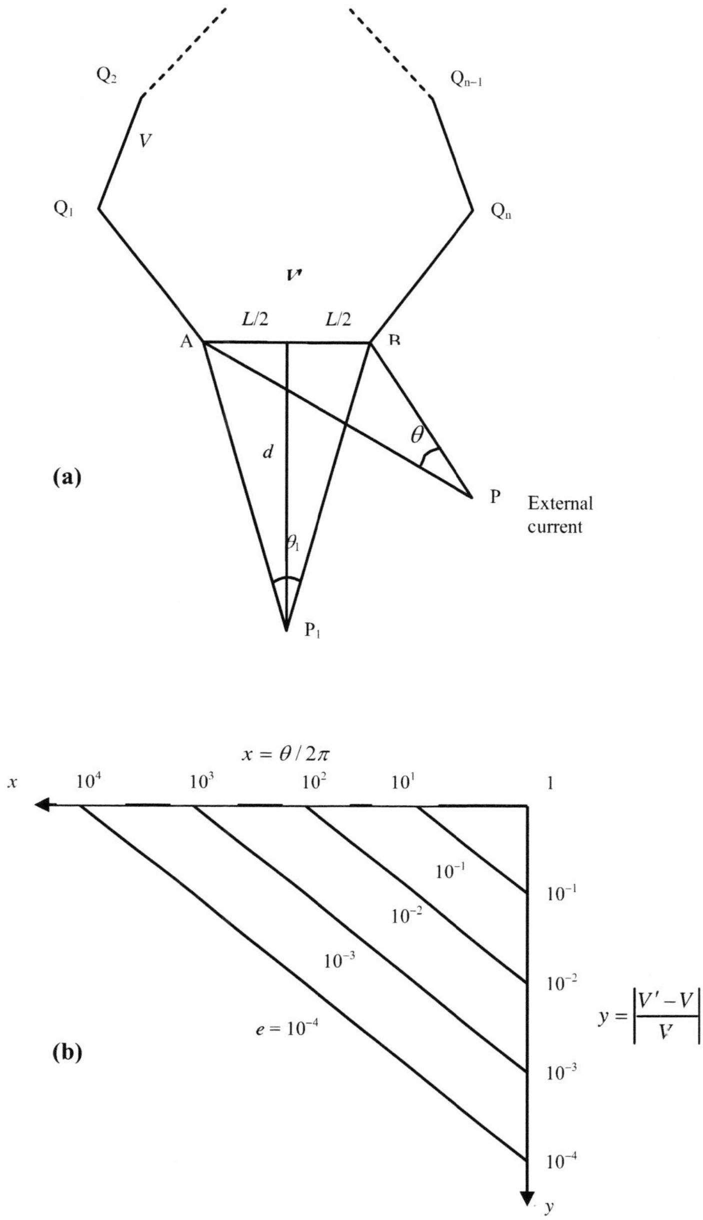

Figure 1. Theoretical value of cross-talk with surrounding current in OCT [3,4], (a) geometry; (b) dependence of cross-talk e on subtended angle θ and Verdet constant inhomogeneity.

Figure 2 shows the experimental demonstration of the cross-talk error due to an open loop [3,4]. In the sensor cell geometry shown in fig. 2(a), the optical path is definitely open without the (4 mm thick) compensating glass plate while the optical path is well-closed with it on the input/output port of the Faraday cell. For both cases, magnitudes of the cross-talk with an external current were measured as a function of the distance of the current from the center of the Faraday cell, and compared with each other in fig. 2(b), together with the theoretical values of eq. (3). Figure 2(b) clearly indicates that, in real OCTs, the cross-talk with close currents becomes a significant error if the optical path is not properly closed.

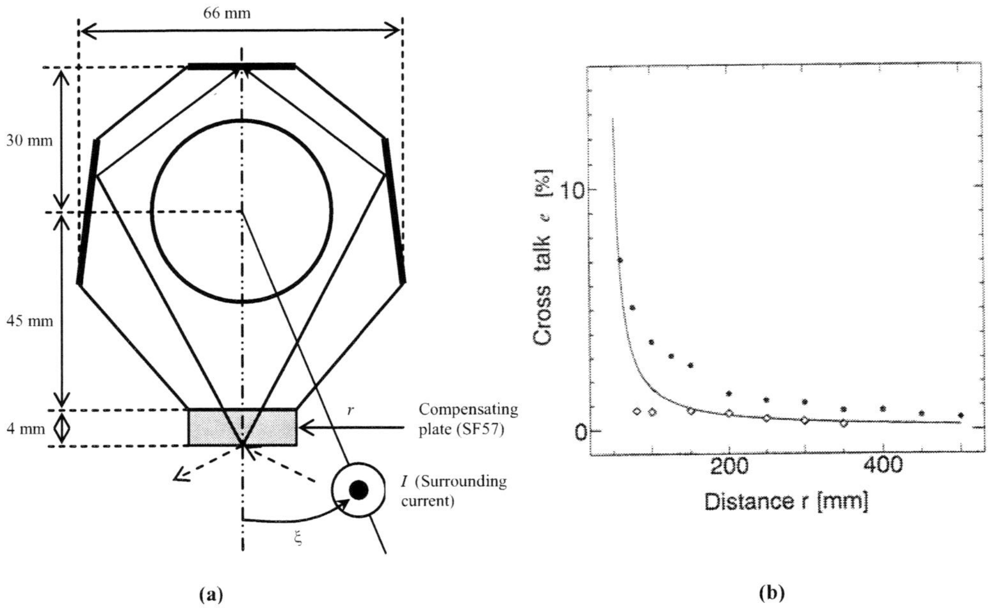

Figure 2. Demonstration of cross-talk with surrounding current in OCT [3,4]; (a) experimental arrangement, (b) dependence of cross-talk e on external current distance r at $\xi = 0°$ with (◦) and without (•) compensator. The solid line represents the calculation by eq. (3).

2.3. Signal Interrogating Schemes

Faraday signal can be detected by some different interrogating schemes. In general, a polarimetric interrogation, as follows, is employed. A linearly polarized light is incident on the Faraday material, the output light is received by two polarizers which are oriented at ±45° to the incident polarization direction, and the degree of polarization of the output light is detected as the OCT signal. As to the light source, a low coherent light is preferable to laser light to avoid its interference noises, and SLD (superluminescent laser diode) is especially used for the coupling to single mode fiber (SMF).

Because the Faraday rotation originates from the magnetically-induced circular birefringence of a substance, interferometric detection of the relevant optical phase shift yields the OCT signal. Though various types of interferometers (e.g., Fabry-Perot, Mach-Zender types) are, in principle, applicable to OCT, a ring interferometer (i.e., Sagnac interferometer) is the most suited one for the Faraday signal interrogation from a closed optical path in particular. The phase difference between the clockwise and the counter-clockwise traveling light beams in the interferometer yields twice the Faraday rotation [5], and then the undesirable reciprocal phase shifts due to temperature changes or vibration are cancelled. Similarly, a ring laser, or an active Sagnac interferometer, can also be used for the Faraday signal interrogation. Interference of the clockwise and the counter-clockwise traveling light beams going out of the ring laser generates an optical beat whose frequency is proportional to the Faraday rotation angle. The Faraday rotation angle is then interrogated in the frequency domain with a linear scale within the free spectral range of the laser [6,7].

In general, the Faraday signal detection is accompanied by optical noises. For dc current sensing, in particular, the drift of light power becomes a serious noise. To reduce it, the heterodyne detection using two frequencies of laser light [5,8] or a phase-modulated (or ac) Sagnac interferometer is very much suited.

3. All-Fiber Type OCT

An SMF made of silica is an ultra-low loss, thin, long, flexible, cheap and intrinsically isotropic optical waveguide, and therefore, its application to OCT started immediately after the invention of the fiber [9,10]. In reality, however, it took a long time to arrive at practical OCTs mainly because their development was much hindered by the involvement of linear birefringence (LB) in real fibers.

3.1. Magneto-Optical Properties of SMFs

The general theory for Faraday effect in isotropic SMFs is presented in ref. [11]. In the simultaneous presence of LB and Faraday effect in a medium, when a linear polarization is input into to medium along the principal axes of LB, the amplitude of the electric field of the output light is given by [12]

$$F = E_0 HVL\mathrm{sinC}\{[(\delta/2)^2 + VH]^{1/2} L\}, \tag{4}$$

where δ is the linear retardation per meter, L is the optical path length, H is the applied magnetic field, E_0 is the amplitude of the electric field of the input light, and $\mathrm{sinC}x = \sin x/x$. Obviously, for $\delta = 0$, eq. (4) becomes $F = E_0\sin(VHL)$, which is the known relationship for Faraday effect in isotropic mediums. In the case that $|\delta| >> |VH|$, eq. (4) becomes $F = E_0 VHL\mathrm{sinC}(\delta L/2)$, so that $|F| \leq 2E_0|VH/\delta|$. For silica fibers subjected to a bending of e.g. 50 mm radius of curvature, $|\delta|$ amounts to the order of 1 radian/m while $|VH|$ is typically 10^{-5} I rad.A^{-1}m^{-1}; it then follows that $|HV/\delta| \leq 10^{-5}$ I A^{-1}, and hence, the field conversion efficiency $|F|/E_0$ is smaller than the order of 10^{-5} I A^{-1}, which is usually very small except for especially heavy currents.

Therefore, some approaches to eliminate LB in silica fibers were conducted. One of them is to induce circular birefringence (CB) by providing a post-drawing fiber with twist [13,14] or an in-drawing fiber with spin (i.e., spun fiber) [15]. For such a fiber, eq. (4) becomes

$$F = E_0 H(VH + \delta_T/2)L\mathrm{sinC}\{[(\delta/2)^2 + (VH + \delta_T/2)^2]^{1/2} L\}, \tag{5}$$

where δ_T is the provided CB/m. In the case when CB is much larger than LB (i.e., $|\delta_T| >> |\delta|$), eq. (5) becomes

$$F = E_0 \sin[(VH + \delta_T/2)L], \tag{6}$$

which indicates that the effect of LB is neglected against CB, and F becomes the same as that for the fiber without LB case except involving a constant bias polarization rotation. When a twist rate of 1 turn/m is applied to a post-drawing silica SMF, $|\delta_T|$ amounts to 1 rad/m.

The study of the thermal relaxation of fiber stress has been attempted, and it has been shown that the bend-induced LB in silica fiber could be reduced by annealing [16]. The annealed fiber, however, is still sensitive to newly applied stress so that, in its practical use, a careful remedy is needed to avoid the application of new stress.

Naturally, it is most desirable to invent an SMF which has inherently no photo-elastic effect. Such fiber was developed by using flint glass as the fiber material [17]. Compared with silica, flint glass has about 100-times smaller photo-elastic constant (Brewster coefficient) and, of additional advantage, its Verdet constant is 6-times larger ($V = 1.5\times10^{-5}$ rad/A at a wavelength of $\lambda = 850$ nm) and little temperature dependent. The transmission loss is, however, pretty high. The flint glass SMF is now commercially available (HOYA Co., Tokyo, Japan).

One more point to be noted in the application of SMF for OCT sensors is the geometrical rotation of the polarization plane of the transmitted light from the fiber. As a general feature of light transmission in optical fibers, if the fiber is not in-plane, the light is then subjected to a purely geometrical rotation of the polarization plane, depending on the geometry of the fiber path [18]. In all-fiber OCTs, this geometrical polarization rotation can cause polarization instability. Fortunately, this geometrical polarization rotation is a reciprocal effect so that it can be cancelled by making the light beam to go forth and back in the same fiber (i.e., by the use of the reflection scheme) [19]. In the transmission type OCTs, however, this geometrical rotation can introduce uncertainty in measurement due to polarization instability. As such, the sensing fiber coil needs to be fixed on a rigid plane in a stable manner.

3.2. Practical All-Fiber OCTs

Various types of all-fiber OCTs have been developed so far and several of them have been put into either test or practice for the electric power systems including power generators in power plants, railway systems, nuclear devices etc.

Regarding flint-glass fiber OCTs, both the transmission and reflection types with polarimetric interrogation have been developed [20]. The transmission-type, in which a flint glass fiber coil is fixed on a durable metallic film, could measure currents up to 300 kA (which corresponds to a Faraday rotation of 45°) within a measurement error of 2 kA to meet the Japanese industrial standard [20]. Figure 3(a) shows the optical diagram of the reflection-type OCT. A low coherent light from SLD ($\lambda = 1550$ nm) is remotely sent (via a polarization maintaining fiber) to the sensing part, and coupled into the flint glass fiber after being passed through an optical unit that serves as both a polarizer and a polarization bias element. The reflected light from the sensing fiber is again passed through the same optical unit and ±45° polarization components of the emerging light are received (via an SMF) by two photo detectors 1 and 2. Figure 3(b) shows the application of the reflection-type OCT for fault current detection. The sensing fiber could, on account of the reflection-type, freely encircle the three phases of electric cables, and accurate phase current detection (summation of the three phase currents) is carried out [21].

The annealed-silica-fiber OCTs with polarmetric interrogation have been put into for testing. In order to reduce the effects of thermal expansion and vibration, the annealed fiber coil is installed in the glass epoxy compound package [22]. A Sagnac interferometer OCT using annealed silica SMF has also been tested for applications, and such OCTs are found to have a measurement accuracy of better than 0.5% for the environmental temperature range from −35 °C to 85 °C [23]. A Sagnac interferometer OCT using spun silica SMF has also been developed for the application in D.C. railway transportation system [24].

All of the all-fiber OCTs reported so far, unfortunately, did not describe about the closing feature of the sensing fiber loop so that there are ambiguities about the possible cross-talk errors with very close surrounding currents.

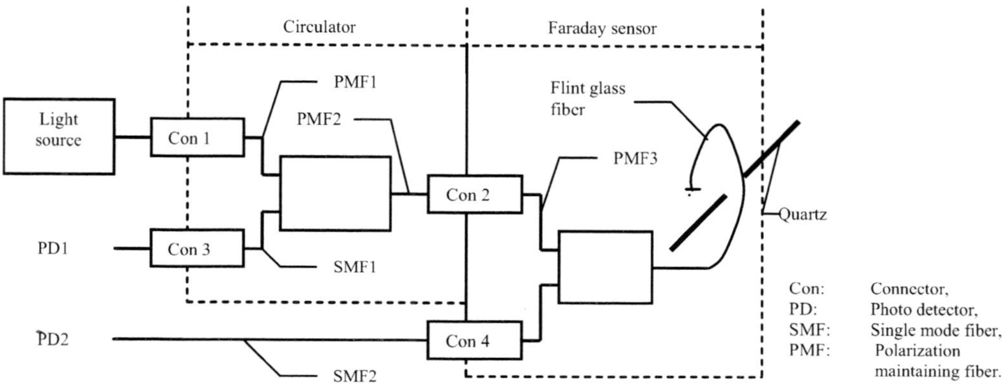

(a)

(b)

Figure 3. Practical flint glass OCT of the reflection type [21] ; (a) optical diagram, (b) fault section locating system for 22 kV power cable lines.

4. Bulk-Optic OCT

In order to get rid of the problem of fiber LB or to make it possible to use a variety of Faraday materials, it is important to study the use of non-fiber bulk-optic materials as the Faraday element. In this type of OCT (or the bulk-optic type of OCT), the light beam travels in a zigzag path within a Faraday cell. The key requirement for the bulk-optic OCT is to achieve the polarization-maintaining 100% reflectance of the traveling light at each reflecting surface of the Faraday cell. To this end, the oldest scheme applied the crossed right-angle total-reflection glass prisms for the corner reflection [25,26] but it, in principle, could not well fulfill the HICOC condition. In order to fulfill the HICOC condition precisely, new schemes have been developed, which applied the thin-film coating technology. Two different approaches have been conducted. The first approach extended the well-known quarter-wavelength multilayer coating technology to perform the polarization-maintaining 100% reflectance. The theory required the specific conditions of "above Brewster's angle and below critical-angle" of light on every boundary of the multilayer thin films. The theoretical requirement was experimentally demonstrated by the use of conventional coating materials. The HICOC condition was thus for the first time very exactly fulfilled [4]. The resulting form of the Faraday cell was a polygon of specific form.

The second approach used the coating of the bi-layer dielectric thin films on the total reflection surface to achieve the polarization-maintaining 100% reflectance for light beam at oblique angle of incidence [27]. The optical design was conducted by the trial-and-error numerical calculation. The sensor system developed is shown in fig. 4(a) [27]. Each of the three reflecting surfaces of a square block Faraday cell made of flint glass was coated by the suitably thick SiO_2 and Ta_2O_5 films to produce no polarization-retardation and 100% reflectance on the internal reflection for 45° oblique incidence of light. The input/output port of the Faraday cell was provided by attaching a low-Verdet-constant (SF7) tiny prism on one uncoated surface of the cell. This Faraday cell had good tolerances with respect to light beam angle, wavelength and film thickness. The current sensing characteristics of this OCT are shown in fig. 4(b). The polarization extinction ratio was better than 2×10^{-4} and high isolation from the surrounding current was demonstrated [fig. 4(c)]. Figure 4(d) shows the ac Faraday signal and the dc light power as a function of applied weight to the cell; it indicates that the OCT is very stable against the mechanical disturbances applied to the Faraday cell.

5. Hybrid Type OCT

The third type of OCT is fabricated by a simple modification of the traditional electric CT. This type of OCT, or the hybrid OCT, is composed of a high-permeability (metallic) magnetic core with a small gap, and a Faraday element inserted in the core gap. In principle, the hybrid type OCT cannot rigorously fulfill Ampere's circuital law, and therefore, cannot be a much exact OCT. Moreover, the use of metallic magnetic cores as the field concentrator reduces the insulating feature of OCT, and saturates the Faraday signal against large currents. Despite these drawbacks, the hybrid type OCT is an attractive one because, compared with the bulk-optic type OCT, the hybrid one allows the use of various Faraday materials, which include crystals and thin films, and does not need specific internal refection schemes of the Faraday cell (unlike the bulk-optic OCT). Also, the hybrid type OCT can be used as a clamp-type OCT, like the electric CT. In order to achieve high isolation from surrounding currents and to improve the accuracy of OCT, it is required that the gap length of the core is possibly small. This is the key requirement for developing high performance hybrid type OCT. Earlier studies developed the multiple-reflection Faraday cell made of a thin FR5 glass plate, as shown in fig. 5, which is suitable for the insertion into a small gap of a ring core [28–30].

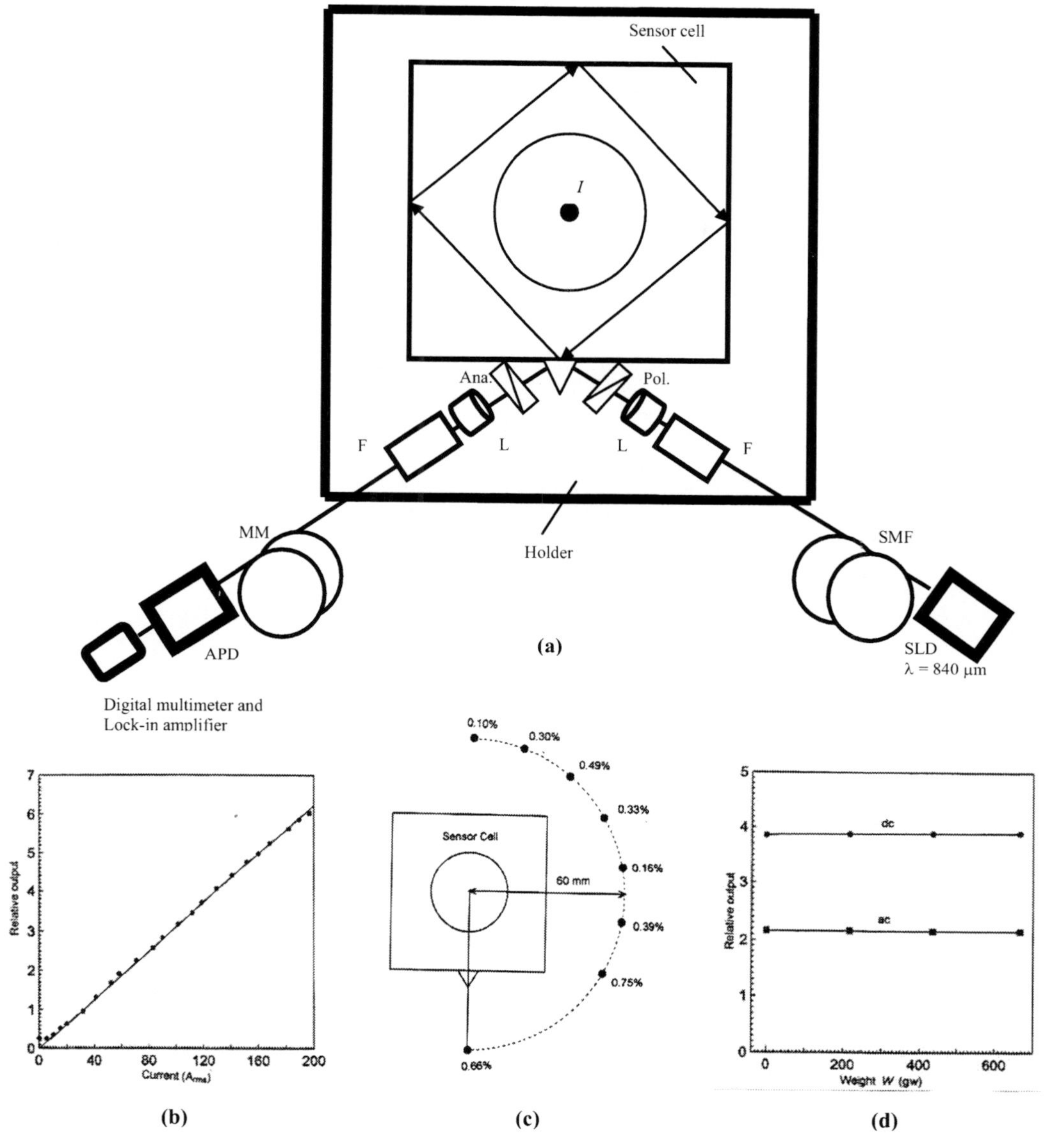

Figure 4. Precise and stable bulk-optic OCT [27]; (a) current sensor system, (b) measured current characteristics, (c) measured cross-talk at different positions of the external current, (d) measured weight characteristics of dc and ac Faraday signals, indicating high mechanical stability of the OCT.

An especially attractive application of the hybrid type OCT is the small current measurement, which is very difficult with the all-fiber or bulk-optic type OCT. The achievable sensitivity is governed by the performances of Faraday materials and magnetic core systems.

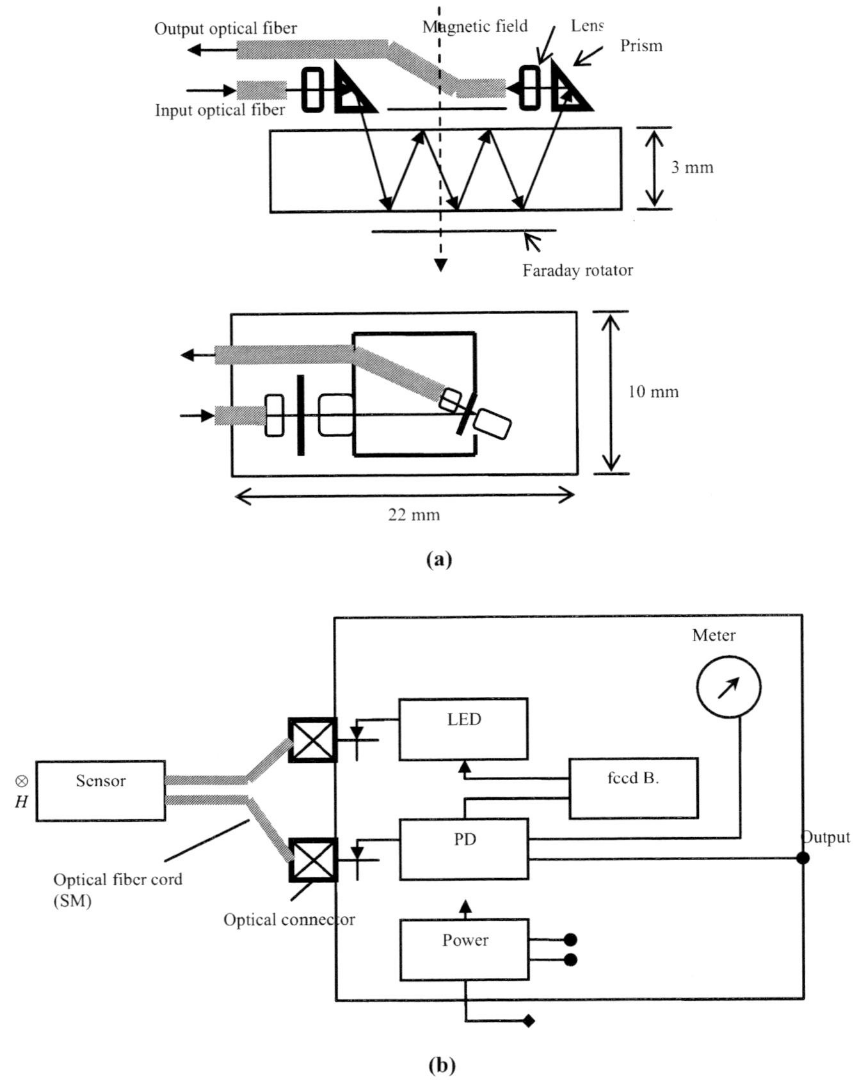

Figure 5. Optical magnetic-field sensor of the multiple-reflection type [30]; (a) sensor head, (b) instrument system.

As for high Verdet constant materials, transparent magnetic materials such as rare earth iron garnet (RIG), e.g., yttrium iron garnet (YIG), are attractive. Improvements on their magneto-optical properties such as temperature characteristics [31], linearity [32] and response speed [33] were conducted. Especially, as for the temperature characteristics, Bi-doped YIG with high temperature stability has been developed, and is now commercially available.

A magnetic material (MM) has the following inherent features as the Faraday material:

i) Verdet constant is typically 100 times higher than non-magnetic materials,

ii) when MM is applied in the hybrid OCT, because of its high permeability, the total magnetic resistance of the core system becomes small, thereby enhancing the current-sensitivity,
iii) it is possible to magnetize MM in the other directions than that of the applied magnetic field by virtue of demagnetization and/or magnetic-anisotropy (such as uniaxial anisotropy) effects peculiar to MM,
iv) the magnetic response of MM is subjected to nonlinearity, hysteresis and saturation against the applied magnetic field, unlike non-magnetic materials,
v) the produced magnetization is not uniform in MM because of the domain structure, which scatters or diffracts the light beam, and hence, causes the measurement uncertainty. This domain effect, however, is averaged and can be reduced by the illumination of a broad-width light beam as the probe light.

It results from the above features of MM that a possibly smaller core gap and a broad-beam illumination are desirable for the hybrid OCT using MM. Usually, in the core gap, the probe light is passed through the Faraday element in the direction parallel to the core gap (longitudinal configuration). Then, because the gap length is very limited, special optical designs such as a core system with holes [34] or a micro-optic system [35] have been introduced.

It has also been demonstrated that the hybrid OCT using MM functions well even in the transverse configuration where the probe light is passed through the Faraday element in the direction perpendicular to the core gap [36,37]. The operating principle of this unusual hybrid OCTs is particularly based on the feature no. (v) of MM as mentioned above.

Figure 6(a) shows the entire system of the hybrid OCT using a YIG bulk crystal Faraday cell in the transverse configuration [36]. The YIG crystal ($3\times10\times10$ mm^3) was tightly inserted in a small gap (of 3 mm) of a permalloy core (of the size 40 mm outer diameter $\times$ 10 mm thickness) to overcome the problem of air gap. Under the illumination of a light beam (from 1.3 μm SLD) on the crystal in the transverse direction to the core, a very high current sensitivity of a modulation depth of 20% per Ampere was obtained, which has been the highest sensitivity of OCT ever reported and about 10^4 times higher than that of the flint-glass fiber OCT. Figure 6(b) shows the typical optical response signal. Good linearity for currents from 0 to 1 Ampere was obtained and the isolation from surrounding currents was very high, as shown in fig. 6(c).

Figure 7 shows the entire system of the hybrid type OCT using a uniaxial Bi-substituted garnet thin film (650 μm thick) in the transverse configuration [37]. The garnet film was obliquely (typically at 45° inclination angle) inserted into the 3.5 mm air gap of a permalloy ring core, and was illuminated by a collimated optical beam (1.3 μm SLD) from the transverse side of the core. The high sensitivity of about 0.08% per Ampere modulation-depth, good linear response from 0 to 1 Ampere and good reproducibility were demonstrated.

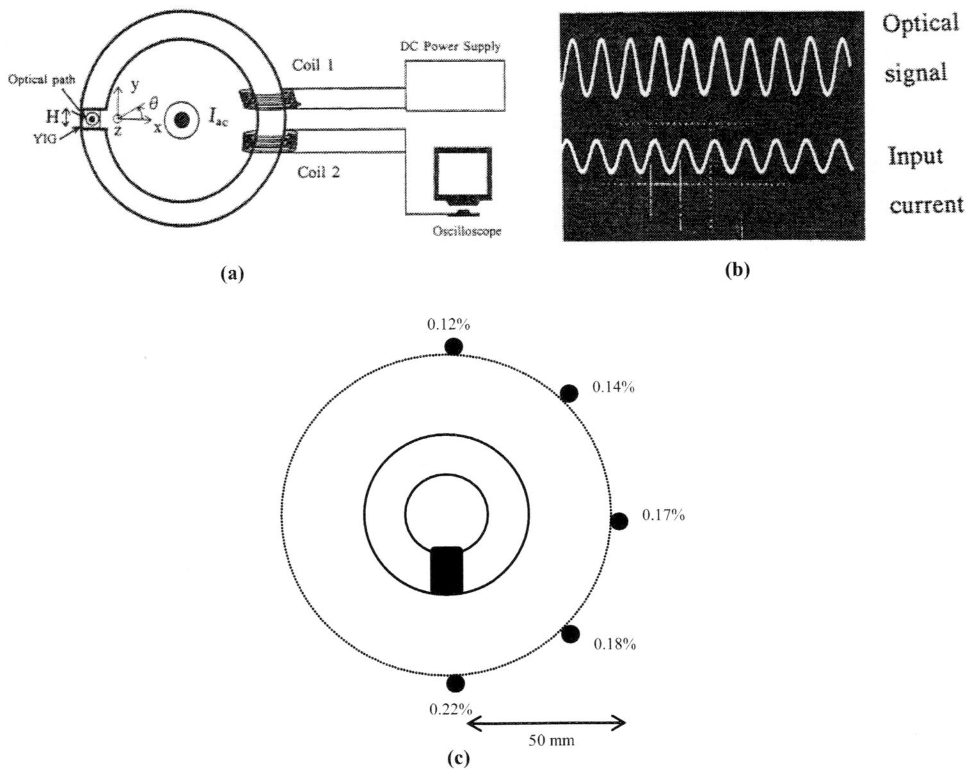

Figure 6. Fiber-linked YIG/ring-core Faraday effect optical current sensor in the transverse configuration [36]; (a) entire system, (b) oscilloscope traces of optical signal for 50Hz-100mA$_{rms}$ current, (c) measured cross-talk from the surrounding current at various angular positions.

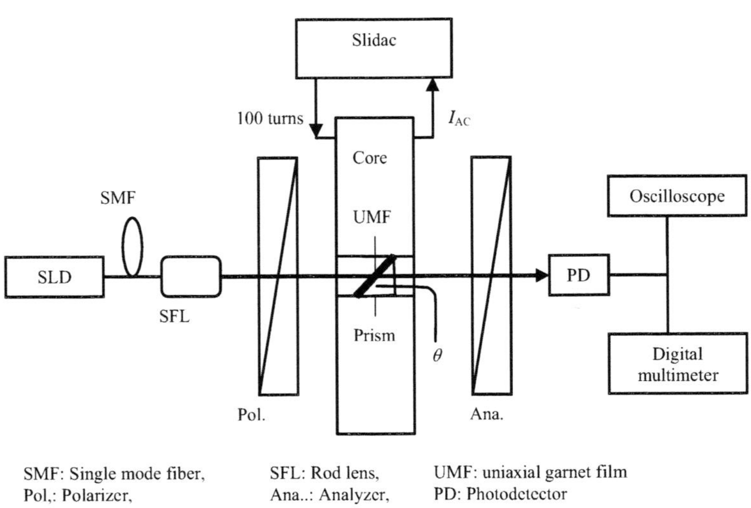

Figure 7. Schematic diagram for magnetic garnet film/ring core Faraday effect optical current sensor in the transverse configuration [37].

6. Other OCTs and Remarks

The OCT technology is closely related to the optical magnetic field sensing technology. Our descriptions are limited to the OCT which basically relies on the Ampere's circuital law for the sensing principle. However, if there are no other currents than a measuring current, and moreover, the magnetic field distribution due to the measuring current is a priori known, then other techniques than the ones based on the optical Ampere's law can also be applied for the current sensing. One of the useful techniques for this purpose is the introduction of a solenoid coil to transform a current into a uniform magnetic field within the solenoid coil, as long as the resistance and reactance of the solenoid do not matter. Various types of solenoid-combined OCTs were studied in connection with the optical sensing of electric power, and their actual performances are described in refs. [38–41].

7. Conclusion

The OCT based on Faraday effect has the invaluable features of non-contact, highly insulating and high-speed measurements. Studies on OCT have been made to fulfill the HICOC condition satisfactorily. The currently-known most effective HICOC technologies use ultra-low photo-elastic coefficient fibers, dielectric-thin film coatings and magnetic Faraday materials in the transverse configuration, for the all-fiber, bulk-optic and hybrid type of OCTs, respectively. Such OCTs can provide very simple and flexible sensor head, accomplish the fulfillment of the HICOC condition very precisely and achieve very high current sensitivity. OCTs are now mostly applied for the electric power industries, but can find other wide applications too in both engineering and science. It is very much expected that the OCT technology will improve more with the progress of optoelectronic materials and devices.

References

[1] Yoshino, T. *Proc. SPIE* 1987, 798, 258–266.

[2] Day, G.W.; Rose, A.H. *Proc. SPIE* 1988, 985, 138–150.

[3] Yoshino, T.; Takahashi, Y.; Gojyuki, M. *Opt. Rev.* 1997, 4, 108–110.

[4] Yoshino, T.; Gojyuki, M.; Takahashi, Y.; Shimoyama, T. *Appl. Opt.* 1997, 36, 5566–5573.

[5] Arditry, H.J.; Bourbin, Y.; Mapuchon, M.; Puech, C. *Tech. Dig. of the 3rd Int. Conf. on Integrated Optics and Optical Fiber Communication (IOOC'3, San Francisco)* 1981, 128–129.

[6] Kueng, A.; Nicati, P.A.; Robert, P.A. *Opt. Rev.* 1997, 4, 56–57.

[7] Takahashi, Y.; Yoshino, T. *J. Light. Tech.* 1999, LT-17, 591–597.

[8] Yoshino, T.; Hashimoto, T.; Nara, M.; Kurosawa, K. *J. Light. Tech.* 1992, LT-10, 503–513.

[9] Smith, A.M. *Appl. Opt.* 1978, 17, 52–56.

[10] Rapp, A.; Harms, H. *Appl. Opt.*1980, 19, 3729–3747.

[11] Yoshino, T. *J. Opt. Soc. Am. B* 2005, 22, 1856–1860.
[12] Yoshino, T. (ed.) Optical fiber sensor technologies (2nd ed.), Dai-ichi International Co.: Tokyo, 1986, 15–160.
[13] Ulrich, R.; Simon, A. *Appl. Opt.* 1979, 18, 2241–2251.
[14] Rashleigh, S.C.; Ulrich, R. *Appl. Phys. Lett.* 1979, 34, 769–770.
[15] Laming, R.I.; Payne, D.N.; *J. Light. Tech.* 1989, 7, 2084–2094.
[16] Tang, D.; Rose, A.H.; Day, G.W., Etzel, S.M. *J. Light. Tech.* 1991, 9, 1031–1037.
[17] Kurosawa, K.; Yoshida, S.; Sakamoto, K. *J. Light. Tech.* 1995, 13, 1378–1384.
[18] Ross, J.N. *Opt. Qunatum Electron.* 1984, 16, 455–461.
[19] Yoshino, T.; Iwama, M. *Opt. Lett.* 1999, 24, 1626–628.
[20] Kurosawa, K. *Opt. Rev.* 1997, 4, 38–44.
[21] Kurosawa, K.; Shirakawa, K.; Saito, H.; Itakura, E.; Sowa, T.; Hiroki, Y. Kojima, T. *Proc. of the 16th Int. Conf. on Optical Fiber Sensors* (OFS'16, Nara) 2003, 316–319.
[22] Willsch, M.; Bosselmann, T. *Proc. of the 15th Int. Conf. on Optical Fiber Sensors* (OFS'15, Boulder) 2002, 407–410.
[23] Bohnert, K., Gabus, P.; Brandle, H., *Proc. of the 16th Int. Conf. on Optical Fiber Sensors* (OFS'16, Nara) 2003, 752–755.
[24] Hayashiya, H.; Kumagai, T.; Hino, M.; Endo, T.; Ando, M.; Negishi, H. *Proc. of the 32nd Meeting on Lightwave Sensing Technology (Jpn. Soc. Appl. Phys., LST'32, Tokyo)* 2003, 141–146.
[25] Saito, S.; Fujii,Y.; Yokoyama, K.; Hamasaki, J.; Ohno, Y. *IEEE J. Quantum Electron.* 1968, QE-2, 255–259.
[26] Kanai, N.; Takahashi, G.; Sato, T.; Higashi, M.; Okamura, K. *IEEE Trans. Power Del.,* 1986, PWRD-1, 91–97.
[27] Yoshino, T.; Yokota, M.; Aoki, K.; Yamamoto, K.; Itoi, S.; Ohtaka, M. *Appl. Opt.* 2002, 41, 5963–5968.
[28] Yoshino, T. *Jpn. J. Appl. Phys.* 1978, 19, 745–749.
[29] Yoshino, T.; Ohno, Y. *Fiber and Integrated Opt.* 1981, 3, 391–399.
[30] Yoshino, T.; Ohno, Y.; Kurosawa, K. *Tech. Dig. of the 2nd Int. Conf. on Optical Fiber Sensors (OFS'2, Stuttgart)* 1984, 55–58.
[31] Inoue, N.; Yamasawa, K. T. *IEE Japan* 1995, 115-A, 1114–1120.
[32] Numata, T.; Tanakaike, H.; Inokuchi, S.; Sakurai, S. *IEEE J. Mag.* 1990, 26, 1358–1360.
[33] Rochford, K.B.; Rose, A.H.; Deeter, M.N.; Day, G.W. *Opt. Lett.* 1994, 19, 1903–1905.
[34] Yoshino, T.; Hara, H.; Sakamoto, N. *Extended Abstracts of the Spring Meeting of Jpn. Soc. Appl. Phys.*, 1988, 860.
[35] Itoh, N.; Minemoto, H.; Ishiko, D.; Ishizuka, S. *Tech. Dig. of the 11th Int. Conf. on Optical Fiber Sensors (OFS'11, Sapporo)* 1996, 638–641.
[36] Yoshino, T.; Minegishi, K.; Nitta, M. *Meas. Sci. Tech.* 2001, 12, 850–853.
[37] Yoshino, T.; Torihata, S.; Yokota, M.; Tsukada, N. *Appl. Opt.* 2003, 42, 1769–1772.
[38] Li, Y.; Li, C.; Yoshino, T. *Appl. Opt.* 2001, 40, 5738–5741.
[39] Li, C.; Cui, X.; Yoshino, T. *IEEE Tans. Instrum. Meas.* 2001, 50, 1375–1380.
[40] Li, C.; Yoshino, T. *Appl. Opt.* 2002, 41, 5391–5397.
[41] Li, C.; Cui, X.; Yoshino, T. *J. Light. Tech.* 2003, 20, 843–849.

Toshihiko Yoshino was born in Tokyo, Japan, on April 27, 1939. He received the B.S., M.S., and Ph.D. degrees in Applied Physics from Tokyo University (Tokyo) in 1963, 1965 and 1968, respectively. From 1968 to 1988, he was with the Institute of Industrial Science, Tokyo University as a Research Associate, Lecturer and successively an Associate Professor. From 1988 to 2002, he was a Professor in the Department of Electrical Engineering, and also, from 2000 to 2002, the Director of the Venture Business Laboratory, Gunma University (Kiryu, Japan). Since 2002, he has been the Principal of the Kaisei Academy (Arakawa-ku, Tokyo). His research area of interest has been Applied Optics. From 1971 to 1973, he was a Guest Researcher with the Hanover University (Germany), where he worked in the area of Laser Spectroscopy. In 1985, he was a Visiting Professor with the IBM Research Laboratory (San Jose, CA), where he worked on optical sensing. He is a Consulting Professor of the North China Electric Power University (Beijing, China).

Dr. Yoshino is a member of the Japan Society of Applied Physics, the Optical Society of America, and a Fellow of Institute of Physics, UK.

Masayuki Yokota was born in Gunma, Japan, on January 31, 1967. He received the B.S., M.S., and Ph.D. degrees in Electronic Engineering from Gunma University in 1989, 1991, and 1998, respectively. Currently, he is a research associate at the Department of Electronic Engineering, Gunma University, where he has been working on optical engineering.

Dr. Yokota is a Member of the Japan Society of Applied Physics (JSAP) and the Optical Society of America (OSA).

ABOUT THE EDITORS

Pankaj Kr. Choudhury started his research career in 1989 from the Department of Applied Physics, Institute of Technology, Banaras Hindu University (BHU), Varanasi (India), and received Ph.D. degree in Physics in 1992. From 1992 to 1997, he worked as a Research Associate in the Department of Electronics Engineering at BHU. Also, during 1995–1996, he was a Guest Lecturer in Physics at the Faculty of Science and Technology, Mahatma Gandhi Kashi Vidyapith, Varanasi (India). In 1997 he joined the Department of Physics at Goa University, Goa (India) as a Lecturer, where he was until 1999. In 1999 he joined the position of Researcher at the Center of Optics, Photonics and Lasers (COPL), Laval University, Quebec (Canada), where he worked on the fabrication and characterization of ferroelectric optical waveguides. During 2000–2003 he was a Researcher at the Faculty of Engineering, Gunma University, Kiryu (Japan). Presently he is a Professor at the Faculty of Engineering, Multimedia University (Malaysia). His current research interests lie in the theory of optical waveguides and EM wave propagation, which include fiber and integrated optics, fiber-optic devices and sensors. He has authored and co-authored over eighty five research publications (including three book chapters) in these areas. He is a Senior Member of IEEE and a Regular Member of the Optical Society of America.

Onkar N. Singh received his M.Sc. and Ph.D. degrees, respectively, in 1963 and 1967 in Physics (Spectroscopy) from the Banaras Hindu University (BHU), Varanasi (India). Joining the BHU faculty in 1963, he became a Professor of Applied Physics in 1986, and served as the Head of the Department of Applied Physics, Institute of Technology (BHU) from 1991 to 1993. He has worked as a visiting/research scientist at the International Center for Theoretical Physics (Trieste, Italy), FOM Institute for Atomic and Molecular Physics (Amsterdam, The Netherlands), Max-Planck Institute for Aeronomy (Lindau, Germany), Ludwig Maxmilian University (Munich, Germany), and the Technical University of Munich (Germany). He has published over 100 research papers and book chapters on molecular spectroscopy, electromagnetic fields, lower stratospheric pollution and climate change, and was co-editor of Vol. 4/E of the *Handbook of Environmental Chemistry* (Springer, Heidelberg, 1999). He was the representative of India to the Atmospheric Chemistry of Commission of IUPAC (1989–1995) and was also a member of the National Committee of World Climate Research Programme (1997–2000). He also co-edited the book *Electromagnetic fields in unconventional structures and materials* (Wiley, USA, 2000). He is Member/Fellow of several learned bodies.

INDEX

A

access, 213, 236
accounting, 112
accumulation, 92, 255
accuracy, 196, 279, 280
acetone, 105
acetonitrile, 105
achievement, 244
additives, 132, 142, 153, 155
adhesion, 252
adjustment, 83
aerospace, 235, 258, 269
aesthetics, 154, 155
affect, 2, 75, 140, 149
ageing, 104
air pollutants, 259
algorithm, 2, 6
alternative, 77, 99, 137, 151
aluminium, 74, 83, 86, 133, 137, 152, 262
ambient air, 77, 80, 262
ammonia, 75
ammonium, 218, 262
amplitude, 1, 2, 3, 6, 24, 27, 41, 90, 162, 163, 167, 172, 177, 223, 277
anisotropy, 33, 131, 283
annealing, 101, 102, 103, 106, 110, 278
antibody, 253
antigen, 253
aqueous solutions, 256
argument, 178
artery, 242
Asia, 128
assessment, 259, 263
assumptions, 26, 35, 227
asymmetry, 181
athletes, 128
atoms, 99, 102, 193
attention, 23, 24, 47, 129, 154
Australia, 125, 128, 147, 158
availability, 74, 146

B

backscattering, 133, 142, 143, 144, 146, 150
bandgap, 194
bandwidth, 59, 60, 61, 118, 172, 192, 204, 205, 213
beams, 147, 150, 217, 239, 251
behavior, 255
Beijing, 287
bending, 76, 88, 238, 244, 252, 254, 277
bias, 118, 119, 120, 163, 167, 168, 170, 172, 173, 180, 181, 182, 185, 188, 192, 195, 197, 208, 251, 278
biomedical applications, 236
birefringence, 18, 48, 251, 276, 277
blackbody radiation, 240
bleeding, 1, 8, 9, 14, 15, 16, 17, 19
blocks, 32, 34, 127
blood, 242, 243
blood flow, 242, 243
body, 107, 131, 132, 133, 153, 240, 242, 263
boundary value problem, 2
breakdown, 119, 120
breathing, 251
broadband, 173, 245, 247
buffer, 75, 174, 176, 184
building blocks, 235
buildings, 126, 128, 129, 131, 133, 151, 153

C

cables, 147, 278
Canada, 271, 289
carbon, 205, 258, 261
cardiac surgery, 243

carrier, 1, 2, 3, 6, 7, 8, 9, 10, 11, 12, 13, 14, 15, 16, 17, 19, 80, 84, 92, 105, 109, 114, 137, 138, 164, 166, 170, 182, 183, 188, 192, 193, 194, 196, 197, 203, 249
cast, 146
catheter, 242, 263
causality, 3
cell, 38, 104, 132, 141, 240, 261, 275, 280, 281, 283
channels, 78, 84, 169, 170, 182, 183, 185, 186, 188, 189, 194, 195, 196, 198, 202, 203, 205, 208
chemical bonds, 99
chemical etching, 218
chemical kinetics, 256, 257
chemical reactions, 101, 256, 257, 259
China, 71, 287
chlorine, 260, 261
circulation, 185, 186, 188
cladding, 74, 75, 76, 80, 87, 91, 92, 110, 218, 223, 224, 225, 226, 227, 231, 242, 251, 252
cladding layer, 74, 75, 76, 80, 91
classes, 124, 125, 126, 214, 262, 268
cleaning, 129
climate change, 272, 290
clustering, 101, 109, 140
clusters, 101, 138
CO2, 125
coatings, 127, 132, 153, 249, 285
coherence, 252
collaboration, 63
collisions, 138, 150, 151
combustion, 101
communication, 191, 192, 208, 213
communication systems, 213
community, 262
compensation, 163, 164, 166, 172
complexity, 79
components, 5, 9, 23, 24, 25, 26, 27, 29, 31, 41, 48, 83, 86, 88, 127, 144, 145, 149, 161, 162, 163, 164, 168, 169, 170, 172, 183, 184, 186, 213, 214, 215, 234, 236, 237, 244, 278
composites, 125, 141
composition, 74, 100, 121, 125, 144, 218, 242
computation, 7
computer simulations, 150
computing, 97
concentrates, 2
concentration, 102, 104, 130, 134, 138, 144, 145, 146, 147, 148, 151, 218, 242, 243, 251, 253, 254, 255, 256, 258, 259, 260, 261
conditioning, 128
conduct, 18
conduction, 107, 118, 119, 194, 195, 198
conductivity, 176
conductor, 129, 132, 134, 137, 164
configuration, 38, 174, 193, 205, 208, 214, 215, 216, 224, 226, 227, 228, 229, 230, 231, 238, 239, 241, 249, 262, 283, 284, 285
confinement, 75, 92, 103, 111, 112, 114
conflict, 126
conformity, 89
construction, 226
consumers, 213
consumption, 236
control, 74, 78, 91, 101, 103, 116, 125, 126, 127, 128, 131, 138, 140, 142, 143, 145, 149, 154, 155, 158, 159, 160, 162, 170, 172, 184, 252, 258, 259, 263
conversion, 41, 78, 134, 159, 163, 168, 188, 191, 192, 203, 204, 211, 214, 277
cooling, 125, 127, 128, 131, 136, 137, 153, 154
copper, 151
correlation, 52
couples, 84, 223
coupling, 73, 76, 79, 80, 81, 82, 83, 84, 85, 86, 87, 88, 90, 117, 119, 120, 134, 149, 161, 205, 213, 214, 215, 218, 219, 220, 221, 222, 223, 224, 225, 226, 227, 228, 229, 230, 231, 232, 236, 238, 239, 245, 262, 276
crystallites, 102
crystallization, 103
crystals, xiii, 2, 23, 24, 67, 68, 70, 71, 103, 115, 116, 235, 280
CVD, 74, 76, 101
cycles, 15

D

damage, 107
data communication, 213
DCA, 167, 168
debt, xiii
decay, 241, 244
defects, 24, 50, 51, 52, 53, 58, 59, 60, 61, 62, 64, 79, 101, 116
definition, 29, 32
deformation, 180, 252
demand, 126, 128, 213
density, 73, 74, 76, 102, 103, 104, 105, 112, 113, 114, 115, 137, 141, 144, 148, 194, 205, 244, 245, 255, 259
density fluctuations, 244
deposition, 1, 2, 18, 73, 74, 75, 76, 79, 81, 82, 83, 86, 88, 89, 101, 141, 152, 260
deposition rate, 2, 76
deposits, 155
depth perception, 150

derivatives, 5
designers, 127
detection, 172, 241, 247, 249, 252, 253, 258, 259, 260, 261, 262, 276, 277, 278
developed nations, 235
deviation, 143, 148, 162, 183, 223, 231, 261
diaphragm, 237
dielectric constant, 137, 138, 140
dielectrics, 100, 101
differential equations, 4
differentiation, 4
diffusion, 74, 80, 91, 142, 144, 180, 198, 200, 203
diffusivity, 101
dimensionality, 97, 98, 100
diodes, 73, 79, 82, 84, 85, 97, 123, 213, 236
direct measure, 145
discretization, 5
dispersion, 25, 26, 35, 37, 38, 92, 103, 138, 145, 146, 164, 172
displacement, 24, 25, 238, 250, 258
distillation, 128
distribution, 25, 120, 125, 127, 132, 141, 143, 144, 148, 151, 152, 154, 155, 175, 176, 177, 178, 179, 180, 193, 196, 285
distribution function, 144
divergence, 150, 214
division, 78, 247
domain, 1, 2, 3, 7, 8, 9, 14, 15, 17, 19, 20, 21, 159, 160, 164, 168, 182, 187, 188, 230, 244, 251, 276, 283
domain structure, 283
dominance, 97
dopants, 125
doping, 103, 104, 134, 139, 144, 146, 147, 148, 149, 158, 193, 194, 205, 218, 245
double bonds, 99, 113
drinking water, 260, 261
duration, 1, 2, 7, 8, 15

E

earth, 129, 236, 244, 282
economic growth, 126
electric field, 24, 25, 35, 41, 108, 117, 118, 164, 167, 176, 177, 180, 181, 191, 193, 195, 196, 197, 198, 202, 203, 205, 216, 219, 277
electrical fields, 194
electricity, 128, 155
electrodes, 92, 159, 160, 164, 170, 172, 173, 174, 176, 177, 178, 179, 180, 181, 183, 203, 237
electroluminescence, 107, 108, 109, 196
electrolyte, 103
electromagnetic, 3, 5, 6, 23, 24, 61, 74, 78, 79, 80, 91, 236, 258, 262, 272, 290
electromagnetic fields, 3, 272, 290
electronic structure, 86, 98
electrons, 73, 98, 107, 108, 109, 119, 193, 196, 197, 198, 203, 210
emission, 73, 97, 100, 105, 106, 107, 111, 112, 113, 114, 124, 134, 146, 147, 148, 158, 191, 192, 193, 194, 196, 197, 198, 199, 200, 201, 202, 204, 205, 208, 213, 227, 238, 240, 259, 260
emitters, 191, 192, 193, 210
energy, 8, 9, 14, 15, 46, 57, 98, 99, 100, 102, 105, 106, 107, 108, 112, 113, 117, 118, 125, 126, 127, 128, 129, 130, 131, 132, 137, 138, 141, 142, 144, 147, 149, 150, 151, 152, 153, 154, 155, 158, 194, 195, 196, 197, 198, 202, 203, 205, 225, 231, 236
energy efficiency, 127, 132, 137, 142, 150, 152, 155, 158
environment, 104, 105, 124, 126. 127, 154, 155, 235, 258, 259
environmental control, 128
equilibrium, 193, 194, 195, 196
equipment, 76
erbium, 249, 251
erosion, 100
etching, 73, 74, 75, 81, 82, 83, 84, 86, 87, 88, 89, 103, 115, 116, 202, 218, 222
ethanol, 101
European Commission, 211
European Union, 71
evanescent waves, 242, 252, 257
evaporation, 1, 101
evidence, 112, 114, 119, 120, 124
evolution, 2, 14, 102, 256, 262, 263
excitation, 19, 107, 108, 113, 114
exciton, 106
experts, 124
exploitation, 97
exposure, 107
expression, 28, 222, 245, 246
extinction, 140, 162, 163, 164, 165, 166, 183, 280
extraction, 90, 100, 127

F

fabrication, 88, 101, 103, 104, 124, 163, 192, 193, 236, 245, 258, 259, 262, 271, 289
fabrication errors, 163
family, 24
Faraday effect, 250, 251, 273, 274, 277, 284, 285
feedback, 250
FEM, 180
Fermi level, 194

fiber bundles, 146
fiber couplings, xiii
fiber optics, 255, 260
fibers, 73, 146, 147, 149, 158, 164, 213, 214, 215, 218, 226, 236, 237, 238, 239, 244, 245, 246, 249, 251, 253, 261, 262, 263, 277, 285
film thickness, 280
films, xiii, 21, 73, 74, 88, 101, 128, 129, 132, 133, 249, 280
finite element method, 176, 180, 181
flatness, 59, 83
flexibility, 127, 146, 262
flight, 101, 193
fluctuations, 151, 252
fluid, 155, 242, 255, 259
fluorescence, 158, 238, 241, 251, 253
focusing, xiii, 226, 227, 228, 231
foils, 136
fossil, 126
France, 266
freedom, 215
Fresnel coefficients, 41
FTIR, 251
fuel, xiii
fulfillment, 273, 274, 285
funding, 211

G

gadolinium, 24
gallium, 24, 160
gasoline, 252
gate oxidation, 83
generation, 1, 84, 108, 159, 160, 162, 167, 172, 173, 182, 186
germanium, 245
Germany, 70, 73, 95, 272, 287, 290
goals, 127, 128
gold, 137, 141, 152, 174
grants, 155
graph, 222, 224, 230
gratings, 247, 248
GRIN, 226
groups, 127
growth, 18, 57, 86, 106, 124, 126, 128, 146, 154, 194, 195, 213, 263
guidance, 18

H

handedness, 2, 4, 18, 258
harvesting, 153
haze, 132, 137, 138, 140
health, 127, 259, 260, 263
heat, 107, 127, 128, 129, 131, 132, 133, 154, 155, 194
heat loss, 128
heating, 142, 193, 196, 197, 201, 224, 226, 252
heavy metals, 262
height, 76, 224, 231
hemisphere, 227
host, 99, 100, 102, 125, 138, 142, 241
hot carrier effects, 194
hot spots, 147
House, 93, 263, 269
human activity, 126
hybrid, 259, 273, 280, 281, 283, 285
hydrocarbons, 259
hydrofluoric acid, 218
hydrogen, 106, 253
hysteresis, 283

I

identity, 4
illumination, 103, 125, 142, 146, 149, 150, 151, 208, 240, 247, 283
images, 102
immersion, 258
implementation, 111, 236, 262
incidence, 23, 25, 33, 34, 35, 36, 38, 39, 40, 41, 43, 47, 48, 51, 52, 55, 63, 127, 130, 131, 132, 133, 136, 143, 148, 280
India, 19, 271, 272, 289, 290
indicators, 237
indices, 25
indirect effect, 100
indium, 138
induction, 24, 25, 196
inductor, 176
industry, 235, 236, 259
inefficiency, 107
inequality, 55
infancy, 127
infinite, 35, 36, 98, 99
influence, 63, 100, 106, 137
information processing, 236
inhomogeneity, 275
input, 52, 54, 55, 56, 57, 61, 84, 146, 148, 149, 150, 155, 160, 161, 167, 168, 169, 175, 177, 182, 183, 184, 185, 187, 188, 204, 217, 227, 231, 238, 239, 240, 244, 247, 249, 263, 275, 277, 280
insertion, 281
inspiration, xiii
instability, 278

instruments, 128, 236, 263
insulation, 132, 137, 153, 154
integrated circuits, 123
integrated optics, xiii, 73, 74, 76, 88, 92, 93, 95, 214, 235, 271, 289
integration, 73, 74, 75, 76, 78, 79, 83, 84, 86, 87, 88, 89, 90, 92, 125, 127, 150, 192, 205, 214, 251, 263, 273, 274
integrity, 262
intensity, 77, 79, 84, 88, 90, 92, 111, 112, 113, 114, 117, 119, 120, 121, 129, 131, 149, 150, 159, 160, 162, 163, 164, 165, 166, 167, 169, 170, 172, 182, 185, 187, 192, 197, 204, 208, 229, 235, 236, 237, 238, 239, 242, 243, 245, 249, 250, 251, 252, 253
interaction, 53, 123, 146, 148, 150, 159, 238, 245, 255, 261
interactions, 138, 141, 148, 159
interest, xiii, 73, 74, 91, 115, 126, 130, 131, 132, 137, 138, 140, 141, 151, 153, 192, 205, 210, 213, 214, 237, 287
interface, 8, 9, 10, 12, 78, 107, 110, 111, 113, 193, 194
interference, 9, 36, 152, 232, 236, 239, 249, 250, 258, 262, 276
internet, 213
invariants, 37
inversion, 37, 111, 112
iodine, 256
ions, 253
iron, 24, 152, 282
isolation, 78, 81, 86, 274, 280, 283
Italy, 97, 124, 272, 290

J

Japan, 159, 264, 267, 271, 273, 278, 286, 287, 289

K

knowledge, 17, 256
Korea, 23, 63, 71

L

labeling, 160, 182, 185
lack of control, 103
lasers, 111, 205, 210, 211, 213, 214, 248, 249, 250, 251
layering, 152
lead, 63, 73, 97, 149
leakage, 80, 84, 86
LED, 108, 109, 124, 125, 146, 150, 151, 260
legislation, 263
lens, 76, 213, 214, 215, 218, 220, 221, 223, 224, 225, 226, 227, 228, 229, 231
lifetime, 92, 100, 105, 113, 114, 241
light beam, 250, 253, 276, 278, 280, 283
light scattering, 115
light transmission, 278
light transmittance, 133
limitation, 99, 104, 226, 244
linear dependence, 25, 197
links, 159
liquid crystals, 16, 19
liquids, 253
lithium, 160
living standards, 125, 126
localization, 105, 119
location, 127, 137, 138, 237, 238
long distance, 134, 146
longevity, 150
luminescence, 103, 106, 107, 111, 112, 113, 193, 238, 241, 259
Luxemburg, 23
lying, 261

M

magnetic field, 24, 25, 41, 55, 63, 176, 180, 273, 274, 277, 283, 285
magnetic moment, 47
magnetic properties, 5
magnetic structure, 57
magnetization, 24, 25, 29, 30, 33, 39, 40, 55, 61, 63, 273, 283
Malaysia, 210, 213, 234, 271, 289
management, 125, 129, 130, 131, 159, 160
manipulation, 160
manufacturing, 263
mapping, 238
market, 154, 213, 235, 236, 258, 259
mass, 98, 99, 101, 130, 137
materials science, 155
matrix, 3, 5, 6, 24, 26, 27, 28, 29, 30, 32, 33, 34, 36, 37, 41, 46, 66, 67, 97, 99, 100, 101, 105, 109, 116, 117, 118, 120, 138, 143, 144, 147, 213, 215, 217, 219, 220, 223, 224, 227, 229, 231
Maxwell equations, 3, 24, 25
meanings, 3
measurement, 88, 117, 144, 204, 208, 222, 224, 225, 229, 230, 231, 232, 236, 237, 241, 242, 243, 244, 246, 250, 251, 252, 253, 274, 278, 279, 281, 283
measures, 112, 113, 253
mechanical properties, 73, 262
mechanical stress, 88

media, 24, 26, 27, 29, 32, 34, 41, 128, 158
median, 143
melting, 252
membranes, 88, 89, 90, 91
memory, 6
metals, 1, 133, 137, 138
methanol, 251, 252
microcavity, 115, 116, 117, 124, 205
microelectronics, 88, 97
micrometer, 193
microscope, 103
microscopy, 229
microspheres, 253
microstructure, 3
milk, 251
mixing, 125, 127, 134, 149, 150, 151
mobility, 193, 203
mode, 27, 36, 38, 39, 40, 41, 42, 47, 48, 49, 50, 51, 52, 53, 54, 55, 57, 58, 59, 60, 61, 62, 64, 74, 110, 112, 115, 117, 141, 160, 162, 165, 172, 180, 181, 192, 200, 201, 205, 206, 208, 213, 214, 215, 223, 225, 231, 245, 276
models, 2, 125, 130, 141, 147, 231, 236, 258
modules, 214
modulus, 178
molecular beam, 101
molecules, 141, 255, 258, 260, 261, 262
momentum, 106, 193
monitoring, 241, 242, 244, 248, 252, 258, 259, 261, 262, 263
monolayer, 249, 252
morphology, 2, 141, 258, 269
motion, 47
motivation, 97, 105
MOVPE, 194, 200, 205
MRS, 156
multimedia, 213

N

nanocrystals, 98, 105, 114, 124
nanometer scale, 75, 95
nanoparticles, 125, 129, 132, 134, 135, 137, 139, 140, 141
nanostructure, 155
nanostructured materials, 129
nanostructures, 124, 128
nanotechnology, 9
nanowires, 1, 2, 3, 18
NATO, 211
Nd, 249, 250
needs, 7, 76, 103, 126, 128, 129, 140, 227, 278
neglect, 138, 167
Netherlands, 71, 156, 265, 272, 290
network, 169
NIR, 129, 132, 133, 137, 138, 139, 140, 141, 151, 152, 153, 155
nitrogen, 109
noise, 169, 205, 277
nucleation, 101
nucleus, 151
numerical aperture, 148, 227, 236, 242

O

observations, 114, 198
OFS, 263, 265, 266, 268, 269
oil, 235, 250, 251
one dimension, 111, 115
optical activity, 2, 18
optical anisotropy, 129
optical fiber, 97, 147, 148, 149, 184, 213, 214, 215, 234, 236, 237, 241, 242, 244, 245, 246, 251, 252, 253, 258, 261, 262, 265, 278
optical properties, 19, 24, 50, 63, 70, 97, 98, 109, 112, 113, 127, 129, 131, 142, 252, 282
optical pulses, 1, 2, 14, 15, 16, 18, 21
optical systems, 128, 132, 159, 163
optimization, 54, 63
optoelectronics, 124, 210
organic polymers, 261
organic solvents, 255
organizations, 263
orientation, 61, 152
oscillation, 119, 160
outline, 24
output, 41, 57, 86, 88, 89, 90, 91, 92, 127, 131, 146, 147, 148, 149, 150, 151, 158, 160, 161, 162, 163, 164, 167, 168, 169, 170, 171, 172, 173, 182, 183, 185, 186, 187, 188, 192, 200, 201, 207, 214, 217, 226, 227, 236, 242, 243, 247, 249, 274, 275, 276, 277, 280
oxidation, 74, 81, 82, 92
oxide thickness, 80, 82, 83, 92, 152
oxides, 1, 152, 252, 262
oxygen, 86, 87, 107, 243
ozone, 89

P

Pacific, 155
packaging, 214, 232, 234, 262
paints, 125, 133, 151, 152

parameter, 7, 14, 46, 52, 53, 58, 59, 60, 62, 90, 109, 131, 143, 144, 146, 147, 148, 216, 220, 224, 227, 228, 231, 232, 237, 242, 274
particles, 98, 100, 105, 117, 137, 138, 140, 141, 142, 143, 144, 148, 150, 151, 153
passivation, 83, 106
passive, 74, 128, 214, 236, 237, 247, 263
patents, 124, 147, 158
penalties, 172
percolation, 140
periodicity, 46, 250
permeability, 3, 23, 24, 25, 38, 280, 283
permittivity, 3, 23, 24, 25, 35, 38, 176, 258
pH, 237, 253, 255
phase shifts, 247, 249, 276
phenol, 253
phenomenology, 2, 17
phonons, 106, 193, 245
phosphorous, 91
photoelastic effect, 246
photographs, 199
photoluminescence, 107, 112, 193
photonics, xiii, 24, 63, 97, 109, 121, 122, 124
photons, 78, 97, 114, 119, 203, 245
photosensitivity, 245, 246
physical fields, 237
physical mechanisms, 131
physics, 95, 122, 125, 129, 155, 210
pitch, 2, 245, 246, 252
plane waves, 1, 2, 8, 9, 15, 258
planning, 63
plants, 259
plasma, 19, 74, 86, 87, 88, 92, 101, 137, 226
platinum, 262
PMMA, 132, 133, 134, 142, 143, 145, 151
polarity, 168, 177, 192, 195, 198, 199, 200, 202, 251
polarization, 1, 2, 7, 8, 9, 15, 18, 24, 25, 36, 38, 46, 53, 57, 63, 205, 236, 238, 251, 273, 274, 276, 277, 278, 280
pollution, 272, 290
polyacrylamide, 253
polymer composites, 158
polymers, 1, 128, 129, 132, 133, 146, 153
poor, 150, 154, 214, 236
population, 111, 112
porosity, 101, 103, 115
ports, 167, 169, 182
potassium, 245, 256
power, 8, 9, 78, 84, 97, 99, 107, 110, 112, 113, 114, 126, 127, 150, 163, 164, 169, 170, 171, 177, 180, 184, 192, 201, 204, 205, 207, 208, 215, 224, 232, 235, 236, 238, 239, 240, 242, 244, 247, 251, 253, 260, 277, 278, 279, 280, 285
power plants, 278
prediction, 141, 198
preparation, 106
pressure, 73, 74, 88, 89, 90, 101, 237, 238, 240, 252
principle, 46, 70, 92, 130, 160, 162, 168, 244, 247, 250, 251, 252, 253, 276, 280, 283, 285
probability, 100, 114, 143
probability density function, 143
probe, 111, 236, 238, 239, 242, 243, 253, 254, 255, 256, 257, 260, 283
process control, 259
production, 152, 193, 205, 262
productivity, 127
program, 122, 148
promoter, 262
propagation, 1, 4, 15, 17, 19, 25, 27, 29, 32, 37, 73, 74, 75, 76, 77, 84, 110, 111, 112, 114, 115, 173, 175, 176, 178, 184, 186, 187, 219, 227, 231
pulse, 1, 2, 3, 6, 7, 8, 9, 10, 11, 12, 13, 14, 15, 16, 17, 19, 116, 118, 119, 188, 192, 201, 207, 250
pyrolysis, 101

Q

quality of life, 259
quantitative technique, 17
quantum confinement, 100, 105, 106, 107
quantum dots, 100
quantum mechanics, 98
quantum well, 99, 117, 191, 196, 198, 205, 210
quartz, 112

R

radiation, 23, 73, 76, 107, 110, 126, 127, 128, 129, 130, 131, 134, 137, 153, 155, 240, 251, 259
radio, 174
radius, 76, 114, 143, 144, 148, 216, 218, 220, 223, 224, 227, 228, 252, 253, 254, 277
rain, 155
Raman spectra, 251
Raman spectroscopy, 193
range, 2, 58, 73, 74, 75, 78, 91, 92, 106, 116, 117, 121, 125, 131, 137, 138, 142, 149, 153, 175, 181, 200, 211, 227, 231, 235, 242, 244, 248, 252, 258, 260, 261, 262, 263, 276, 279
reaction rate, 256
reagents, 256
reality, 110, 277
reciprocity, 37
recombination, 100, 105, 113, 114, 198, 203, 206
recombination processes, 105

reconstruction, 99, 101
reduction, 100, 125, 138, 142, 259, 262
reflection, 2, 9, 19, 23, 36, 41, 79, 80, 134, 142, 149, 151, 153, 181, 183, 184, 185, 186, 187, 205, 206, 232, 239, 250, 252, 259, 265, 278, 279, 280, 281, 282
reflectivity, 115, 117, 153, 176, 179, 194, 200, 205, 239
refraction index, 47
refractive index, 23, 27, 63, 74, 97, 125, 160, 194, 218, 239, 257
regulations, 127
rejection, 152, 154
relationship, 218, 220, 224, 242, 255, 277
relaxation, 113, 138
reliability, 259
remembering, 48
replacement, 31, 86
resistance, 193, 261, 283, 285
resolution, 77, 115, 116, 117, 169
resonator, 249, 262
resources, 155
response time, 91, 236
retardation, 274, 277, 280
returns, 240
rewards, 262
risk, 126, 150, 225
rods, 128, 134, 149, 150
room temperature, 201, 207
roughness, 74, 75, 84
routing, 182

S

safety, 128, 259, 263
Sagnac effect, 250
sample, 105, 106, 107, 111, 113, 114, 115, 116, 117, 118, 119, 120, 193, 203, 251, 253, 261
sampling, 147, 169
saturation, 283
savings, 144, 154, 155
scaling, 9, 99, 126
scaling law, 99
scatter, 132
scattering, 79, 110, 111, 117, 125, 127, 129, 130, 137, 138, 139, 140, 141, 142, 144, 146, 147, 193, 242, 244, 245, 253, 254
schema, 112
school, 124
search, 35
segregation, 84
selectivity, 131, 132, 136, 140, 153, 253
self, 103, 141, 155, 202, 214, 218
semiconductor, 92, 97, 98, 124, 134, 191, 192, 193, 208, 210, 214, 240, 249, 251, 262
semiconductors, 73, 118, 193, 209, 210
sensing, xiii, 89, 90, 141, 235, 236, 237, 238, 241, 242, 243, 244, 245, 246, 247, 249, 251, 252, 258, 259, 260, 262, 263, 273, 277, 278, 279, 280, 285, 287
sensitivity, 89, 90, 104, 130, 172, 237, 242, 248, 249, 251, 253, 254, 256, 257, 258, 262, 281, 283, 285
sensors, 74, 92, 154, 234, 235, 236, 237, 238, 239, 240, 241, 242, 243, 244, 245, 246, 247, 248, 249, 250, 251, 252, 253, 256, 258, 259, 260, 261, 262, 263, 269, 271, 273, 278, 289
separation, 53, 82, 100, 101, 102, 103, 109, 174, 193, 198, 199, 215
series, 35, 119, 174, 191
shape, 1, 2, 7, 9, 14, 75, 137, 138, 192, 214, 215, 253
shaping, 1, 2, 15, 16, 17
Si3N4, 74, 76, 77, 92, 99, 109, 110
sign, 34, 55, 109, 151
signalling, 128
signals, 159, 160, 161, 162, 163, 164, 167, 169, 170, 172, 182, 183, 186, 236, 237, 241, 261, 281
silica, 99, 103, 107, 109, 244, 246, 251, 252, 255, 260, 277, 278, 279
silicon, xiii, 73, 74, 75, 76, 78, 79, 80, 81, 82, 83, 84, 86, 87, 88, 89, 90, 91, 92, 93, 95, 97, 98, 100, 101, 102, 103, 104, 105, 106, 107, 108, 109, 110, 115, 116, 121, 122, 124, 133, 136, 214, 262
silicon nanocrystals, 97, 105
silver, 137, 152
simulation, 18, 19, 73, 116, 130, 143, 144, 150, 154, 169
SiO2, 74, 76, 77, 83, 99, 100, 101, 102, 103, 109, 110, 132, 133, 134, 153, 176, 183, 280
skeleton, 103
skin, 129, 261
SOI, 74, 91, 110
solar cells, 128
sol-gel, 152
solvents, 105
spectral component, 168, 182, 186, 188
spectroscopy, 71, 124, 210, 242, 251, 253, 257, 260, 261, 272, 290
spectrum, 8, 98, 99, 100, 106, 107, 112, 113, 115, 117, 119, 120, 129, 131, 162, 167, 169, 183, 192, 196, 200, 201, 207, 208, 240, 241, 247, 251, 252, 253, 254, 256, 258, 259
speed, xiii, 92, 159, 160, 163, 164, 170, 172, 173, 176, 183, 186, 193, 213, 262, 273, 282, 285
speed of light, 176, 186
spin, 277
stability, 6, 14, 215, 236, 281, 282

standards, 126
statistics, 148
steel, 133
storage, 261
strain, 99, 245, 246, 248, 249, 252
strategies, 127
strength, 53, 98, 108, 203, 241, 249
stress, 278
students, xiii, 155
substitution, 27, 31, 35, 65, 67
substrates, 2, 18, 73, 133, 151, 183, 194
summer, 128
Sun, 129
superlattice, 103, 117, 118, 119, 121
supply, 126, 129, 235, 243
suppression, 164, 166, 188
surface layer, 50
susceptibility, 25
Sweden, 158
switching, 92, 97, 109, 163, 166, 170, 172, 182, 184, 186, 187
Switzerland, 124
symbols, 173, 220
symmetry, 26, 27, 29, 30, 33, 34, 37, 40, 47, 67
synthesis, 100, 101
systems, 115, 117, 121, 123, 125, 126, 128, 129, 132, 133, 134, 138, 141, 142, 147, 149, 151, 153, 155, 159, 160, 169, 173, 174, 182, 183, 184, 202, 210, 211, 214, 217, 235, 236, 237, 238, 240, 241, 244, 249, 259, 260, 261, 262, 263, 278, 281

T

Taiwan, 210
tanks, 261
tantalum, 252
technology, xiii, 2, 18, 73, 74, 80, 88, 90, 95, 97, 126, 127, 129, 147, 154, 155, 214, 234, 235, 236, 242, 249, 258, 259, 261, 262, 263, 265, 273, 280, 285
telecommunications, 2, 18, 97, 159, 213, 265
temperature, 75, 86, 101, 102, 103, 106, 107, 110, 128, 153, 154, 196, 197, 200, 201, 237, 238, 240, 241, 244, 245, 246, 248, 251, 252, 259, 262, 276, 278, 279, 282
temperature dependence, 244, 252
terminals, 159
theory, 70, 71, 91, 110, 152, 173, 227, 258, 259, 271, 277, 280, 289
thermal energy, 154
thermal evaporation, 210
thermal oxidation, 74, 75, 82
thermal relaxation, 278
thermal treatment, 76
thin films, 1, 18, 19, 71, 128, 129, 132, 133, 137, 158, 252, 258, 269, 280
threshold, 104, 201, 204
time, 1, 2, 3, 5, 6, 7, 8, 9, 14, 15, 17, 19, 21, 55, 91, 101, 104, 110, 113, 114, 118, 152, 159, 160, 169, 170, 172, 182, 187, 188, 193, 198, 203, 210, 222, 241, 244, 256, 257, 258, 262, 277, 280
timing, 159, 184, 187
tin, 135, 138, 262
tin oxide, 135, 138
titanium, 137, 180
toluene, 105
total energy, 2
total internal reflection, 132, 144
tracking, 129
trade, 170, 254
trade-off, 170, 254
trajectory, 227, 230, 231
transducer, 237, 242, 258
transformation, 8, 25, 228
transformation matrix, 25
transistor, 81, 83, 86, 90, 108, 109, 193
transition, 77, 79, 80, 81, 83, 84, 85, 98, 99, 100, 105, 262
transmission, 2, 19, 23, 41, 46, 47, 48, 49, 50, 51, 52, 54, 55, 56, 57, 58, 59, 63, 73, 92, 97, 103, 112, 113, 115, 117, 119, 120, 121, 127, 144, 163, 164, 172, 173, 213, 216, 236, 237, 252, 256, 258, 259, 278
transport, 126, 129, 132, 134, 146, 147, 148, 149, 158, 191, 192, 193, 198, 203, 205, 209, 210
transportation, 279
trend, 259
trial, 280
tunneling, 97, 107, 108, 109, 119, 120, 121, 124, 196
Turkey, 210, 211

U

UK, 19, 123, 155, 191, 210, 265, 287
Ukraine, 23, 63, 64, 70, 71
uncertainty, 274, 278, 283
underlying mechanisms, 151
uniform, 18, 24, 32, 36, 49, 71, 125, 127, 141, 142, 149, 150, 151, 152, 158, 180, 218, 244, 252, 283, 285
universities, 210
USSR, 70
UV, 129, 132, 155, 245, 252
UV light, 245

V

vacuum, 1, 3, 7, 8, 9, 10, 12, 14, 16, 17, 50, 51, 128
valence, 107, 119, 194, 195, 198
validity, 257
values, 7, 27, 31, 33, 37, 39, 40, 46, 47, 51, 52, 54, 55, 57, 67, 103, 109, 138, 143, 148, 152, 181, 213, 223, 227, 230, 231, 253, 254, 260, 275
vapor, 1
variable, 111, 138
variables, 7, 28, 217
variation, 60, 98, 100, 115, 145, 153, 218, 232, 244, 245, 254, 257, 260
VCSEL, 191, 192, 193, 197, 200, 201, 204, 207, 208, 211
vector, 3, 4, 5, 6, 7, 8, 9, 14, 15, 16, 17, 24, 25, 27, 29, 30, 32, 39, 41, 55, 61, 63, 274
vehicles, 133
velocity, 176, 242, 243, 245
versatility, 127
vessels, 242, 263
vibration, 250, 276, 279
vision, 214

W

water, 105, 125, 127, 128, 129, 153, 154, 155, 251, 252, 260
water heater, 128
wave number, 35, 40, 43
wave propagation, 271, 289
wave vector, 25, 26, 27, 35, 36, 38, 40, 41, 46, 50, 61
wavelengths, 2, 3, 4, 7, 8, 9, 74, 129, 131, 132, 134, 140, 144, 145, 151, 152, 161, 204, 205, 240, 243, 244, 251, 253, 260
weakness, 154
wells, 117
windows, 125, 128, 129, 131, 155, 158, 169, 192, 208
wires, 83, 99, 103
words, 202
work, xiii, 2, 3, 21, 33, 63, 97, 109, 128, 130, 141, 142, 147, 150, 153, 155, 158, 192, 205, 208, 224, 227, 231, 234, 248, 249
workers, 122

X

X-axis, 46

Y

yield, 74, 79, 84, 149, 249, 251
yttrium, 24, 282

Z

zinc, 262
ZnO, 132, 133